T0413215

Laser growth and processing of photonic devices

Related titles:

Biomaterials, artificial organs and tissue engineering

(ISBN 978-1-85573-737-2)

Biomaterials are materials and devices that are used to repair, replace or augment the living tissues and organs of the human body. The purpose of this wide-ranging introductory textbook is to provide an understanding of the needs, uses and limitations of materials used in the human body and to explain the biomechanical principles and biological factors involved in achieving the long-term stability of replacement parts in the body. This book examines the industrial, governmental and ethical factors involved in the use of artificial materials in humans and discusses the principles and applications of engineering of tissues to replace body parts. The approach necessarily incorporates a wide range of reference material because of the complex multidisciplinary nature of the fields of biomedical materials, biomechanics, artificial organs and tissue engineering. There is an accompanying CD-ROM providing supplementary information and illustrations to support the book.

Surfaces and interfaces for biomaterials

(ISBN 978-1-85573-930-7)

This book presents our current level of understanding on the nature of a biomaterial surface, the adaptive response of the biomatrix to that surface, techniques used to modify biocompatibility, and state-of-the-art characterisation techniques to follow the interfacial events at that surface.

Shape memory alloys for biomedical applications

(ISBN 978-1-84569-344-2)

Shape memory metals are suitable for a wide range of biomedical devices including applications in dentistry, bone repair, urology and cardiology. This book provides a thorough review of shape memory metals and devices for medical applications. The first part of the book discusses the materials, primarily Ti–Ni based alloys; chapters cover the mechanical properties, thermodynamics, composition, fabrication of parts, chemical reactivity, surface modification and biocompatibility. Medical and dental devices using shape memory metals are reviewed in the following section; chapters cover stents, orthodontic devices and endodontic instruments. Finally, future developments in this area are discussed including alternatives to Ti–Ni based shape memory alloys.

Details of these books and a complete list of titles from Woodhead Publishing can be obtained by:

- visiting our web site at www.woodheadpublishing.com
- contacting Customer Services (e-mail: sales@woodheadpublishing.com; fax: +44 (0) 1223 832819; tel.: +44 (0) 1223 499140 ext. 130; address: Woodhead Publishing Limited, 80, High Street, Sawston, Cambridge CB22 3HJ, UK)
- in North America, contacting our US office (e-mail: usmarketing@woodheadpublishing.com; tel.: (215) 928 9112; address: Woodhead Publishing, 1518 Walnut Street, Suite 1100, Philadelphia, PA 19102–3406, USA

If you would like e-versions of our content, please visit our online platform: www.woodheadpublishingonline.com. Please recommend it to your librarian so that everyone in your institution can benefit from the wealth of content on the site.

Woodhead Publishing Series in Electronic and Optical Materials:
Number 27

Laser growth and processing of photonic devices

Edited by
Nikolaos A. Vainos

WOODHEAD
PUBLISHING

Oxford   Cambridge   Philadelphia   New Delhi

Published by Woodhead Publishing Limited,
80 High Street, Sawston, Cambridge CB22 3HJ, UK
www.woodheadpublishing.com
www.woodheadpublishingonline.com

Woodhead Publishing, 1518 Walnut Street, Suite 1100, Philadelphia,
PA 19102–3406, USA

Woodhead Publishing India Private Limited, G-2, Vardaan House, 7/28 Ansari Road,
Daryaganj, New Delhi – 110002, India
www.woodheadpublishingindia.com

First published 2012, Woodhead Publishing Limited
© Woodhead Publishing Limited, 2012
The authors have asserted their moral rights.

British Library Cataloguing in Publication Data
A catalogue record for this book is available from the British Library.

Library of Congress Control Number: 2012939793

ISBN 978-1-84569-936-9 (print)
ISBN 978-0-85709-622-7 (online)
ISSN 2050-1501 Woodhead Publishing Series in Electronic and Optical Materials (print)
ISSN 2050-151X Woodhead Publishing Series in Electronic and Optical Materials (online)

The publisher's policy is to use permanent paper from mills that operate a sustainable
forestry policy, and which has been manufactured from pulp which is processed using
acid-free and elemental chlorine-free practices. Furthermore, the publisher ensures that
the text paper and cover board used have met acceptable environmental accreditation
standards.

Typeset by Newgen Publishing and Data Services, India

Contents

v

x   Contents

Contributor contact details

(* = main contact)

Editor and Chapter 1

Nikolaos A. Vainos
Department of Materials Science
University of Patras
Patras 26500
Greece

and

Theoretical and Physical Chemistry
Institute -T.P.C.I.
National Hellenic Research
Foundation-N.H.R.F
48 Vas. Constantinou Ave.
11635, Athens
Greece

E-mail: vainos@upatras.gr

Chapter 2

Robert W. Eason,* Timothy C.
  May-Smith, Katherine Sloyan,
  Rossana Gazia, Mark Darby and
  Alberto Sposito
Optoelectronics Research Centre
University of Southampton
Highfield
Southampton
SO17 1BJ
UK

E-mail: rwe@orc.soton.ac.uk

Chapter 3

Artur Medvid'
Riga Technical University
Faculty of Material Science and
  Applied Chemistry
14, Azenes Str.,
Riga, LV 1048
Latvia

and

National Academy of Science of
  Ukraine
Institute of Semiconductor Physics
41, Pr. Nauki
Kyiv, 03028
Ukraine

E-mail: medvids@latnet.lv

Chapter 4

Quan-Zhong Zhao
Shanghai Institute of Optics and
  Fine Mechanics
Chinese Academy of Sciences
390 Qinghe Road
Jiading District
Shanghai 201800
China

E-mail: zqz@siom.ac.cn

Chapter 5

John T. Fourkas
Department of Chemistry &
    Biochemistry
Institute for Physical Science and
    Technology
Center for Nanophysics and
    Advanced Materials
and Maryland NanoCenter
University of Maryland
College Park, MD 20742
USA

E-mail: fourkas@umd.edu

Chapter 6

Volker Schmidt
Joanneum Research
    Forschungsgesellschaft mbH
MATERIALS – Institute for
    Surface Technologies and
    Photonics
Franz-Pichler-Straße
30 A-8160 Weiz
Austria

E-mail: volker.schmidt@joanneum.
    at

Chapter 7

Loukas Athanasekos* and Nikolaos
    A. Vainos
Department of Materials Science
University of Patras
Patras 26504
Greece

and

Theoretical and Physical Chemistry
    Institute -T.P.C.I.
National Hellenic Research
    Foundation-N.H.R.F
48 Vas. Constantinou Ave.
11635, Athens
Greece

E-mail: athanasekos@eie.gr;
    vainos@upatras.gr

Stergios Pispas
Theoretical and Physical Chemistry
    Institute -T.P.C.I.
National Hellenic Research
    Foundation-N.H.R.F.
48 Vas. Constantinou Ave.
11635, Athens
Greece

Chapter 8

Changhai H. Wang
School of Engineering & Physical
    Sciences
Heriot-Watt University
Edinburgh
EH14 4AS
UK

E-mail: c.wang@hw.ac.uk

Chapter 9

John Canning*
Interdisciplinary Photonics
    Laboratories (iPL)
222 Madsen Building F09
University of Sydney
NSW 2006
Australia

E-mail: john.canning@sydney.edu.au

Somnath Bandyopadhyay
Fibre Optics Laboratory
Central Glass and Ceramic
    Research Institute (CGCRI)
Council of Scientific & Industrial
    Research
Kolkata – 700032
India

Chapter 10

M. Ams,* D. J. Little and
    M. J. Withford
Centre for Ultrahigh bandwidth
    Devices for Optical Systems
    (CUDOS)
MQ Photonics Research Centre
Department of Physics &
    Astronomy
Macquarie University
NSW 2109
Australia

E-mail: martin.ams@mq.edu.
    au; douglas.little@mq.edu.au;
    michael.withford@mq.edu.au

Chapter 11

Shane M. Eaton* and Roberto
    Osellame
Istituto di Fotonica e
    Nanotecnologie (IFN)-CNR
Politecnico di Milano
Piazza Leonardo da Vinci 32
20133 Milan
Italy

E-mail: shane.eaton@gmail.com

Giulio Cerullo
Dipartimento di Fisica
Politecnico di Milano
Piazza Leonardo da Vinci 32
20133 Milan
Italy

Chapter 12

Stavros Pissadakis
Foundation for Research and
    Technology – Hellas (FORTH)
Institute of Electronic Structure
    and Laser (IESL)
N. Plastira 100
Vassilika Vouton
Heraklion 70013
Greece

E-mail: pissas@iesl.forth.gr

Dedication

To my wife Ioanna and my daughter Aria
Inspiring and understanding my efforts

Woodhead Publishing Series in Electronic and Optical Materials

xvii

Preface

Laser-induced photorefractivity discovered and studied in the late 1970s was the prelude to emerging laser-based materials growth and device processing technologies in photonics. Tailoring the refractive index of optical media by intense short wavelength laser radiation paved the way for periodic Bragg structures used as selective reflectors, optical fibre filters and sensor devices, all available today as off-the-shelf components for industry and telecoms. These prime developments have been followed by a rich portfolio of science and technology, with thousands of scientific articles and a plethora of patents addressing materials fundamentals and technical solutions to long-standing industrial problems. Novel materials compositions and phases become manipulated by laser radiation effects and drive this now blooming technology to its nanoscale limits, by taking advantage of ultrashort, femtosecond, laser pulses and their extremely high peak power. Internal structure manipulation is achieved with extreme precision, enabling tailoring of photonic crystal fibres and other complex devices with novel functional properties.

Pulsed laser deposition by use of high power excimer lasers achieved transparent epitaxial films of ferroelectrics ($BaTiO_3$) and paraelectrics ($Bi_{12}GeO_{20}$) in the late 1980s and the start of the 1990s. Such success reinforced with optimism those first apprehensive steps and pointed to the capacity for epitaxial materials growth of complex compositions with high optical quality. The successful growth of waveguide lasers (Ti:Sapphire and Nd:GGG) became reality after a couple of years' effort. 'Lasers-make-Lasers' fabrication approaches revealed a unique potential for the development of new materials for photonics applications demanding extreme structural and functional qualities. It was not only the congruent nature of laser ablative deposition that enabled hard-to-achieve structures and advanced stoichiometries. Mastering the laser-growth techniques allowed in-growth structural manipulation, yielding active photonic nanocomposites and multi-layered structures with functionalities not available by other means.

Novel functional properties come about from miniaturization and new dramatic nanoscale effects unravel a palette of new functionalities for photonics. Along these lines, tailoring photonic interfaces by laser light would permit the evolution of yet unknown phenomena and light–matter interactions at the molecular level. Highly energetic photons 'hammer' the interphase neatly, with no contamination, building micro- and nano-structures in direct ablative or, alternatively, subtle reactive processing modes. Ultraviolet laser micro-etching and micro-printing approaches are now complemented by multiphoton lithographic and other photoreactive processing approaches for surface tailoring and photonic structure fabrication. Taking advantage of the materials' optical nonlinearities and the good beam quality, the processing resolution at sub-wavelength levels offers an unsurpassed potential for three-dimensional nano-structuring.

Laser growth and processing of photonic devices thus emerges as a new global technological approach addressing both materials and device concepts. This book aspires to provide a comprehensive outline of the field, an overview of fundamentals and a focus on the latest developments and future trends. It is within its scope to assist a deeper understanding of laser–matter interactions and related phenomena aiming to trigger and support new frontier research lines. Novel functional photonics structures and devices offer a horizontal technology platform targeting applications of multidisciplinary interest, from information processing and sensing to bio-photonics and energy harvesting. Even though overall not at the level of maturity and the compatibility required by microelectronics industrial foundries, I am confident that, with its unique qualities and penetrating character, this technology will find its way to industrial implementation and the marketplace.

The book comprises a balanced content of science fundamentals and technology. It highlights important aspects and currently open issues, also pointing to the emerging and future trends. The introductory chapter overviews the fundamental phenomena and the basic processes driving this emerging field. By providing some review and pioneering references it aims to guide the reader through the contents without exhausting the subject. Some more background details on laser ablative processing are given to cover the topic. The following three main parts are presented by eminent contributors in the field, respectively addressing materials and surface structuring, three-dimensional structures and materials structure tailoring. In Chapter 2, Eason *et al.* detail the latest developments in the field of pulsed laser deposition for photonics applications, presenting novel approaches and hardware which will make feasible materials composition and structural tuning. In Chapter 3, Medvid' reviews the fundamentals and the applications of nanocone formation on semiconductor interfaces by laser-induced self-assembly, pointing to a potential future sustainable technology in the microelectronics environment. In Chapter 4, Zhao discusses the deployment of femtosecond laser multiple-beam interference methods for the fabrication of photonic nanostructures, enabling a range of unique functions. In Chapter 5 Fourkas elaborates a detailed discussion on fundamentals and three-dimensional fabrication by nonlinear, multiphoton effects using ultrashort pulses, addressing lithographic and alternative photochemical methods, as well as novel materials for micro- and nanostructure fabrication. In Chapter 6 by Schmidt, the laser becomes a versatile lithographic and rapid prototyping tool for multi-dimensional photonics, giving practical components and devices of high potential and prospective impact in several areas. In Chapter 7, a novel approach of three-dimensional microstructuring by applying laser radiation gradient forces is introduced by Athanasekos *et al.*, clarifying its reversibility and its distinct differences to 3D lithographic processing. In Chapter 8, Wang discusses laser assembly of photonic devices, a range of versatile laser-based fabrication techniques, including microwelding and packaging, methods of direct industrial relevance. In Chapter 9, Canning and Bandyopadhyay introduce materials structural modifications for photonics, detailing thermal glass processing in the nanoscale with recent results unravelling a unique potential and fabrication flexibility. In Chapter 10, Ams *et al.* thoroughly address laser-induced refractive index manipulation, from the fundamentals to novel concepts and a complete range of photonic devices produced by these methods. In Chapter 11, Eaton *et al.* present in detail the thermal writing methods for photonic device fabrication in glass and polymers and investigate the role of laser exposure parameters in refractive manipulation for waveguide formation, a topic of crucial practical importance. Finally in Chapter 12, Pissadakis reviews exhaustively laser processing of optical fibres, presenting thoroughly index engineering and microstructuring approaches and concluding with emerging topics relating to tailoring of photonic crystal fibres and applications.

I am hopeful that this book will become a useful resource for the academic and the industrial reader, an aid to their future research and endeavours.

The constant guidance, encouragement and support of Woodhead Publishing's editorial staff throughout this project have been invaluable. I am especially grateful to the Commissioning Editor Ms Laura Pugh, Senior Project Editor Ms Cathryn Freear and Publications Coordinators Ms Rachel Cox and Ms Anneka Hess. Their devotion to our targets and their patience with my, unavoidably, delayed responses has been remarkable. Many thanks also go to my colleagues Miltos Vasiliades, Loukas Athanasekos and Dimitris Alexandropoulos, for their useful discussions, support and contributions in the preparation of this volume.

Finally, I would like to express my deepest and sincerest gratitude to all authors of this book. Without their authority and eminent contributions this work would not have been realised.

Nikolaos A. Vainos

1

Laser growth and processing of photonic structures: an overview of fundamentals, interaction phenomena and operations

N. A. VAINOS, University of Patras, Greece and National
Hellenic Research Foundation, Greece

Abstract: Laser materials processing meets some innovative applications
for advanced photonic technologies. The use of lasers in the fabrication
of unique devices unavailable by other means opens up a field rich of
science and engineering and promises many benefits in the years to come.
This chapter illustrates the fundamental effects and the concepts behind
the applications addressed in this book. It overviews the phenomena
and the underlying mechanisms and sets the basis for understanding the
advanced topics discussed in the following chapters. It further continues
with a more detailed presentation of laser ablation methods in materials
growth and processing. This part concludes summarizing the trends and
prospects of emerging laser-based technologies in the fabrication of
photonics devices.

Key words: laser–matter interactions, laser materials processing, photonic
device processing, pulsed laser deposition of photonic materials, laser-
microfabrication.

1.1 Laser processing concepts and processes: an introduction

Laser radiation emerged in the 1960s as a very promising alternative tool
in materials science and technology (Beesley, 1978). Novel sources of
coherent radiation had the ability to provide a large amount of directional
energy which could be handled and manipulated in free space with remarkable convenience using optical systems. Significant advantages are drawn
from the high intensity, directionality and wavelength specificity of radiation. Radiation energy can thus be efficiently coupled into the workpiece
and results in spatial localization of laser–matter interactions, enabling the
achievement of superior quality processing results and providing novel
means for advanced materials processing and engineering (Ion, 2005).

Primarily addressing cutting and welding in the engineering workshop,
this technology progressed steadily to several industrial applications, from
the heavy shipbuilding industry to the niche micro-engineering foundry

1

(Steen and Mazumder, 2010). The growth of photonic materials, the fabrication of novel optical systems and the tailoring of complex waveguiding circuits are some first steps addressed in this book aiming at the realization of the future miniaturized three-dimensional (3D) photonics and hybrid multifunctional nanosystems.

Initial applications of lasers in materials processing have mainly relied on thermal effects produced by the use of high power carbon dioxide (CO_2) lasers. This most energy efficient laser source emits 10.6 μm radiation in the long wave infrared (IR) region. Its ability to deliver kilowatt power beams in both the continuous wave (CW) and pulsed emission modes, at considerably low capital and running costs, makes it a favoured choice for materials processing. Depending on the optical properties of materials, energy is effectively transferred from the laser beam to the workpiece and it is coupled into a rather small volume near the surface, resulting in the rapid localized increase of temperature in the interaction region. Thermal properties take up and become responsible for the evolution of consequent effects. In the 1970s materials cutting, drilling and welding, followed by surface processing, such as thermal annealing and alloying, were all successfully adopted by the industry (Cline and Anthony, 1977).

Laser material interactions are usually initiated by focused laser radiation. Depending on the rate of energy deposition and the energy out-diffusion from the interaction region to the bulk, a number of associated effects, from phase changing and melting to evaporation and ablation, are possible. The properties of the laser radiation and the nature of the material determine the highly interdependent physical and chemical effects which yield processing of the material (Bäuerle, 2000).

Thermal processing of a metallic sample irradiated by a laser beam is an excellent example to illustrate the range of operations:

(a) At low laser intensity heating at a suitably high temperature causes phase transformation and leads to the alteration of the mechanical, electrical, optical or other properties of the material.
(b) Increasing the energy deposition produces melting and consecutive re-solidification in a spatially localized region. This may result in phase transformation while more intense effects of crystallization or amorphization and vitrification can be produced depending on the conditions. In this mode, alloying and impurity doping by in-diffusion are also possible yielding considerable structural modifications and improving materials performance.
(c) Higher laser intensities can produce local vaporization of the material and mass removal from the interaction region under thermodynamic equilibrium. This process can be quite slow at relatively low intensities. The resulting heat diffusion yields an extended re-solidification and heat affected

zone. Materials vaporization and removal in this case becomes a distillation process yielding elemental separation and phase transformations.

(d) Using very high-intensity nanosecond pulses, the above vaporization process evolves rapidly under non-equilibrium conditions. The energy is deposited in such a very short time scale which effectively does not allow heat diffusion through the bulk. A small superheated volume is explosively vaporized. Violent materials ejection occurs usually associated with the production and evolution of plasma. This is the so-called pulsed laser ablation (PLA) process. It is associated with minimization of the heat-effected zone and the simultaneous congruent removal of the materials constituents, in a mixture of atoms and ions, molecules, clusters and micro-particulates.

Laser technology advanced very rapidly (Siegman, 1986). The advent of lasers emitting at shorter wavelengths offered novel potential for materials processing due to the high energy photons absorbed near the surface. In addition, the stronger focusing and higher resolution imaging at short wavelengths set the basis of the laser micro-processing technology. Original developments in the field concerned thermal processing by use of pulsed Nd:YAG lasers emitting from a few milli-joules to several joules per pulse at 1.06 μm in the near infrared (NIR). Such systems are currently deployed in industrial operations for laser welding, cutting and drilling achieving micron scale accuracy appropriate for the micro-engineering and micro-electronics industries. Solid state laser sources emitting higher harmonics, as for example 2ω:532 nm, 3ω:355 nm and 4ω:266 nm of the Nd:YAG, as well as metal-vapour high power lasers, offered new tools based on visible and ultraviolet (UV) wavelengths, improved the accuracy and enhanced the coupling of radiation with superior results.

Significant advances are recorded with the availability and deployment of excimer lasers emitting mid and deep-UV radiation in nanosecond duration pulses (Laude, 1994). The intense beams of highly energetic photons produced by the metastable dimmers (emitting at) XeCl (308 nm), KrF (248 nm), ArF (193 nm) and more recently, F_2 (157 nm) lasers, are responsible for a number of distinct processing operations unavailable by other means. First, the absorption coefficient at these short wavelengths is large for most materials and this reduces significantly the penetration depth (defined to $1/e^2$ intensity point) of the incident radiation. The result is the accumulation of energy in a very shallow interaction volume near the irradiated surface of the sample. In effect, a considerable amount of energy is deposited typically in a few nanoseconds in a very limited material volume. The material becomes superheated and ablated. Second, the photon energy may be capable of direct molecular photo-dissociation, or can trigger violent photochemical reactions and photomechanical effects. Overall such effects yield superior processing quality with minimal thermally affected zones. Third, materials ablation is a congruent process as it produces highly energetic multi-component plasma owing to the minimization of the inter-action region, heat transport and loss through the bulk. The deposition of the ablated material on solid substrates opened up the field of pulsed laser deposition (PLD) we address here (Chrisey and Hubler, 1994).

The availability of intense UV laser sources made possible a number of indirect non-destructive processes primarily relating to photo-polym-erization, which to date has proved of the utmost importance in micro-fabrication technologies (Jain, 1990). The operations take advantage of the narrow line-width spatially incoherent UV laser sources, to yield short exposure times and improved imaging and irradiation procedures. These features have enabled accurate processing methods in micro-fabrication production lines and they have been integrated as an efficient industrial standard technology for micro-engineering, micro-electronics and photon-ics. Further photophysical and reactive photochemical processing methods for surface treatment and laser-assisted vapour deposition by use of pre-cursor compounds for applications in micro-engineering, aerospace and other niche areas have been exhaustively investigated (Bäuerle, 2000). Direct ablative processing also attracts great interest as a viable alterna-tive to conventional multi-step lithographic methods. The method has been developed and applied as a single step processing tool for micro-optics and multifunction device fabrication exhibiting great flexibility and universal-ity as discussed in this work.

In recent years ultra-short laser pulses became available, with the Ti:sapphire laser system attracting major interest, due to its capability to emit sub-picosecond pulses, from about 200 femtoseconds, to less than 50 fs at 800 nm. Pulse energies range from the nJ/pulse produced by the high repetition rate (100 MHz) oscillators, to mJ/pulse low repetition rate main frame amplifier systems yielding from GW to TW range peak intensities. In this range the absorption of materials reaches the nonlinear regime. Materials transparent at the irradiating wavelength are becoming opaque due to nonlinear absorption with severe consequences in their response.

Ultra-short pulsed laser radiation defines the current research trends in laser materials processing and fabrication (Haglund, 2006). Although sev-eral well-established operations employing IR and UV lasers represent today's industrial standards, multiphoton processing offers new potential and unique advantages. Multiphoton absorption reflects on further localiza-tion of the deposited energy in the high-intensity region of the beam. This is also assisted by the high quality Gaussian beams produced, which allow efficient beam control and delivery to the target workpiece. High quality direct ablative and/or photophysical and photochemical reactive processing are obtained.

This chapter provides an overview of the fundamental concepts and laser interaction effects, emphasizing mechanisms and parameters in the context of materials growth and processing. It aims to establish the appropriate background and guide the reader through the main body of this book. Acquaintance with the underlying basic physics and optics is assumed. Emphasis is placed on the physical notions of the various processes, rather than a complete mathematical treatment of the various quite complex topics, most of which are subjects of current research.

The growth and processing of photonic materials adds a new dimension to the traditional laser materials processing technology, owing to further stringent requirements imposed by the optical quality and the special properties of the final product. Pointing to the important aspects and parameters involved, the second section of this chapter summarizes the concepts behind the real processing systems and includes aspects of laser beam propagation and radiation coupling. A more detailed overview of the fundamental processes is found in section three, commencing from the absorption of radiation and elaborating the energy coupling and macroscopic processing effects. The fourth section aims to introduce the most important materials processing operations for photonics, focusing on cases of ablative operations and device fabrication. The final section concludes the discussion by summarizing the trends of this emerging technology and gives the floor to the eminent contributors of this book.

1.2 Laser radiation, propagation and delivery

1.2.1 Laser radiation, properties and sources

Laser radiation today covers an extended section of the electromagnetic spectrum, from the deep-UV to the far-infrared (FIR), while free electron systems extend to the X-ray region. CW and pulsed laser systems are available, delivering from a few mW to several kW optical power and laser pulses of millisecond to femtosecond duration at pulse energy levels ranging from nJ to several Joules for industrial systems, or kJ for systems developed for fusion and other high energy applications.

The principal properties of laser radiation, as contrasted to incoherent radiation in the same spectral regions, are briefly outlined here:

(a) The coherence of laser radiation is the result of the stimulated emission process in the atomic or molecular system. It relates to the narrow spectral bandwidth into which all available energy is channelled, thus achieving a very high spectral density. In addition to the tremendous properties of interference and diffraction, this property allows a wide range of applications including spectral pumping and tuning, efficient nonlinear propagation, selective absorption and activation and others.

(b) The directionality of the radiation by the formation of Gaussian beams is the result of the optical resonance necessary to achieve and sustain laser oscillation. Laser beams have increased spatial coherence and exhibit minimal deterioration due to diffraction, thus enabling efficient energy transmission and delivery on the target under processing.

(c) Both the above aspects are responsible for the celebrated high intensity of the laser beam, which together with the spatial and temporal coherence, as well as the polarization control, are crucial in materials processing operations.

A limited number of laser types have been proved suitable for and are efficiently deployed in materials processing. While laser radiation covers a very broad spectrum, from the X-ray or deep-UV to the FIR region, a number of significant appropriate parameters and requirements must be fulfilled, including:

(i)   *Wavelength of radiation, propagation in free space and absorption by the workpiece under processing.* Selection of specific wavelength is important in many cases. It determines the beam delivery technology and the nature of operations.

(ii)   *Emission at CW and/or pulsed operation.* It determines the nature and range of operations, as well as the technical and financial aspects of the deployed technology.

(iii)   *Efficiency of operation in technical and economic terms and power efficiency in reference to electrical-to-optical power conversion.* It affects the operational costs and the market value of the final products.

(iv)   *Reliability of operation, technology complexity and automation.* They are very important in the industrial floor.

(v)   *Acquisition and operational costs, production efficiency and final product price.* They affect primarily the industrial production and also research operations.

The selection of laser types currently available and used in the materials processing operations includes:

(a) Carbon dioxide laser: CO_2 laser emitting at 10.6 μm from the molecular transitions of carbon dioxide. Highly efficient (~ 30%) system offering both pulsed and CW operations from mW to kW CW output power levels and pulses of mJ to kJ energy. Systems widely deployed in industrial production, from heavy industry to microelectronics.

(b) Excimer lasers produce UV radiation by the metastable excited molecular dimers ArF (193 nm), KrF (248 nm), XeCl (308 nm) and the more recently developed F_2 (157 nm). They only emit pulsed operation and are pumped by high current discharges or electron beams. Their efficiency varies depending on their type. They deliver mJ to Joule pulses, of a few nanoseconds' duration. Special hybrid dye-excimer systems offer sub-picosecond radiation. Excimer lasers offer highly energetic photons and enable unique operations, even though their beam coherence and spatial quality are reduced and limited by the electric discharge effects.

(c) Nd:YAG, Nd:Glass and related rare earth solid sate laser systems. Nd-based solid state sources have been and are widely used owing to their high efficiency operation at the NIR and the potential of visible and UV emission by use of frequency doubling and optical parametric oscillator systems. By these means they cover the range from the UV to the mid-IR, although at varying laser efficiencies. In addition to the fundamental frequency ω: 1064 nm, significant wavelengths deployed are 2ω: 532 nm, 3ω: 355 nm and 4ω: 266 nm. The use of various laser hosts and ions give new possibilities with prime candidates the Er^+ (1.5–1.6 μm) and the Ho^+ (2.1 μm) laser systems. Solid state technology has become widespread and enables highly efficient operations by the use of spectrally selective pumping by diode lasers. Systems deployed in CW at power levels of less than 1 mW to kW and pulsed modes emitting from the quasi CW to picoseconds pulses of nJ to tens of J energy are well established in industry. The wide range of free space systems is now enriched by the production of all-fibre lasers which deliver picosecond and femtosecond pulse trains in the NIR.

(d) Semiconductor diode lasers are deployed owing to the achievement of considerably high power levels delivered by laser diodes arrays. They operate in both pulsed and CW modes, achieving power levels approaching the 1 kW in the NIR range from ~ 800 nm to ~ 1600 nm. They are small, compact and reliable sources of very robust construction and in most cases offer fibre delivery alleviating the hazards of invisible radiation.

(e) Ti-Sapphire, Ti:Al$_2$O$_3$ laser system emitting broadband radiation at ~ 800 nm. This is one of the latest solid state laser developments which operate in the CW and pulsed modes. The broad gain bandwidth allows efficient mode locked operation producing 150 fs pulse trains at ~ 100 MHz repetition rates and above. By use of pulse compression and amplification techniques, these lasers offer currently sub-50 fs pulses of mJ energy, yielding extremely high peak power levels and several TW/cm^2 intensities on target. These systems offer an extended wavelength selection by use of optical parametric oscillators (UV to NIR). Their use defines the current trends in several fields.

1.2.2 Laser beam formation and propagation

Optical feedback is fundamental in the operation of the laser oscillator. The optical cavity containing the active laser medium redirects the photons needed for stimulating the emission and thus determines the formation of the output laser beam. The feedback operation sets the boundaries for the propagation of the otherwise plane wave. It limits the spatial and temporal characteristics of the oscillating field by requiring a self-consistent regeneration of the optical field. The latter established in the resonator yields the specific spatial and spectral content of the emitted radiation.

The high intensity of the laser beam is the result of two effects. First, the stimulated emission is by nature a coherent process yielding high intensity and spectral purity. Second, the paraxial localization of propagating energy, with the formation of Gaussian beams in the laser cavity, limits the cross-sectional area and results in high intensities. Coherent laser radiation thus exhibits a narrow spectrum which can be tuned, filtered, modulated and generally manipulated in space and time.

Central to the laser oscillation is the formation of the fundamental Gaussian beam in the laser resonator (Yariv, 1986). It is the field mode determined by wave propagation under the slowly varying envelope approximation and represents the diffraction limited case. The beam reproduces itself in the resonator self-consistently and 'survives' after infinite round trips.

Figure 1.1 presents the typical geometry of the time-independent Gaussian beam propagating along the z-axis with wave vector, k, given by equation 1.1 as a function of various significant parameters defined in Table 1.1. The beam extends laterally to infinity. The spatial limits shown in Fig. 1.1 represent the envelope to $1/e$ points of the maximum electric field at z-position, or equivalently to $1/e^2$ intensity points, with respect to the on axis, $z = 0$, maximum value. The hyperboloid bounds physically the major part of the power transmitted along the z-direction.

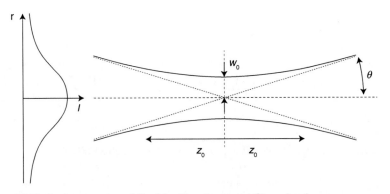

1.1 Typical geometry of the ideal fundamental Gaussian beam.

Table 1.1 Notation and physical content of the fundamental Gaussian beam and related parameters

1	w_0	Beam waist; the radius of the beam spot at focus; a measure of the diffraction limit.
2	$w(z)$	Spot size; radius of the spot at the specified position z of the propagation axis.
3	$q(z)$	Complex radius of the beam at position z.
4	$R(z)$	Real radius of curvature of the wavefront at position z.
5	$\eta(z)$	Phase parameter describing the on axis departure of the wavefront from the phase of the planar wave.
6	z_0	The distance from focus ($z = 0$) at which the cross-sectional area of the beam is doubled. Also referred as the Rayleigh length and defines the confocal parameter or depth-of-focus (DOF) of the beam, $b = 2z_0$, the region in which the beam is considered to remain in focus.
7	θ_0	Divergence of the Gaussian beam; defines the diffraction limit and the speed of intensity decrease along propagation.
8	θ_M	Actual beam divergence.
9	M^2	M^2-factor; determines the departure from the ideal Gaussian form; a measure of beam quality.

Eq. [1.1] is an analytical expression of the fundamental Gaussian field:

$$E(x,y,z) = E_0 \frac{w_0}{w(z)} \exp\left\{-i\left[kz - \eta(z)\right] - i\frac{kr^2}{2q(z)}\right\}$$

$$= E_0 \frac{w_0}{w(z)} \exp\left[-i\left(kz - \eta(z)\right) - r^2\left(\frac{1}{w^2(z)} + \frac{ik}{2R(z)}\right)\right]. \qquad [1.1]$$

where the various significant parameters are defined in Eq. [1.2] and their significance is outlined in Table 1.1:

$$w^2(z) = w_0^2\left[1 + \left(\frac{\lambda z}{n\pi w_0^2}\right)^2\right] = w_0^2\left(1 + \frac{z^2}{z_0^2}\right)$$

$$R = z\left[1 + \left(\frac{n\pi w_0^2}{\lambda z}\right)^2\right] = z\left(1 + \frac{z_0^2}{z^2}\right)$$

$$\eta(z) = \tan^{-1}\left(\frac{\lambda z}{n\pi w_0^2}\right) = \tan^{-1}\left(\frac{z}{z_0}\right) \qquad [1.2]$$

$$z_0 \equiv \frac{n\pi w_0^2}{\lambda}$$

$$\theta_0 = \tan^{-1}\left(\frac{\lambda}{n\pi w_0}\right)$$

The spatial extent, $w(z)$, of the beam determines the area of the interaction region, which is irradiated by the beam and it is a very important parameter as it defines the intensity, the power and the energy deposited on the target workpiece under processing.

Bearing in mind that in many processing operations focused radiation is used, the spot size at beam waist, w_0, becomes of extreme importance and, effectively, determines the maximum intensity on target and the minimum interaction region. In turn, they reflect on the overall 'intensity profile' available to the processing operation which affects the results.

One important aspect of the Gaussian beam concerns its propagation properties. The Gaussian function transmitted in free space is responsible for the spatial invariability of its form. Diffraction in the far field (Fraunhofer region) is determined by Fourier transformations and preserve the Gaussian shape of the beam. In addition, transmission of the Gaussian field through an ideal lens becomes equivalent to the far-field diffraction and preserves the beam shape. It is important to underline here again the significance of the spot size, w_0. The diffraction limited beam divergence, θ_0, is inversely proportional to the 'tightness' of focus. A strongly focusing beam rapidly converges towards the focal point and then diverges fast on its propagation. The focusing condition determines the interaction volume and the intensity and energy content, which are the most crucial parameters for the processing operations. On the other hand, a weakly focused Gaussian beam having a small divergence is the most appropriate for radiation transmission and delivery, since it results in minimum energy loss through the system, while preserving the main beam characteristics, even in the presence of optical imperfections. Very important to the focusing and propagation properties is the confocal parameter of the beam. It is also referred to as the Rayleigh length, z_0, and describes the depth-of-focus (DOF) of the beam. It is the $\pm z$-point where spot size increases by $2^{1/2}$ with respect to the waist, or equivalently, the spot area becomes doubled, thus representing the best focus region.

In the above context, the importance of a beam quality parameter M^2-factor becomes apparent. A real laser beam usually departs from the ideal diffraction limited case, with the consequence of a larger beam divergence and confocal parameter (depth of focus) for a given spot size. The beam quality parameter, M^2-factor, is defined by Eq. [1.3] in terms of actual beam divergence with respect to the ideal Gaussian as:

$$\theta_M = M^2 \frac{\lambda}{n \pi w_0} \tag{1.3}$$

In addition to the fundamental Gaussian beam discussed above, higher order solutions-modes of the wave equation may exist under certain boundary values imposed by the resonator optics.

The larger spatial extent is also expressed in terms of the M^2-factor. The possible coupling and power between modes, however, should also be mentioned and relates to the delivered laser beam quality on target. This is a crucial parameter affecting directly the final processing quality.

1.2.3   Laser beam delivery and radiation coupling

Laser power is delivered on target either directing the beam in free space using reflecting and focusing elements, such as lenses and mirrors, or by use of fibre optics including solid, hollow or liquid filled lightguides.

The Fourier transformations preserve the Gaussian form in free space propagation and delivery. In effect, natural diffraction transforms the Gaussian beam to itself, by effectively changing the pair (w_0, θ_0), thus preserving the power but amending appropriately the intensity profile along the path. The optical system may take the form of a free space waveguide, formed by use of a series of focusing elements. This implies maximum transmission efficiency, experimentally approaching the diffraction limited performance. Various practical experimental geometries have been developed in the form of table-top optical systems, flat-bed and scanning work-stations, as well as robotic articulated-arm delivery systems.

The use of fibre optics and lightguiding offers significant advantages and convenience of operations covering the range from the deep-UV (> 200 nm) to the extended NIR (3 μm). Speciality silica fibres are used to transmit radiation of wavelength below 250 nm and non-oxide, chalcogenide glass (Ga-La-S, Ag-Se) waveguides transmit in the far-IR wavelengths above 5 μm. In cases requiring high power transmission, hollow and liquid filled light pipes are used, however, at decreased beam quality.

In materials processing operations, laser radiation must be optically coupled to the workpiece via its optical interphase, as depicted in Fig. 1.2. Figure 1.2a presents a typical laser beam, 1, incident on the target, 2. Irrespective of the specifics, the incident field may be approximated by a plane wave of wavelength, λ, and electric field, E, where:

$$E = E_0 \exp i(kr - \omega t) \qquad [1.4]$$

Part of the incident radiation is reflected in a specular mode, 3, and a part, 4, is transmitted and absorbed in the material (Born and Wolf, 1989).

In effect the material exhibits a complex refractive index:

$$n_c = n(1 + i\kappa) \qquad [1.5]$$

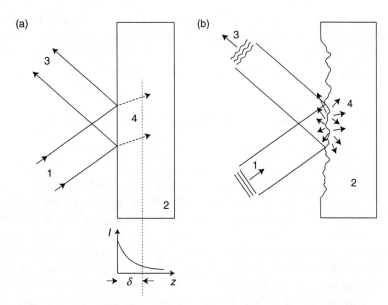

1.2 (a) Specular reflection and transmission geometry by an ideal planar interphase with inset absorption graph and (b) reflection and diffraction by a real interphase exhibiting a finite roughness (drawn exaggerated for clarity). Incident beam (1), sample (2), reflected beam (3), transmitted / scattered and absorbed beams (4).

where n is the real part and κ is the imaginary part describing energy loss. For a planar interface the reflection coefficient for normal incidence is given by:

$$R = \frac{n^2\left(1+\kappa^2\right)+1-2n}{n^2\left(1+\kappa^2\right)+1+2n}$$

[1.6]

Practically the beam transmitted in the z-axis is described by:

$$E = E_0 \exp\left(-n\kappa k z\right)\exp\left[i\left(nkz-\omega t\right)\right]$$

[1.7]

and undergoes absorption. The absorbed fraction of the incident power is used for processing the irradiated material. In the linear absorption regime the absorption coefficient, α, and penetration depth, δ (illustrated in Fig. 1.2), in a first approximation independent of the intensity, given by:

$$\alpha = \frac{2\pi n\kappa}{\lambda} = \delta^{-1}$$

[1.8]

describe the exponential decay of the field along propagation into the bulk workpiece. These parameters are of great significance since, in conjunction with the spatial extent of the irradiating beam, they determine the laser–matter interaction volume.

Depending on the absorption coefficient, α, at the particular wavelength, penetration depth for the deep UV may be in the nanometre scale for many materials and certainly for metals. Figure 1.2b illustrates a more realistic situation. A laser beam approximated by the plane wave, 1, is incident on a nearly planar but textured sample, 2, having a finite roughness as met in most cases. A specularly reflected beam 3 is observed, but the inclusion of surface imperfections modifies significantly the problem. A scattered beam, 4, is observed at the interface and energy is redistributed with a part in free space and another part coupled into the material bulk.

More specifically:

(a) Irrespective of the nature of the material, the workpiece is described by a 3D interphase profile function of the form $n_c(x, y, z)$, which can be analysed in terms of spatial frequencies.
(b) The incident field is diffracted by this interphase in both semi-spaces breaking up into components determined by the particular geometrical and physical characteristics of the problem. The presence of suitable spatial frequencies in the sub-wavelength regime may reduce severely the reflectivity and can effectively produce opacity by presenting an artificial refractive index to the incident wave. Radiation may thus be coupled into the sample and become absorbed, irrespective of the prime natural reflection properties of the material. A totally different absorption response is therefore attained by the otherwise assumed 'planar' but textured interphase.
(c) The existence of localized imperfections produces scattering which leads to local modifications of the field. The ideal planar wave incident on the workpiece produce large local field intensities which can affect the material significantly. Multi-pulse irradiation produces localized damage in the material even at relatively low average intensity. Subsequent pulses may thus find totally different surface properties, which in nominal terms respond quite unexpectedly.

Further to the above, the use of ultra-short pulse high-intensity radiation yields nonlinear refraction and absorption effects. In practice this may be equivalent to the absorption at $\Omega = n\omega$, but the relevant refractive index and absorption coefficients are functions of radiation intensity and the overall absorption along the propagation may be described by:

$$-\frac{dI}{dz} = \alpha I + \beta I^2 + \mu I^3 + \cdots + \xi I^n \qquad [1.9]$$

where the first is the linear term and the second and following terms with nonlinear absorption coefficients $\beta, \mu, ..., \xi$, respectively concern the possible 2-, 3-, ... n-photon processes participating in the interaction with decreasing probabilities of occurrence.

Overall, the delivery of laser radiation and the coupling to the target is a complex problem which traditionally has been overlooked. Experimental evidence shows that the quality of the laser processing results depends primarily on surface properties which affect severely the radiation coupling. This may become of prime importance when working in the photonics miniaturization domain close to diffraction limits, where processing quality and accuracy at the microscopic level are important issues.

1.3   Summary of the interactions of laser radiation with condensed matter

The interactions of laser radiation with solid materials are primarily determined by the nature of the materials and the properties of the laser radiation, which define the parameter space of the individual problem. The high precision required in photonics and micro-engineering usually necessitates specific measures and advanced technological solutions. The requirements for each operation may differ significantly. Even though the basic technology remains the same, the high accuracy needed required our understanding of the fundamentals, in order to master the processing operations.

1.3.1 An outline of fundamental effects

The materials processing operations and observed phenomena are closely associated with a range of fundamental interactions:

(i) *Electronic excitation and de-excitation upon radiation absorption:* Transitions between bands by use of a single photon or many photons (multiphoton), associated with radiation emission (fluorescence, luminescence) and also non-radiative energy transfer via coupling to the lattice which may cause significant structural modifications.

(ii) *Photo-ionization:* Removal of electrons to the vacuum state usually followed by avalanche ionization, which leads to materials breakdown, triggering ablation and plasma formation.

(iii) *Molecular photo-dissociation:* Absorption and excitation to repulsive states leads to the dissociation of molecular bonds and disintegration of the material.

(iv) *Photochemical reactions:* Activation of chemical reactions, synthesis of new molecules or manipulation effects.

(v) *Laser-plasma effects:* formation, plasma post-ionization, resonant effects and superheating.
(vi) *Other atomic and molecular interactions:* radiation-induced forces, optical trapping and atomic cooling.

1.3.2 Range of macroscopic phenomena

Macroscopic phenomena observed are directly or indirectly associated with the above fundamental processes. They are supportive or detrimental to the materials processing operations. More specifically:

(a) Thermal effects are of prime interest and their participation cannot be excluded in any case. They range from the initial phase transformation to melting and re-solidification, evaporation and pyrolysis, processes evolving under thermodynamic equilibrium.
(b) Photomechanical effects relate to the direct or indirect application of mechanical forces. At low light intensities they concern radiation pressure and gradient field forces. In addition, thermomechanical stress effects, acoustic waves and shockwaves can be induced at high intensities, especially upon photo-ablation.
(c) Photochemical effects, including photo-dissociation and photolysis, photo-catalysis, photochemical reactions, photosynthesis and photo-polymerization.
(d) Photo-desorption and photo-ejection from surfaces include molecular, atomic, ionic and electronic ejection by absorption of radiation. This is termed non-explosive materials ablation as it is effectively removal of materials in the atomic/molecular level.
(e) Photo-ionization of atoms and ions, plasma formation and driving of atomic beams.
(f) Explosive photo-ablation is a composite explosive effect embracing most of the above phenomena, developing under non-equilibrium thermodynamic conditions. It is associated with violent ejection of plasma accompanied by neutral species such as molecules, clusters and particulates.

The properties of laser radiation especially relate to the laser intensity and energy density (fluence) on target. In addition, the spectral and polarization characteristics are of supreme importance, affecting both radiation coupling and the nature of the developing process itself.

The following section gives a thorough overview of the microscopic processes involved to explain their utilization in the present context of photonics technology.

1.4   Radiation absorption and energy transfer

1.4.1   Metallic absorption and free electron coupling

The free electron model (Drude, 1900) provides an excellent account for the absorption by metallic surfaces at relatively low photon energies of the long-wavelength range. Free electrons in the conduction band are coupled to the electric field of the optical wave, become accelerated and dissipate their energy. This represents the main source of absorption in the far-IR, mm-wave and μ-wave regions.

In a more generic approach, the classical dispersion Lorentzian model approximates well the complex dielectric constant:

$$\varepsilon_c = n_c^2 = 1 - \frac{4\pi Ne^2}{m}\frac{1}{\omega(\omega - i\gamma)} \tag{1.10}$$

with:

$$\varepsilon_{Re} = n^2\left(1 - \kappa^2\right) = 1 - \frac{\omega_c^2 + \gamma^2}{\omega^2 + \gamma^2}, \quad \varepsilon_{Im} = n^2\kappa = \frac{\gamma\left(\omega_c^2 + \gamma^2\right)}{2\omega\left(\omega^2 + \gamma^2\right)} \tag{1.11}$$

for γ the loss (friction) coefficient and ω_c the plasmon frequency, while such expressions can be generalized for multiple resonances (Roberts, 1955).

In real materials the electron mass should be replaced by the effective mass $m^* = \hbar/\partial^2 E/\partial k^2$ and the time constant $\tau = \gamma^{-1}$ is typically of the order of 10^{-15} to 10^{-13} s and describes the electron relaxation. Collisions with lattice imperfections, including phonons, impurities, vacancies and dislocations, are directly related to this prime cause of absorption and loss. At relatively low frequencies, below the plasma frequency and the relaxation rate ($\omega \ll \tau^{-1}$) the metallic behaviour is evident with $\alpha \sim (2\pi\sigma/\omega)^{1/2}$ and penetration depth α^{-1} diminishes at very high conductivities yielding unity reflectivity, as described by the Hagens-Rubens relation $R \sim 1 - 2\,(2\pi\omega/\sigma)^{1/2}$. However, for the optical frequencies above the plasmon frequency and the relaxation eigenfrequency $\omega \gg \tau^{-1}$, extinction decreases, $\kappa \ll 1$, manifesting a dielectric behaviour for the metallic material. In this regime, however, interband transitions and direct photo-ionization become profound and are the principal causes of absorption in the material. Electrons may be excited to the continuum and transfer their energy to the lattice non-radiatively, resulting in temperature increase. In fact, at room temperature free electrons may have velocities of the order of 10^6 ms^{-1} (Fermi level). Under low radiation intensities, electrons are accelerated under the action of the e/m field of the incident radiation. This is a small perturbation of the random motion and electron collisions act to randomize and restore the distribution effectively with a time constant τ_D. For metals the Fermi level is much larger that kT and in practice the available states are only found close to the Fermi level. Therefore, electrons gaining energy from the optical field tend to occupy these states and thus determine the value of relaxation, τ, for the total distribution. The relaxation time constant, τ_D, in this case refers to the previously defined rapid processes in the time scale of $\tau \sim 10^{-13}–10^{-15}$ s. This random redistribution of electrons having energy larger than kT contributes to the heat content of the material. There is, however, a significant difference between the randomization of electron momentum and the randomization of energy gained through interaction with the optical field, which is certainly slower, and it is represented by the energy relaxation time constant, τ_E. Considering a scattering cross-section σ and a density of scattering centres, N, the mean free path is $l_{free} = (\sigma N)^{-1}$. This parameter determines the efficiency of energy transfer to the lattice, providing energy and momentum conservation.

The increase of light intensity implies the application of a high electric field of several Volts/cm which strongly accelerates the electrons. 'Hot electrons' are thus produced that are moving fast under the influence of the e/m wave. The cloud develops its own high electron temperature T_e. This electron energy gained is exchanged with the lattice and leads to equilibrium described through this energy relaxation time constant, τ_E, which becomes an important parameter. With the exception of aggregates and interfaces in the solid, especially found in alloys and nanocomposites, impurity scattering is quite low at room temperature and phonons are the main scattering centres. The acoustic branches cannot be responsible for a large velocity change associated with energy transfer, since they are limited by the speed of sound in the solid. Therefore, the largest energy loss mechanism is through scattering by optical phonons where the loss per collision is of the order of $k\Theta_D$, where Θ_D is the Debye temperature. The inclusion of structural inhomogeneities results in dramatic energy loss. Overall such processes yield large increase of temperature eventually leading to thermal transformations, melting and evaporation, phenomena fundamental in materials processing. In the case of significant energy excess, beyond the levels of the unperturbed electron gas, the rapid increase of its energy content yields strong electron collisions capable of producing avalanche ionization and explosive ablation of the material with plasma formation.

The evolution of effects in various laser intensity regimes are considered and discussed in the following sections of this chapter.

1.4.2 Absorption and energy transfer in semiconductors and insulators

Semiconductors and insulators constitute another important category exhibiting individual behaviours. The absorption processes are presented

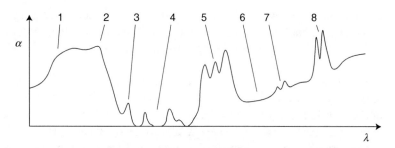

1.3 Absorption processes in a model semiconductor from the X-ray and extreme UV region (1) to beyond the FIR region. The fundamental absorption edge (2) embraces possible excitonic states (3) and is followed by impurity absorption states (4). The free electron IR absorption edge (6) with molecular coupling (5) and direct photon–phonon interactions (7), are followed by long-wavelength spin and cyclotron resonances (8), which are of minimal importance in our case.

schematically in Fig. 1.3 with the typical absorption curve describing the various processes.

The various regions and associated specific processes are as follows:

(i)   In the UV spectral range (1–2) strong absorption is caused by interband transitions from the valence to conduction bands. The fundamental absorption edge, 2, may be in the visible or near-IR for semiconductors. The absorption coefficient is typically above $\alpha \sim 10^5$ cm^{-1}. For larger photon energy in the X-ray region and beyond, this absorption decreases rapidly. The energy and momentum conservation, respectively, $\Delta E = E_f - E_i = \hbar\omega(\beta)$ and $k_f - k_i = \beta$ are valid. Towards the longer wavelengths, the edge 2 defines the limit of the band gap, $E_g = h\nu$, and yields a sharp decrease of absorption. Usually at low temperatures for bulk materials, this region comprises excitonic states, 3. Rayleigh and Mie scattering are always present and result in increased coupling effects.

(ii)   Further to the exciton states, 3, absorption is due to impurity centres in the gap, 4, which become ionized. Intraband absorption concerns also the case of free electrons in the conduction band or holes in the valence band. High intensities may yield reflectivity increase due to free electrons. In contrast to the metallic behaviour, this type of absorption depends on the electron density and is limited. Electron density in insulators is negligible. For photon energies below the energy gap, $h\nu < E_g$, absorption decreases with ω^2, defining the free electron absorption edge, 6, that extends to the lowest limit of the energy gap. Impurity absorption changes considerably the behaviour of the materials and has dramatic effects in the overall response. In the band gap where the material can in practice be fully transparent, ion centres or other imperfections lead to strong resonance absorption which depending on the matrix environment can present a quite broadened response. A linear absorption spectrum is thus observed. Characteristic examples here are the hydroxyl (O-H) bonds in silica, of special interest in optical fibres, which limit the NIR transmission, and the C-H bonds in polymers, with several chromophores producing dramatic absorption in the otherwise transparent materials.

(iii) In the far-IR region (> 20 μm) the interactions between photons and phonons are direct, in region 5. Photons interact with the optical modes due to dipole moment of the lattice and the comparable energy content (0.05–0.02 eV). At high frequencies the probability for the participation of multiphoton processes increases and absorption constants can approach 10^5 cm^{-1} in polar crystals. In covalent crystals, however, smaller values of 10–100 cm^{-1} are observed. Lattice imperfections again play a crucial role and lead to nonlinear coupling. Very low frequency effects, 7–8, such as cyclotron and spin resonances are not important in our case.

(iv) High laser intensities reveal a range of nonlinear processes in solids, owing to non-harmonic driving of the materials polarization as $P \sim \chi E + \chi^{(2)} E^2 + \chi^{(3)} E^3 + \cdots$. Further to the frequency generation processes, nonlinear Raman or Brilluin scattering induce strong radiation coupling to the solid via stimulated processes. Of great significance here is the nonlinear absorption process occurring at high laser intensities, in which the simultaneous presence of multiple photons results in excitation and ionization of atoms and molecules, even by use of low photon energies. The response is analogous to the nth-power of the intensity as, $\Phi \sim I^n$, where n is the process order defined as the number of participating photons.

In the above context, high-intensity laser radiation can be coupled directly or indirectly to the molecular or lattice vibrational modes depending on wavelength and release its large energy content locally in the material bulk. The large temperatures induced can drive pyrolytic reactions which can have a dissociative or a synthetic character. Such processes may be realized not only in solids, but also in the liquid and gas phase, while the involved processes may assume a chemical, a physical or a mixed character. Laser beams are ideal for delivering their energy on the spot and pyrolytic processing takes advantages of this effect. In addition, the thermal properties of the sample under processing must have appropriate values. In particular, thermal diffusivity is crucial for localized high temperature processing. In addition, resonant absorption aids further, by allowing selective and efficient processing.

Photolytic processes are also of particular interest. In this case highly energetic photons are used to activate photo-dissociative reactions which decompose the material by breaking molecular bonds. These operations make use of single or multiple photons of the UV, visible and up to the near-IR region. They may be realized under subtle thermodynamic equilibrium or take an explosive, photo-ablative character, depending on the irradiating intensity and materials nature. This process, however, can result in materials deposition by use of precursors in reactions of the type $MX + hv \rightarrow M^* + X$, where * denotes the possible presence of an excited state. Classic examples are the cases of metal-organic dissociation employed in Laser-assisted Chemical Vapour Deposition (Laser-CVD) by use of alkyl, alkalide, carbonyl and hydride precursors.

In molecular solids, transitions are realized between vibrational levels of the molecule (Frank-Condon principle) and high photon energy may lead directly to repulsive states and bond breaking. Depending on light intensity and nature of the material, in particular the state of impurity content, imperfections, water/analyte level of organic materials, a variety of processes may take place. Synthetic chemical reactions among molecular species are a major processing class, with photosynthesis of organic compounds found widespread in nature. Photo-polymerization is also a principal delegate. Further to conventional UV light and electron beam methods, laser lithography is also a well-established micro-fabrication process and it is widely deployed in micro-electronics and micro-engineering industries, due to the reliable process integration of deep-UV KrF (248 nm) and ArF (193 nm) laser sources. Current trends discussed in this work relate to multiphoton laser lithography which enables stereo-lithographic processing with submicron accuracy, as addressed in the following chapters.

A number of further processes concern molecular photo-adsorption and photo-desorption by relatively low intensity laser radiation (Georgiou et al., 1998). Laser radiation may resonantly drive atomic ensembles and produce large perturbation of their thermal distribution and thus retardation and cooling (Letokhov and Minogin, 1979). In addition the use of enhanced optical near fields in the vicinity of nanostructures such as metal tips leads to selective surface patterning (Jersch and Dikmann, 1996). Such processes are of fundamental interest in surface science leading to selective deposition in the atomic and molecular level with prospective impact in several fields, especially in forthcoming deposition and patterning micro-fabrication applications.

1.5 Materials processing phenomena: appraisal of energy dependencies

Laser processing operations commence upon the absorption of radiation and follow the conversion of energy. Absorption is described by the absorption coefficient, $\alpha(\omega)$, which is a strong function of materials nature and the frequency of light. The individual absorption processes and the transfer and conversion of the energy are characteristic of the specific materials class as analysed in the previous section. Nevertheless, the solid workpiece may differentiate its physical response to the incident radiation, depending on surface properties, structural defect and impurity content. All these factors can differentiate greatly the processes involved and the operations performed even by using the same substance.

To illustrate the dependencies on radiation power properties we refer to Fig. 1.4 presenting a schematic description of typical events evolving by increasing laser intensity. We include in our discussion both cases of CW and pulsed operation, in conjunction with appropriate energy transfer and diffusion issues, as follows:

(i)   At relatively low intensities $< 10^4$ W/cm^2 the energy absorbed can heat rapidly the material, as shown in Fig. 1.4a. Depending on energy deposition rate by beam 1 and heat transfer properties of the sample 2,

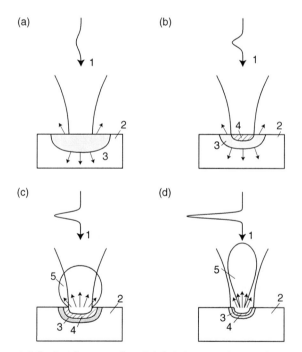

1.4 Scale of processing: (a) Subtle, non-destructive processing by low intensity radiation with physical and chemical effects yielding thermal, synthetic and polymerization operations, (b) higher laser intensity may lead to two zone transformations, (c) increasing intensity produces evaporation of material and may lead to photo-ablation process (d) with directional plasma formation and ejection of particles in a non-Maxwellian distribution.

including heat conductivity and geometry, an elevated temperature profile is established. A high enough temperature is able to produce initially phase changes and transform the material of region 3. Provided appropriately short wavelength laser radiation may induce photochemical reactions in region 3, processes well known in photochemical synthesis and photo-polymerization. Region 3, the heat (or radiation) affected zone (HAZ) of the workpiece, may exhibit upgraded or degraded physical and/or chemical properties (hardening, polymerization, etc.).

(ii)   Irradiating by a beam 1 of intensity above 10^5 W/cm^2, the material, 2, may melt under thermodynamic equilibrium, as shown in Fig. 1.4b. The energy deposition rate must exceed loss by thermal diffusion and the overall energy transfer determines again the established temperature above the melting point. The melt region solidifies after removal of radiation. Heat affected zone 3 and re-solidification zone 4 are observed. This case is relevant to alloying, vitrification and the like processes.

(iii)   The case of higher beam intensities of the order 10^6 W/cm^2 is presented in Fig. 1.4c where melting and evaporation become profound. Vaporized material 5 is ejected in vacuum (or the ambient gas atmosphere) uniformly leaving considerable heat affected (3) and re-solidification (4) zones on the sample surface. As a result of this process a crater is drilled in the material and may be fully penetrating. In metals the prolonged interaction may lead to the formation of a highly energetic plasma-radiation coupled region in the crater, namely a dynamic keyhole effect. This highly energetic keyhole is responsible for deep-penetration materials processing (drilling, cutting, welding). In this intensity regime, the use of pulses of highly energetic photons, for example in the UV, can lead to molecular photo-dissociation and atomic desorption which may also undertake an ablative character.

(iv)   With the use of well above 10^6 W/cm^2 intensities as shown in Fig. 1.4d, the energy of beam 1 is deposited very fast on the material. This regime is reached by use of short (nanosecond) and ultra-short (< 1 picosecond) laser pulses. The interaction with the material thus becomes violent and explosive with a highly energetic plasma plume, 5, produced, which has directional properties. This is the laser ablation regime evolving under unstable non-equilibrium thermodynamic conditions and in practice it is an explosive process. Material is ejected rapidly from the target and undertakes a thermal or a less thermal character depending on wavelength and materials nature. The ejected material is a mixture of plasma species together with molecular clusters and micro-particulates formed on ablation. This is a very complex process caused by several interdependent physical and chemical effects. Here the extent of regions 3 and 4 is smaller and a 'cleaner' cut is etched on the surface, with minimal thermal or radiation affected zones.

Depending on wavelength and pulse duration thermomechanical and chemical effects are induced and drive the process with respective modifications. Pulse energy and duration are both crucial as discussed in the next section. Laser intensity determines the absorption linearity and the energy deposition rate – both of importance in process quality. The fluence and total energy deposited on the workpiece are associated with the extent of the operation. They quantify the ablation zone and the process affected zones on the workpiece.

Laser materials processing is a multifaceted operation rich in science and technology. The complexity of the involved processes and the interdependency of effects make the analysis of the involved processes very difficult. Most topics addressed here are subjects of current research, especially as they relate to photonics and micro-engineering sectors. They are a challenging field with many open problems yet to solve, but many promises to be fulfilled.

1.6   Laser-based materials processing for photonics

Two main processing modes are met and have been applied in the growth and processing of photonic materials and devices. The non-ablative, non-destructive subtle operations concern mainly thermal and photochemical processes. Three-dimensional polymerization and more recent advances relating to further photophysical effects such as the application of radiation forces are fields open to research. The second major category concerns the application of laser ablation in the growth and processing of photonic materials and devices. This section provides a brief outline of the operations and highlights the unique features and the achievements of the original developments, discussed further in the following chapters of this book.

1.6.1   Subtle, non-destructive materials processing

Operations in this category cover a wide range of photophysical and photochemical effects and applications, making use of a wide range of available laser sources. We outline here the main topics in the field and point the reader to the following chapters for a detailed state-of-the art analysis.

Photophysical effects

While laser radiation is an important tool for metalworking and the microelectronics packaging industries, laser-based processing in photonics technology meets niche applications (Basting and Marowsky, 2005). Laser annealing has been widely applied for the phase transformation of semiconductor interfaces such as in thin film transistor (TFT) production, as well as for selective materials doping. Thus, laser offers selective processing in the material bulk (Gamaly *et al.*, 2006) leading to structure modification, crystallization, optical poling glass and polymers, and the induction of nonlinear optical properties Along these lines, photorefractive processes represent a large class of refractive index modification of materials by the application of usually intense optical fields. The refractive index of a material can be tailored by structured light even at milli-Watt power levels. Such changes may relate to the excitation of specific impurity centres or vacancies resulting in localized refractive modifications. In several cases these effects are also associated with strong photochromism owing to the excitation of relatively deep levels which affect the absorption properties, especially by use of UV radiation. Characteristic examples concern silica and other glass bulk and optical fibres with significant impact for fibre grating technology and other widespread photonic applications.

Structural modifications can be induced even by low intensity laser beams in amorphous and glass oxides (Mailis *et al.*, 1998), and also non-oxide chalcogenide glass materials (Gill *et al.*, 1995). The strength and the permanency of these effects depend on materials, laser wavelength and radiation intensity in the interaction region. Dynamic processes may be based on relatively long-lived effects that relate to impurity state excitation. They are followed by electronic or ionic relaxation phenomena which provide a degree of reversibility to the original materials state after removal of the irradiation. Intense irradiation, especially highly energetic deep-UV photons, may yield permanent structural changes usually associated with structural deformations. These effects result in significant variation of the refractive index and provide the means for the fabrication of micro-optics, optical waveguides and other photonic devices. In this case the high-intensity electric field rather than thermal effects has effectively modified the internal structure locally in the bulk in the area of beam focus, inducing large optical nonlinearities. Such effects take advantage of multiphoton processes induced by ultra-short high-intensity pulses, while refractive filaments in silica bulk offer new possibilities for photonic structure fabrication in the future (Haglund, 2006).

Quite different processing operations are possible by the deployment of laser radiation forces (Sigel *et al.*, 2002). The application of structured optical fields leads to the production of reversible structures. They are formed by compressive forces in semi-dilute entangled soft matter causing rapid osmotic extraction of the solvent and subsequent solidification, forming three-dimensional micro-objects, discussed in a later chapter.

Photochemical effects

Some of the most well-known photochemical processes are those met in the reduction of silver in photographic emulsions. The dissociation of silver halide produces silver nanoclusters which darken the emulsion and develop the image, via the reaction $AgX + hv \rightarrow Ag + X$. This process can also be activated by a focused or an interfering laser beam and can produce optical gratings in the bulk or the surface, a toolbox for the advent of holography and its numerous applications. The reduction of metallic compounds by using light is today a source for nanocluster production and it is applied as an alternative to chemical reduction methods.

The exposure of photoresists in lithographic processes use routinely high power UV lasers in the industrial floor. The relatively narrow spectrum of excimer lasers emitting at 248 nm (KrF) and 193 nm (ArF) and the high-intensity radiation allow tuning of the resist properties. In addition pulsed exposure allows fast processing speeds and high industrial yield. Deep-UV exposure has been advantageous in reducing the feature size of lithographic processing to the 100 nm size for contact methods. Resist technology has thus been developed in recent years to cope with lithography towards the extreme UV range. A favourable candidate appeared with the development of high power F_2 excimer laser emitting at 157 nm. Even though this radiation promised to provide reduction of feature size well below the state of the art, two main problems hindered the further deployment of the technology. First, the delivery of beams at 157 nm requires complex vacuum or nitrogen or noble gas purged transmission lines, which increases the complexity of the technology. Second, the highly energetic photon of 7.9 eV produces molecular dissociation of the photoresist and large amounts of debris which is deposited on masks or near field optics thus decreasing the efficiency and the accuracy of the lithographic process.

Current industrial developments in the micro-electronics and photonics industries incorporate high power UV laser lithography for large-scale production and multiphoton lithographic processing currently provides significant feature size reduction. This is caused by the use of the high-intensity section of the focused beam, which can be tuned to reach processing regions sized below the optical diffraction limits. In this context scanning laser lithography by femtosecond lasers leads to 3D structure development, as an evolution of conventional stereolithography. Complementary approaches relating to metallized structures have also been presented.

Further to the photo-polymerization processes, intense UV beams have been used for the deposition of metals and the growth of semiconductors, which can be realized in gas, liquid and solid environments. For example, the use of organometallic compounds has provided selective deposition of metals by photo-dissociative reactions such as $M(-C_xH_y-) + hv \rightarrow M\downarrow + C_xH_y\uparrow$. Metal hydrides have been similarly used with significant examples of amorphous silicon growth through silane decomposition in a series of photoreactions with deep-UV radiation to yield hydrogenated amorphous silicon through: $SiH_4 + hv \rightarrow \{SiH_2 + 2H\} + hv + Si \rightarrow \alpha\text{-}Si{:}H$.

Further applications involve oxidation and reduction processes with important examples being the production of reactive gas such as oxygen and nitrogen roots for oxide and nitride materials growth. The above processes may be realized as part of more complex processes such as physical vapour, molecular beam deposition, or laser ablative processes.

Photoreactions may also be used to achieve selective corrosive processing, while in most cases a suitable chemical equilibrium may be established depending on the application sought.

As a general rule the surface or bulk photochemical process embraces:

- Physical contact of the reactive species with the surface or dilution in the bulk
- Molecular adsorption on the surface of interest or establishment of a suitable composition which enables to develop further reactions
- Activation by laser radiation of suitable photon energy
- Production of reactive species and/or dissociation of products
- Secondary reactions
- Resulting processing effects
- Separation and removal of primary and secondary products.

The localization or not of the targeted process depends exactly on the laser interaction region and the extent and nature of the process achieved via localized or delocalized primary and/or secondary reactions. Effects may take an explosive character under intense illumination, thus denoting a rather 'grey borderline' between subtle and ablative processes.

In its relevant chapters this book provides an exhaustive account of state-of-the-art photophysical and photochemical processing, expanding in detail on the emerging research and future technologies.

1.6.2   Ablative materials growth and processing

Main ablation schemes

Of extreme fundamental and technological interest are the ablative processes realized upon interaction of solid or liquid materials with high-intensity laser pulses. Figure 1.5 depicts the typical interaction geometries implemented to date. In Fig. 1.5a, an intense laser pulse 1 is focused on surface of a solid target 2. The material absorbs the pulse energy and the material is explosively photo-ablated and forms a deposit, 3, on suitable oppositely positioned substrate, 4. In this case a pulsed laser deposition (PLD) scheme is realized, in which a highly energetic plasma plume 5 is formed. Figure 1.5b emphasizes the direct laser ablative processing scheme, where the target is etched with

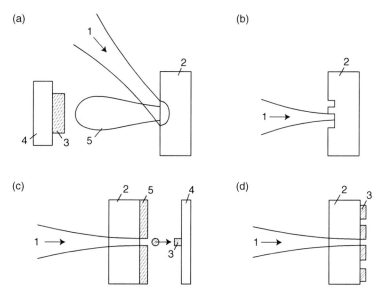

1.5 Ablative processes: (a) backward ablation and materials deposition, (b) resulting surface etching, (c) forward ablation and micro-printing operation by laser-induced forward transfer and (d) selective film etching by forward materials ablation.

high precision by a highly focused beam or a complex pattern imaged on the surface, 2, at high resolution (Vainos *et al.*, 1996). Both operations require laser intensity on target exceeding the characteristic threshold value. This threshold value is materials, surface and wavelength dependent. Below this value photo-desorption effects are possible which may be assisted by the substrate and/or the matrix.

The forward ablation scheme (Bohandy *et al.*, 1986) and the resulting micro-printing are illustrated in Fig. 1.5c. A transparent target substrate embodying a thin or thick film 5 is used. The irradiating pulsed beam is transmitted through the substrate 2 and is absorbed by the film 5 which is thus ablated at its interphase. The film material is expelled by the vaporized layer and forms a deposit 3 on the receiving substrate 4. The accuracy of this micro-printing operation performed by laser-induced forward transfer (LIFT) can reach the submicron levels (Zergioti *et al.*, 1998). The remaining film is etched with high precision as shown in Fig. 1.5d, forming patterns, 3, of complementary design to those being micro-printed.

As we will discuss later in this chapter, both backward and forward ablation modes have achieved quite remarkable results. A wide range of materials, from glass and amorphous to high crystallinity epitaxial films have been achieved, while high precision direct etching has been proved a viable scheme capable for micro-fabrication.

Ablation mechanisms

Depending on the nature of the material, exposure of a local region of the materials interphase a (usually) focused high-intensity beam produces a rapid, abrupt vaporization of the surface material under thermodynamic non-equilibrium conditions. A superheated volume builds up close to the surface, the Knudsen layer, establishing internal temperature much higher than the vaporization temperature and a very high pressure. The density of this layer is proportional to the laser pulse intensity and the very small mean free path aids to equalize instantaneously the internal temperature of the layer.

This superheated layer is expanded from the near surface region expelling the possibly present ambient gas. Depending on pulse intensity this expansion may be subsonic (speed < Mach 1), sonic (speed = Mach 1) or supersonic (speed > Mach 1). In case of Mach 1, the external ambient pressure determines the vaporization. Mach 1 is a limiting case from which mass continuity is satisfied by the production of new vapour that adds on, finally creating a shock wave which propagates in the neighbouring materials.

This is exactly the case which onsets a series of laser interactions relating to the production of plasma as shown in Fig. 1.5. The intense light pulse produces fast ionization and the formation of a highly energetic directional plasma plume ejected as shown in Fig. 1.5a. Depending on the process parameters this plasma can be produced either by radiation heating of the ejected vapour, or by photo-ionization and direct photo-dissociation of the material or their combination. Plasma can be established under local thermodynamic equilibrium (LTE), governed by the Saha equation:

$$\frac{n_e n_i}{n_g} = 2 \frac{Z^+}{Z^0} \frac{(2\pi m k_B)}{h^3} T^{3/2} \exp\left(-\frac{E_i}{k_B T}\right) \qquad [1.12]$$

where respectively: n_g, n_e, n_i are the number densities of atoms, electrons and ions, Z^+ and Z^0 the partial functions of ions and neutral atoms, m_e the electron mass, T temperature, k_B and h are Boltzmann's and Planck's constants respectively, and E_i is the ionization energy.

Plasma extends and expands above the interaction region with a visible plume formed with temperatures reaching above 10 000°K and pressures of the order of 10^4–10^7 Pa. Depending on intensity, particles are ejected at supersonic speeds, above 10^6 cm/s. Depending on the number density and radiation wavelength, plasma may absorb or reflect radiation, with such effects naturally becoming more profound by using IR radiation and long laser pulses.

Bearing in mind the above, it is important to underline that there is not an integrated theory available to date capable of describing well these complex phenomena. The thermal, photophysical, photomechanical and photochemical mechanisms are coupled together and constitute this explosive photo-ablation effect. The participating strength of each mechanism depends on the specific experimental parameters, which determine the evolution of events, and no absolute borders can be defined in all cases.

There are two types of ablation processes demonstrated to date. The backward ablation scheme described above (Fig. 1.5a) leads to a highly directional plasma plume which is characterized by multiple constituent components. The ejected matter comprises electrons and ionic species, neutral atoms, complexes, clusters and micro-particles, all ejected at different speeds from the ablation volume and following different spatiotemporal distributions. This latter effect is of extreme importance and offers significant advantages in applications requiring filtering of specific species. On the one hand the eroded material exhibits a minimal distillation effect, and the remaining part after ablation preserves the original composition of the target, and on the other hand, the ejected material also preserves its overall composition and can be deposited on a substrate achieving the original stoichiometry and in many cases the crystallinity of the target material. While there are exceptions, it has been observed that the less thermal the nature of the ablation process is, the more stoichiometric the expected deposits would be.

Influence of the actual experimental parameters

There are a number of important parameters which influence the ablation process, its nature and individual characteristics. These parameters are interdependent and affect not only the coupling of energy to the target but also the nature of the process itself as follows:

(a) Nature of target material, properties and surface characteristics.
(b) Wavelength of the laser beam and optical properties.
(c) Pulse duration and pulse intensity.
(d) Total pulse energy and its spatial density.

The nature of the target material determines the overall process and it is in full dependence on the laser parameters. The optical and physical properties are decisive in the development and evolution of these effects. Wavelength and surface properties are the main factors that determine the degree of energy coupling into the target, through the absorption property and the diffractive coupling effects induced by the surface texture.

According to our previous discussion, radiation incident on the planar surface is absorbed in the penetration depth, $\delta = \alpha^{-1}$, where α is the absorption coefficient at the particular wavelength. The surface texture of spatial frequency comparable to the wavelength is acting to increase coupling of radiation, while sub-wavelength components are enhancing the most this nominal apparent absorption. The absorption of most materials in the deep-UV region is usually very high and the penetration depth is minimal, with the exception of highly transparent fluorides and related compounds. The penetration depth of UV radiation in metals is a few nanometres, except in some alkali. Infrared radiation can be more penetrating and thus energy is deposited in a larger interaction region than the UV.

The physical properties of the material undertake the next task of transferring this energy to the bulk. The nature of the material then determines the evolution of effects. In a case of solid having finite thermal conductivity, it is the combination of penetration depth δ and thermal diffusion length, L_{th}:

$$L_{th} = 2\sqrt{D_{th}\tau} \qquad [1.13]$$

where, D_{th} is the thermal diffusion coefficient and τ is the laser pulse duration, which determines the interaction. In effect, the energy is deposited in a depth $\delta = \alpha^{-1}$ and is converted to heat. During the laser pulse this heat is diffusing in depth L_{th}. A maximum energy accumulation effect is thus expected when during the irradiation time, τ, all energy deposited in δ cannot diffuse away, but remains in the interaction region with $\delta \sim L_{th}$. Provided the pulse energy is high enough, this fact will lead to efficient ablation.

The intensity of the laser beam thus becomes the most crucial parameter. It relates to the energy deposition rate per unit area. While the pulse energy fluence is an important factor, it is effectively the pulse duration, τ, what governs the overall process. Figure 1.6 presents a range of effects evolving at different pulse length values and the respective intensities. In Fig. 1.6a CW radiation or a relatively long pulse 1, up to the nanosecond regime, irradiates the sample 2. Radiation may be absorbed within a specific volume, 3, by electrons in the valence band or impurity states, 4. This energy is transferred to the lattice via collisions and heats the material bulk. The more than few nanoseconds' duration of interaction greatly exceeds the energy relaxation time (~ picosecond), meaning that interaction extends during the laser pulse irradiation. Region 3 is thermally transformed and melts under laser exposure. Non-destructive chemical effects may also take place.

The use of high intensity, relatively long nanosecond laser pulses produces an electron avalanche inducing strong collisions with the lattice and finally causes materials breakdown by avalanche ionization yielding the photo-ablation effect. In molecular solids such as polymeric materials high energy photons may yield excitation to higher molecular states and enable the reach of dissociative states and molecular breakdown with minimal thermal effects. Even though in the case of polymer ablation thermal action cannot be excluded, the low thermal conductivity may yield a rapid passage to the evaporation state with minimal heat diffusion effects in the non-irradiated volume (Urech and Lippert, 2010).

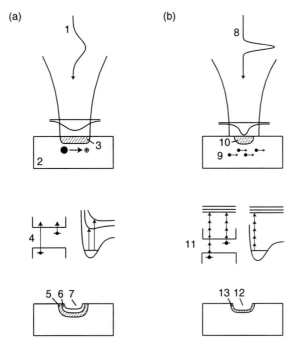

1.6 (a) Long pulse effects from the cw to nanosecond pulse irradiation and (b) ultra-short pulse effects of sub-picosecond duration inducing multiphoton processing.

In the nanosecond regime photo-ablation yields a visible ablated crater, 7, accompanied by a heat affected zone, 5, and re-solidification zone, 6.

Quite different effects are met by using high-intensity ultra-short pulses (Pronko *et al.*, 1995). In Fig. 1.6b a sub-picosecond duration pulse, 8, irradiates the sample, 9. Irrespective of the laser wavelength a relatively high energy is confined in a small laser spot, thus leading to a smaller diameter interaction region, 10. This is caused by two effects: (a) the short interaction time becomes comparable to the energy relaxation time in the material and consequently it confines in space the whole event; (b) electrons are excited and become ionized by multiphoton absorption from the valence or impurity bands, or from a molecular state, as shown in 11 of Fig. 1.6b. The high-intensity pulse induces extreme acceleration and direct photo-ionization of electrons, leading to direct ablation. High laser intensities thus produce ablation with a smaller crater formation, 12, and minimal heat affected and re-solidification zones, 13. It is now proved experimentally that multiphoton effects enable high-resolution processing.

Laser ablation is characterized by intensity (and energy density) threshold effects and specific erosion (etching) rate behaviours. These are strong functions of the material, the wavelength of radiation and the pulse

duration, through the processes outlined above. Multiphoton processing has a smaller ablation threshold and attains linearity of the etching
rate. Radiation is absorbed and interacts in a shallow region near the surface, which yields high quality ablative processing. On the other hand, the
energy density (fluence) value represents the total energy deposited per
unit target volume and determines the extent of the overall etching process. In the long pulse regime, even in the nanosecond scale, the longer
duration of interaction leads to larger ablation volumes, but with inferior
etching quality effects. The ablation threshold is higher and the etching
rate as a function of energy density shows a linear behaviour followed by a
nonlinear saturation region. Figure 1.7 presents typical experimental data

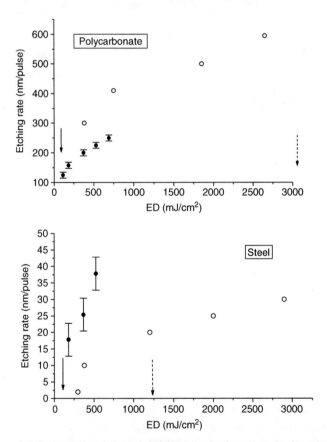

1.7 Typical experimental etching results for planar polycarbonate and
polished steel substrates obtained by 20 ns (open circles) and 0.5 ps
(closed circles) of KrF excimer laser beam at 248 nm, under identical
focusing conditions. Solid arrows indicate single pulse etching threshold using 0.5 ps pulses and dotted arrows indicate the single pulse
threshold using 20 ns pulses, both at 248 nm. (After Mailis et al., 1999.)

for threshold and etching (erosion) for polymer and metal substrates as a function of energy density. It is noted that the beam intensity is about 50 000 times higher in the ultra-fast case, though the energy deposited per unit area remains a quite important parameter. The material dependence is in any case distinguishable with the metal alloy showing clear improvement of the erosion rate and lowering of ablation threshold. It is worth noting here the difference between single and multi-pulse experiments. In the latter case the gradual deterioration of the structure results in lowering the apparent threshold and increasing the overall etching rates. Characteristic also is the difference between sub-picosecond and 20 ns pulses where the great difference in peak intensity leads to severe reduction of the etching threshold (Mailis *et al.*, 1999).

Considering the ablation process the intensity affects the plasma distribution leading a narrow cone of plume and matter ejection. Energy density also affects significantly the products of ablation. While femtosecond laser pulses were initially thought to yield highest quality ablation products and a smaller particulate content this has not been verified experimentally and nanosecond pulses are still seen to produce superior results in PLD experiments.

1.7   Specific laser processing schemes for photonics applications

Following our discussion of fundamental effects and processes we focus here on two specific laser processing schemes used in the fabrication of materials and devices, both relating to ablative processing. The reader is directed to the following chapters of this book for a presentation of schemes relating to lithographic and photorefractive processing.

Pulsed laser deposition is a prime application of congruent laser ablation having developed numerous devices and applications in the last decades (Eason, 2007). The method has been applied for the growth of photonic materials since the late 1980s, by exploring the nature of the process in the case of highly complex epitaxial oxide structures (Vainos *et al.*, 1998), and proceeding to date with highly sensitive novel nanocomposites (Grivas *et al.*, 1998). In all cases laser methods offer unique tools especially for materials that have proved hard to grow at high quality by conventional methods.

Micro-ablation effects are explored in micro-fabrication by etching and tailoring the surfaces and bulk of optical materials to produce photonic structures by use of nanosecond and femtosecond laser pulses. This latter scheme offers an alternative to well-established lithographic processing methods, with unique advantages in terms of single step, generic surface processing and in-the-bulk tailoring not available by other means. In addition, micro-printing applications by laser-induced forward transfer (Fogarassy *et al.*, 1992) are discussed here in the context of photonics.

1.7.1 Pulsed laser deposition (PLD) of photonic materials

Appraisal of methods and apparatus

The growth of photonic materials imposes stringent requirements on the nature of the ablation process, the materials used and the conditions applied. Photonics require extreme quality with respect to optical properties such as transparency, bulk homogeneity and surface quality, in addition to specific individual properties, such as composition, doping or other physical properties.

To comply with these requirements, extreme control of the process is required and a huge effort has been made in this respect. These initial efforts have been directed to ferroelectric and paraelectric oxides, such as respectively barium titanate ($BaTiO_3$) and bismuth silicate ($Bi_{12}SiO_{20}$) as holographic photorefractive materials (Youden *et al.*, 1991). This progress has been followed by further materials including garnets and sapphire, which led to the achievement of the first laser actions by PLD grown in Ti-Sapphire, Nd:YAG and Nd:GGG optical waveguides (Anderson *et al.*, 1997a, 1997b). More recently, nanostructured thin film oxide materials for photonic sensor applications have been developed (Mazingue *et al.*, 2005). The overall methodology is developed and presented in reference to the typical PLD experimental configuration presented in Fig. 1.8a.

A high or ultra-high vacuum (HV or UHV) reactor vessel is especially designed and constructed to incorporate optical windows and feed-through ports. The ambience of the vessel can be modified by gas injection and monitored by, for example, a quadrupole mass spectrometer (not shown). A pulsed laser beam 1 is directed through a transparent window and is focused on the target material 2. The target from which material is to be ablated may be in a solid crystalline, glass or sintered ceramic form. The target is usually attached on a rotating or translating holder platform, to ensure refreshing of the material undergoing ablation. This material is laser sputtered and grown as film, 3, on an oppositely placed substrate, which is usually a glass or crystalline plate, or wafer. A train of laser pulses, usually delivered at a repetition rate of a few Hz to some tens of Hz, ablate the target material producing a plasma plume, 4, which is directional and flushes the substrate. The ejected material is thus deposited on the substrate and grows a film, 3, on the substrate, while measures are taken to ensure uniform coverage in film growth. In Fig. 1.8b and 1.8c two distinct examples of photonic structures

(a)

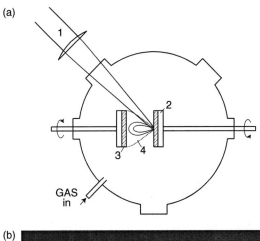

(b)

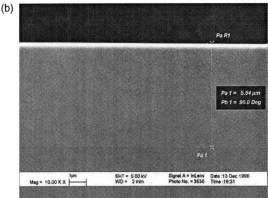

(c)

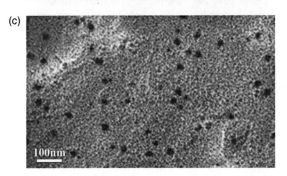

1.8 (a) Typical configuration of a PLD system. Includes a focused laser beam (1), target material (2), substrate with grown film (3), plasma formation and ablated material ejection (4). (b) Example of high quality laser glass composite grown for optical telecom applications and (c) indium oxide nanocomposite structure for light activation photorefractive effects.

are shown: a highly compacted waveguide optical amplifier (Vainos *et al.*, 1999) and indium oxide nanocomposite comprising nanocrystals and metal clusters in the amorphous matrix (Grivas *et al.*, 2002).

While the overall laser sputtering process appears to be simple in its principle and provides a large flexibility in its operation, there are several interdependent parameters which need to be tuned for the successful growth of a specific structural composition. Even though there are various commercial systems available which operate reliably for specific relatively simple coating processes, the specific individuality and the interdependency of parameters make PLD a not yet industrially standardized process.

Interdependency of PLD processing parameters and methodologies

Several important issues determine the PLD methodology, with specific aspects of the scheme concerning:

(i)   *The nature of the target material:* This determines the composition of the ablation plume and the stoichiometry of the deposit. The atomic and molecular composition and its constituents are determined by the density of the target and its form. There are several types of targets to be used: (a) solid crystalline target of material, (b) glass compound material, (c) sintered ceramic target and (d) liquid target used in specific cases. The more concrete and dense target produces better results in terms of particulate content and stoichiometry. In addition the rotation of the target is very important in refreshing the material and producing a relatively uniform composition over the duration of growth, the latter ranging from a few minutes to several hours. It is noted here that upon irradiation the target is usually severely eroded by ablation. Two effects are observed: (i) the composition of the target material is altered and in effect this differentiates the products of ablation over time and consequently the composition of the grown film; (ii) the morphology of the target is changed and this affects the direction of the plasma plume and the overall ejection of species, which consequently affects the growth process and the final composition and morphology of the grown film.

(ii) *Laser parameters and ablation products:* These are fundamental in plasma production and deposition and determine the nature of ablation. Following the previous discussion on ablation and plasma formation they determine the ablation process as follows:

   (a) The laser wavelength determines the initialization regime and the type of ablation. Longer wavelengths produce thermal ablation effects, while the use of deep-UV laser pulses, such as those delivered by excimer lasers, is seen to produce advantageous results in terms of preserving the target stoichiometry in the ablated plume and reducing heat effects (congruent ablation). This preservation of stoichiometry is crucial in producing the desired composition on the film grown. Furthermore, high photon energy values are usually associated with a decrease of threshold intensity, with penetration depth also decreasing.

(b) The laser beam profile and the pulse intensity are crucial and determine the nature of the process and the energy of the ejected material. They affect the ablation threshold and the specific plasma plume formation and the ablation products distribution. The latter parameters are significant for establishing an efficient PLD configuration, in terms of substrate–target distance and geometry, the use of atomic, ionic, cluster or particulate ejects is crucial in the overall growth quality. The intensity of laser pulses determines the energy of the ejected products and the composition of electrons, ions, atoms, molecules, clusters and particulates, which subsequently travel in free space, vacuum or gas ambient and are incident on the target to grow the film.

(c) The total energy density on target determines the quantity of the ablated material and the overall deposition rate, by controlling the mass ejected per incident pulse and consequently the incremental thickness growth.

(iii) *Experimental geometry and ambient atmosphere:* The target–substrate configuration in particular is crucial in the growth of the thin film. The composition of the ablation plume is not uniform over the angular range. In addition the velocity profile of the ejected particles also varies in angle and depends on the nature of the particulates. The angular and velocity profiles are strong functions of the intensity profile on target and the specific geometry on the irradiated spot. Furthermore, the ambient atmosphere affects significantly the velocity profile and also the deposition through reactions of the ejected products upon collisions with the gas molecules. In effect, the presence of oxidizing and reducing gases affects greatly the growth process and can yield completely different compositions. Oxides and composites have been grown with variable stoichiometry (Vainos *et al.*, 2004), as well as nitride materials, starting from metallic targets. These parameters are tuned in relation to the laser and other specifics of the scheme, in order to achieve the final growth goals.

(iv) *Substrate temperature and physical properties:* The growth of most inorganic materials and structures, especially in the form of crystalline epitaxial films necessitate deposition on substrates that are heated at high temperatures. The temperatures are characteristic of the specific material to be grown and for most oxides are above 600°C and exceed 1100°C for refractory oxides such as sapphire (Al_2O_3). More stringent requirements apply in the growth of epitaxial materials, in which case the choice of a suitably lattice matched substrate is necessary, in addition to the elevation at high temperatures. In all growth cases the highly energetic plasma plume is highly beneficial because the kinetic energy of the ablation products is transferred on to the substrate and assists to establish the proper thermodynamic conditions for growth. The net effect is that even by depositing at room temperature the produced films are quite compacted and exhibit higher adherence to the substrate, thus differing in strength and properties to those of similar composition grown by other sputtering methods. Several heating configurations have been employed in the growth of photonic materials commencing from conventional heating wires and filaments and progressing to ceramic elements. In these cases the contact of the heating plate with the substrate has been found to produce non-uniform temperature profiles which affect severely materials growth, in terms of uniformity, composition and crystallinity. Carbon dioxide laser heating methods have been developed in the last decades making use of beam-folding-mirror and light-pipe homogenizers as well as raster scanning methods coupled to temperature profile monitors by thermal imaging, all targeted to producing uniform temperature profiles in the substrate (Barrington and Eason, 2000).

(v) *Process optimization – Film quality:* While significant progress in the growth of materials has been made, the strong interdependency of parameters necessitates detailed optimization to be applied for each material or at least materials class. Furthermore, the spatially narrow and non-uniform distribution of ablation products creates difficulties and inhibits deposition on large size substrates with good uniformity. Typical deposition areas of the order of 1–2 cm² are most usual. Special approaches using eccentric or planetary motion of substrate and target are explored to extend to larger substrates of the order of 5 cm diameter for generic growth. It should be underlined here that thermal laser ablation, for example by IR lasers, of single element materials, such as carbon or silicon or metal, may indeed provide appreciable results over relatively large areas of 5–10 cm diameter wafers, owing to reduced directionality and simple elemental composition. Nevertheless, this is not generally applicable, especially when using high-intensity UV pulses forming multi-component plasma to enable multi-element and complex compound growth. One very important issue in PLD growth is the presence of cluster and particulates in the ablated material, which is directed and incident on the substrate becoming integrated in the deposited film. Two effects are worth mentioning here: (a) the cluster and submicron particulates which become embedded in the film may act in various ways. First, they act beneficially assuming the role of growth centres assisting the creation of crystallinity. Second, they may act disadvantageously since they create scattering inhomogeneities especially by incorporating particles sized in the tens of nanometres range and larger. However, in other applications the incorporation of such particulates has also been seen to act advantageously in creating nanocomposite materials, in which an amorphous and finely structured matrix hosts crystalline nanoparticles which have been produced by the ablation process, (b) It is usual to observe relatively large particles on the surface of the film. Especially those sized several hundred nanometres to micrometres create significant problems in photonics applications producing considerable optical scattering and waveguide losses, the latter being deleterious in highly demanding laser applications. There have been several schemes proposed aiming to reduce particulates by use of gas streams, velocity filters in the form of mechanical shutters, electromagnetic filters to steer the plasma and others. However, in view of the above discussed central use of nanoparticulates in crystalline growth and the formation of nanocomposites it is the specific application and the target operation what will dictate the appropriate methodology.

(vi) *Advanced multi-target operations:* The development of PLD systems is to date in a quite advanced state, although full automation to the degree of industrial standardization of the scheme has not yet been achieved. Systems available to date offer significant flexibility and considerable ability for generic operations by use of multi-target systems which allow multi-compound growth, multi-layers and complex dopants, in amorphous and crystalline growth results. Those use advanced environmental control and target and substrate motion protocols for growth control. Relevant issues and current advances are presented in the following chapters.

In conclusion, PLD has proved to be a quite versatile growth tool which can be adapted to the application needs. While the system is quite flexible, the very nature of the process has not yet allowed industrial standardization and reliability for large-scale production. The limited area and uniformity of deposition, together with the difficult control of parameters in large-scale and long-run operations are some present impediments. The latter yields an overall incompatibility with the existing micro-electronics and photonics foundry, thus limiting wide exploitation of the scheme.

Advanced schemes for photonics are discussed in a later chapter.

1.7.2   Micro-ablative processing operations

Laser micro-etching operations

The use of laser beams finely focused in the micron scale produces materials micro-ablation. This operation is limited in the micron or submicron size laser spot area and requires relatively low pulse energy in the range of a few milli-joules. In this regime the strong focusing in micron size spots of nanosecond pulses leads to extremely high intensities in the GW/cm^2 range corresponding to energy density values of several J/cm^2. These values are well above the ablation threshold of most materials including the highly transparent glass and crystalline compounds. The micro-ablation process thus leads to a series of processing operations allowing direct materials treatment spatially. These are now considered viable as alternatives to well-established lithographic methods, offering specific advantages and unique features as analysed here (Mailis *et al.*, 1999).

Figure 1.9 provides an outline of a direct laser micro-fabrication station. The beam, 1, is delivered by a laser source (not shown) and is transmitted through a beam-shaping optics system, 2. The system performs beam expansion, filtering and possibly homogenization. The processed laser beam is directed through a folding mirror, 5, into a focusing objective, 3, which delivers the radiation onto the workpiece, 4, for processing. This workpiece is positioned on a platform allowing controlled x-y-z motion to enable focusing in the z-direction and x-y surface processing. It is noted here that depending on the specific configuration folding mirror 5 may be replaced by

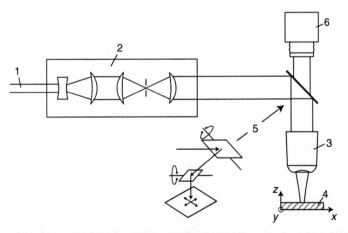

1.9 Schematic of a direct laser micro-fabrication station. Includes laser beam (1), beam-shaping optics (2), focusing optics (3), workpiece for processing (4), folding system (5) ocular observation and monitoring system (6).

a raster scanning system, especially to be used with long focal length focusing. The overall processing area is monitored by ocular system, 6, which may include video imaging. We note here that such a system may be used in projection lithography in which case the laser intensity and power throughput needs may be reduced in favour of resolution.

The operation of the system is assisted by the controlled motion of the platform especially by use of high precision nanomotion-motorized stages controlled by position encoders currently available with sub-100 nm accuracy. This allows precise processing of the workpiece in raster scanning and/or vector scanning modes. In addition, the beam-shaping system may comprise a suitable imaging mask or diaphragm which is projected onto the surface of the workpiece for direct etching of complex patterns such as holographic patterns and images, in conjunction with beam homogenizers.

The above imaging operation mode is considered to be most appropriate as it allows precise intensity control to be applied in the system through calibration procedures. It allows high precision ablative processing via high-resolution imaging, yielding controlled intensity profile and pattern transfer onto the workpiece. This scheme, combined with absolute motion control, enables high precision direct processing. It is worth underlining that the use of a focused beam is not always appropriate due to the Gaussian or other highly peaked beam profile at the waist which may lead to inhomogeneous processing. Nevertheless, in some cases this allows a type of super-resolved processing in the sub-wavelength regime (previously shown in Fig. 1.6b), utilizing only the high-intensity part of the beam, while the beam shoulders have intensity below the required ablation threshold.

Figure 1.10 depicts the first developed examples of micro-fabrication of computer-generated holographic optics etched directly on the surface of acrylic and polished stainless steel plate with their optical reconstructions having appreciable finesse though non-optimal configurations have been applied. Fig. 1.10c shows a Fresnel micro-lens of about 500 μm diameter. Fig. 1.10d shows a polymer optical fibre processed for sensor applications with long light coupling slot. This unique operation proves the generic processing capability of the scheme. Such operations benefit from the nature of ablation which is able to cope with different materials of composite nature without concern for differential erosion or preferential etching effects, provided the intensity is well above the ablation threshold. The quality and the speed of processing depend clearly on the laser parameters and the physical properties of the material. All issues discussed previously concerning the surface quality and the ablation mode established for processing also apply here. Bearing in mind the sensitive and the precise nature of the operation, the laser parameters must be tuned exactly in relation to the physical properties of the workpiece and the application sought in order to achieve high quality results. It is clear that excessive intensity may cause severe damage

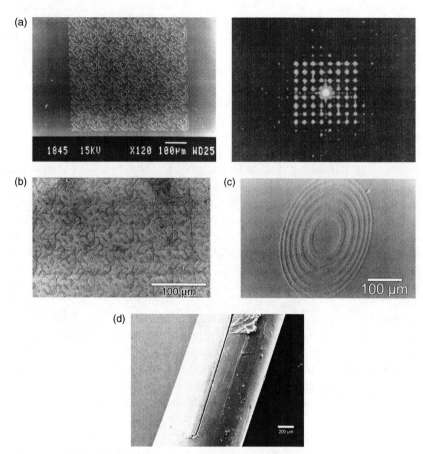

1.10 Direct micro-fabrication by UV laser radiation: Computer-generated holograms etched by direct ablation on (a) acrylic using 0.5 ps excimer laser pulses at 248 nm and optical reconstruction of interconnect matrix (b), stainless steel plate etched with computer-generated hologram using 20 nanosecond excimer laser pulses at 248 nm, (c) Fresnel lens element etched on polymer deposited on fused silica and (d) processed polymer optical fibre for sensor applications (unpublished). (After Vainos *et al.*, 1996, Mailis *et al.*, 1999 and unpublished data.)

to the sample, while reduced intensity may lead to gradual deterioration of surface quality, making the required process impossible.

Specific examples of photonics-related processing include the production of surface relief patterns, the production of micro-lens arrays, the processing of optical fibres and the tailoring of waveguide structures, as well as photonics packaging applications. The main advantages of the direct laser processing operation are drawn from its materials generic nature and the flexibility of application on surfaces of arbitrary geometry, practically on every solid surface. The scheme has been recently explored successfully in the fabrication of authenticity micro-holograms for security applications in the frame of the EU project initiative 'HOLAUTHENTIC', which resulted in a laser-based authentication system.

Laser micro-printing approaches

A complementary operation has been developed for photonics applications by use of the typical micro-fabrication station (Fig. 1.9). The material to be transferred is deposited on a transparent 'carrier' substrate as previously presented in Fig. 1.5c. An ultra-short laser pulse irradiates the film after transmission in the carrier plate and ablates the material at the substrate–film interphase. The net result is the ejection of the material in free space, propelled by the ablation gas products, and its deposition in the nearby positioned substrate. Figure 1.11 presents the dynamics of this forward ablation process by Schlieren imaging of the ejected material in free space (Zergioti *et al.*, 2002). The velocity of the ejected material is a function of intensity of the laser pulse and determines the kinetic energy which is in turn deposited on the substrate. It is worth noting here that this operation can be performed either in vacuum or in gas atmosphere, while velocities in the range of 1 Mach have been recorded. It is noted that in this particular case the produced ablates are expanding in flight, while in other configurations the solid or liquid ejects are propelled and deposited

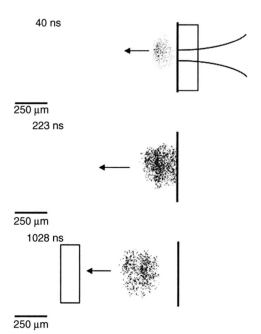

1.11 Dynamics of forward micro-ablation process. (After Zergioti *et al.*, 2002.)

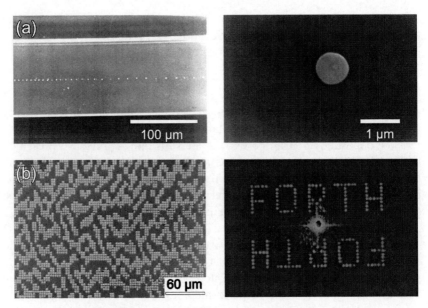

1.12 Micro-printing of (a) submicron chromium dots on the planar surface of D-optical fibre and (b) computer-generated hologram with its optical reconstruction by a HeNe laser. (After Mailis *et al.*, 1999 and Zergioti *et al.*, 1998.)

on the substrate. This is the case with either metal targets or inorganic deposition by use of dynamic release layers, which rapidly photo-dissociate and detonate upon absorption of laser radiation.

Figure 1.12 presents the first laser micro-printing operation in the submicron regime by using a single 0.5 ps laser pulse at 248 nm. Series of 700 nm diameter metallic Cr dots deposited on glass, silica and silicon substrates show excellent adherence, a fact which verifies the energetic nature of the process. By using a modified deposition cell micro-printing on optical fibres has been possible, as shown in the examples in Fig. 1.12b for dot series on D-fibre and high curvature etched cylindrical fibre. Such operations may lead the way to photonic sensor applications, enabling enhanced optical coupling and advanced interrogation.

While the first application of laser-induced forward transfer concerned the repair of photomasks, the reduction of printed patterns and the high quality of deposits led the original developments in the field of photonics. Micro-printing operations thus first addressed the fabrication of photonic micro-structures in the form of computer-generated holograms of metals and metal-oxides.

Since these original developments several techniques have been developed for photonics applications, with most recent advances relating to the fabrication of waveguide devices by depositing metallic Ti stripes on lithium niobate for subsequent in-diffusion for waveguide fabrication. Further developments involve the use of polymeric dynamic release layers (Banks *et al.*, 2008), which upon photo-dissociation produce a considerable amount of gas that enables efficient propulsion of the transfer layer, used especially in cases of transferring inorganic oxides, such as GGG and YAG laser garnets. It is worth mentioning that the method has so far allowed the deposition of submicron metal nanodroplets, achieved currently in the range of 500 nm and below to 300 nm diameter, by using amplified femtosecond Ti-Sapphire laser pulses in forward and backward transfer modes. Metallic films show a most suitable behaviour for decreased feature size, because the film melts locally releasing nanodroplets due to the unstable hydrodynamic conditions induced that enable propelling of matter to the oppositely placed substrates.

A case of global laser-based micro-fabrication

In Fig. 1.13 an example of a global laser-based device fabrication approach is presented (Koundourakis *et al.*, 2001). The process commences with the PLD of metal oxide. In this specific case a non-stoichiometric mesoscopic indium oxide (InO_x) nanocomposite is grown by reactive PLD using metallic indium target as shown in Fig. 1.13a. The material is deposited by use of nanosecond laser pulses at 248 nm on fused silica substrate. Its structure comprises several defects in the form of oxygen vacancies and metal nanoclusters, which enable charge transfer processes to evolve leading to variations of its electrical and optical properties. The latter effects lead to holographic recording in the material due to localized photorefractive effects induced by interfering UV laser beams.

This thin-film-on-carrier-plate structure is used in the laser processing station to micro-print a diffraction grating on a transparent glass plate by 0.5 ps 248 nm pulses, as presented in Fig. 1.13b. This grating is fabricated by repetitive transfer of grating lines sized in the micron range. We note that the build-up of the grating is achieved by precise relative positioning of the carrier plate and the receiver substrate, with the distance between film carrier and substrate kept in the few microns range.

The functionality of the deposited structure has been investigated by diffraction experiments under UV illumination at different temperatures. Fig. 1.13c depicts the typical experimental configuration used for functionality testing. The inset Fig. 1.13d shows an atomic force microscope image of the final diffractive structure. A UV laser beam is turned on and off and the diffraction efficiency of the micro-printed grating is altered within a certain response time determined by the participating charge transfer processes. In a manner identical to the parent InO_x nanocomposite material, the refractive index of the grating is altered by the illuminating UV beam. This produces changes

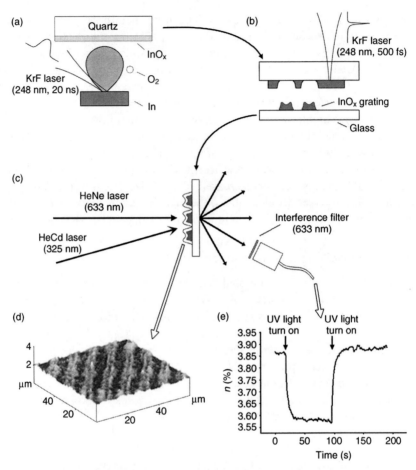

1.13 Exclusively laser-based fabrication process producing active diffractive indium oxide diffractive elements. (a) PLD of the InO_x parent film on carrier plate, (b) micro-printing of InO_x grating, (c) functionality testing by diffraction experiments under the activation of a UV beam, (d) AFM image of a part of the InO_x grating, (e) response of diffraction efficiency enhancement upon UV illumination. (After Koundourakis *et al.*, 2001.)

in the observed diffraction efficiency as shown in Fig. 1.13e in a similar time scale to that observed in the original 'parent' material.

It is worth mentioning that the alternative approach would be the lithographic processing of the planar film in order to produce the desired diffraction grating with expected similar results. Further to the complexity of this multi-step processing operation, a unique advantage of the laser-based scheme is drawn from its potential to produce micro-printed structures of arbitrary design on practically every given surface of arbitrary morphology, even of high curvature or discontinuous geometry, with no restriction. This is the significant advantage of the scheme in addition to the generic and inert processing nature of laser ablation. Indeed, as discussed above, lithographic approaches involve steps of wet deposition of masks, possibly followed by chemical or reactive etching which affects the nature of sensitive materials.

1.8   A suite of emerging concepts driving future trends

Direct laser processing schemes have emerged as viable alternatives to well-established fabrication and processing methods. Among the several open fields, we may focus on two regimes and distinguish them with respect to the intensity of participating processes corresponding to subtle and explosive operations.

Further to the widespread conventional laser lithography and laser surface annealing applications, relatively low power radiation induces photophysical and/or photochemical effects which directly modify the nature of the materials and yield photonic structures of extreme importance, such as waveguide networks, diffractive elements and switches or three-dimensional photonic structures for advanced coupling, filtering and sensing operations.

The second category embraces high-intensity operations associated with ablative materials photo-dissociation or explosive reactions leading to materials synthesis. Significant advances in this category concern the growth of photonic materials by PLD methods, in effect verifying the suitability of the method for the production of high optical quality and low loss materials suitable for photonic devices such as laser sources, waveguide circuits, switches and holographic storage.

Laser-based materials growth and processing exhibits unique advantages which concern the direct and non-contact nature of the operation. The required effects are induced usually by single-step operations and evolve simultaneously with laser irradiation. This offers high speed and flexibility in processing. In addition, the nature of the effects is quite generic, and, in effect, tuning of a set of parameters can suit to several materials and processing operations, with no further assumptions. The highly energetic photons of UV radiation, for example, induce photorefractivity, permanent or dynamic, in a variety of materials and has already led to Bragg grating and switching devices, as well as direct written optical circuits, tailored photonic crystal fibres and others.

Along the above lines, micro-ablative processing offers novel tools for direct etching and printing of materials in a single processing step, with no assumptions on materials nature or further compromises on the morphology and nature of the substrate or workpiece under processing. By these means, the etching of thin optical fibres or the printing of structures on a high curvature cylinder surface becomes possible, as does the internal processing of structures such as photonic crystals or multi-layer composites. In addition, large-scale ablation yields the production of multi-component plasma which is appropriate for the growth of complex multi-element epitaxial crystals or multi-layer stacks and nanocomposites exhibiting performances not available by other means.

Emerging technologies are multi-sectoral with novel concepts covering an interlaced network of techniques, materials and applications. The PLD growth of complex materials and devices is expected to follow niche routes leading to advanced photonics functionalities. While these systems may not be readily industrially acceptable, trends point to the exploration of their novel fabrication potential for advanced multi-layer and nanocomposite structures primarily of inorganic dielectrics and polymer-based structures. Laser devices, waveguide structures and sensors are current targets. The nanostructuring of semiconductor devices by laser radiation adds a new potential and exhibits characteristics compatible and integration ready with semiconductor processing foundries. The growth and the manipulation of devices will be realized by well-controlled PLD and surface processing techniques developed for specific classes of compounds. To this effect, it is yet unclear whether the full flexibility and the generic character of PLD will be exploited in the near future and technical limitations and specific features will prevail. However, there is scope and it is feasible to achieve appreciable size of growth by applying well-controlled multi-target planetary deposition approaches in perfectly mastered ambience.

Direct laser processing in the submicron scale is true and leads the way to nanoscale operations. In prime position are found deep-UV lasers but an important role will also be played by the available ultra-short lasers emitting high-intensity pulses at variable repetition rates. Solid state laser systems are favoured due to their compact size, reduced maintenance needs and overall ease of deployment, even though their cost remains high. The trend to ultra-short pulses is aiming to enhance performance based on multiphoton effects. The simultaneous interaction of two or more photons with the material enables the excitation of high energy states by use of low energy photons emitted by NIR lasers such as, for example, Ti-sapphire, Er-doped fibre lasers (EDFL) and the like. Furthermore, the attainment of super-resolution is realized by utilizing only the high-intensity 'peak' section of the beam profile, the lateral size of which can be smaller than the wavelength of light used when strong focusing with large numerical aperture objectives is applied.

Index manipulation in bulk devices and fibres also has great potential due to its novel character, while drawing benefits from multiphoton processing. The routine fabrication of Bragg fibre gratings with broad applications is a contemporary example of this rich area. Unique operations produced exclusively by laser light such as photorefractive and other light-induced structure modification effects are expected to offer enhanced performances combined with new materials and devices. Optical poling and phase changes, as well as localized thermal manipulation in glass and polymers are leading to advanced diffractive devices for telecommunications and sensing. The exploitation of these methods in speciality fibres and photonic crystal fibre structures represents new routes to follow in the future.

On the horizon for industrial application, ultra-short laser pulses for multiphoton lithographic processing and structure manipulation are important. Nano-fabrication will benefit by the use of ultra-short pulses in rapid prototyping and the direct construction of specialized three-dimensional devices by multiphoton stereolithography techniques enabling processing in the sub-wavelength scale. In this context structured fields come to play a very important role for the realization of innovative architectures. Of special interest are optical interference fields produced by femtosecond laser beams potentially enabling processing in a small fraction of the wavelength. Equally important are related multiphoton technologies which enable metallization, localized activation and micro-moulding processing, presenting a complete suite of multiphoton micro/nano-fabrication methods.

The following chapters of this book expose the richness of 'Laser Growth and Processing for Photonics' in a set of thorough reviews. The flexible methodologies developed in this emerging field are based on numerous scientific and technical achievements of the contributing authors and are expected to impact in several fields. Even though it remains unclear whether these methods will gain full reception and integration in the current industrial fabrication facilities, they hold a great potential and a living promise for new science and innovation in the years to come.

1.9   Acknowledgements

The original developments described in this chapter have been the result of successful research initiatives implemented in the frame of EU and national funded projects. In particular, the development of PLD schemes for photonics has been supported by the HCM EU-HCM ERB 4050 PL 92 1612: 'Advanced Optical Waveguide Components and Systems' and the INCO-Copernicus: 'Inter-European PLD Network'. Diffracted optics and sensing systems have been developed through the support of EU-ESPRIT 6863: 'Parallel Optical Processors and Memories', the EU-GROWTH 'Holographic Authenticity sensors (HOLAUTHENTIC)', the EU-IST 2001 'Nanostructured photonic sensors (NANOPHOS)' and various national PENED and ENTER projects funded by the Hellenic General Secretariat for Research and Technology, as well as the current HERACLITUS initiatives. The support of the ESF-COST Actions P2, P8, and MP0604 on advanced photonic materials is gratefully acknowledged. Many thanks go to Miltiadis Vasileiadis, Loukas Athanasekos and Dimitris Alexandopoulos for useful discussions and help with this chapter. The author remains grateful to all his collaborators who participated in the above research efforts – without their eminent contributions and hard work these developments could not have been realized.

1.10   References

Anderson, A. A., Bonner, C. L., Shepherd, D. P., Eason, R. W., Grivas, Ch., Gill, D. S. and Vainos, N. A. (1997a), 'Low loss (0.5 dB/cm) Nd: $Gd_3Ga_5O_{12}$ waveguide layers grown by pulsed laser deposition', *Optics Communications*, **144**, 183–186.

Anderson, A. A., Eason, R. W., Hickey, L. M. B., Jelinek, M., Grivas, Ch., Gill, D. S. and Vainos, N. A. (1997b), 'Ti:Sapphire planar waveguide laser grown by pulsed laser deposition', *Optics Letters*, **22**, 1556–1558.

Banks, D. P., Kaur, K., Gazia, R., Fardel, R., Nagel, M., Lippert, T. and Eason, R. W. (2008), 'Triazene photopolymer dynamic release layer-assisted femtosecond laser-induced forward transfer with an active carrier substrate', *EPL (Europhysics Letters)*, **83 38003**, 1–6.

Barrington, S. J. and Eason, R. W. (2000), 'Homogeneous substrate heating using a feedback controlled raster scanned c.w. CO2 laser with temperature monitoring', *Review of Scientific Instruments*, **71**(11), 4223–4225.

Basting, D. and Marowsky, G. (2005), *Excimer laser technology*. Berlin: Springer-Verlag.

Bäuerle, D. (2000), *Laser processing and chemistry*. Berlin: Springer-Verlag.

Beesley, M. J. (1978), *Lasers and their applications*. London: Taylor and Francis.

Bohandy, J., Kim, B. F. and Adrian, F. J. (1986), 'Metal deposition from a supported metal film using an excimer laser', *Journal of Applied Physics*, **60**, 1538–1539.

Born, M. and Wolf, E. (1989), *Principles of optics*. London: Pergamon.

Chrisey, D. B. and Hubler, G. K. (eds.) (1994), *Pulsed laser deposition of thin films*. New York: Wiley.

Cline, H. E. and Anthony, T. R. (1977), 'Heat treating and melting with a scanning laser or electron beam', *Journal of Applied Physics*, **48**, 3895–3900.

Drude, P. (1900), 'Zur Elektronentheorie der Metalle', *Annalen der Physik*, **1**, 566–613.

Eason, R. W. (ed.) (2007), *Pulsed laser deposition of thin films: Applications-led growth of functional materials*. Hoboken: Wiley.

Fogarassy, E., Fuchs, C., de Unamuno, S., Kerherve, F. and Perriere, J. (1992), 'High Tc superconducting thin film deposition by laser induced forward transfer', *Materials and Manufacturing Processes*, **7**, 31–51.

Gamaly, E., Luther-Davies, B. and Rode, A. (2006), 'Laser–matter interaction confined inside the bulk of a transparent solid'. In Misawa, H. and Juodkazis, S. (eds.), *3D laser microfabrication: Principles and applications*. Weinheim: Wiley, pp. 5–36.

Georgiou, S., Koubenakis, A., Lambrakis, J. and Lassithiotaki, M. (1998), 'Formation and desorption dynamics of photoproducts in the ablation of van der Waals films of chlorobenzene at 248nm', *Journal of Chemical Physics*, **109**, 8591–8601.

Gill, D. S., Eason, R. W., Zaldo, C., Rutt, H. N. and Vainos, N. A. (1995), 'Characterization of Ga-La-S chalcogenide glass thin optical waveguides fabricated by pulsed laser deposition', *Journal of Non-Crystalline Solids*, **191**, 321–326.

Grivas, Ch., Gill, D. S., Mailis, S., Boutsikaris, L. and Vainos, N. A. (1998), 'Indium oxide thin-film holographic recorders grown via excimer laser reactive sputtering', *Applied Physics A: Materials Science & Processing*, **66**, 201–204.

Grivas, Ch., Mailis, S., Eason, R. W., Tzamali E., and Vainos, N. A. (2002), 'Holographic recording mechanisms of gratings in indium oxide films using 325 nm helium-cadmium irradiation', *Applied Physics A: Materials Science & Processing*, **74**, 457–465.

Haglund Jr., R. F. (2006), 'Photophysics and photochemistry of ultrafast laser materials processing'. In Misawa, H. and Juodkazis, S. (eds.), *3D laser microfabrication. Principles and applications*. Weinheim: Wiley, pp. 139–175.

Ion, C. J. (2005), *Laser processing of engineering materials: Principles, procedure and industrial application*. Oxford: Elsevier Butterworth-Heinemann.

Jain, K. (1990), *Excimer laser lithography*. Bellingham: SPIE.

Jersch, J. and Dikmann, K. (1996), 'Nanostructure fabrication using laser field enhancement in the near field scanning tunneling microscope tip', *Applied Physics Letters*, **68**, 868–870.

Koundourakis, G., Rockstuhl, C., Papazoglou, D., Klini, A., Zergioti I., Vainos, N. A. and Fotakis, C. (2001), 'Laser printing of active optical microstructures', *Applied Physics Letters*, **78**, 868–870.

Laude, L. D. (ed.) (1994), *Excimer lasers*. Dordrecht: Kluwer Academic Publishers.

Letokhov, V. S. and Minogin, V. G. (1979), 'Cooling, trapping, and storage of atoms by resonant laser fields', *Journal of the Optical Society of America*, **69**, 413–419.

Mailis, S., Anderson, A. A., Barrington, S. J., Brocklesby, W. S., Greef, R., Rutt, H. N., Eason, R. W., Vainos, N. A. and Grivas, Ch. (1998), 'Photosensitivity of lead germanate glass waveguides grown by pulsed laser deposition', *Optics Letters*, **23**, 1751–1753.

Mailis, S., Zergioti, I., Koundourakis, G., Ikiades, A., Patentalaki, A., Papakonstantinou, P., Vainos, N. A. and Fotakis, C. (1999), 'Etching and printing of diffractive optical microstructures by femtosecond excimer laser', *Applied Optics*, **38**, 2301–2308.

Mazingue, T., Escoubas, L., Spalluto, L., Flory, F., Socol, G., Ristoscu, C., Axente, E., Grigorescu, S., Mihailescu, I. N. and Vainos, N. A. (2005), 'Nanostructured ZnO coatings grown by pulsed laser deposition for optical gas sensing of butane', *Journal of Applied Physics*, **98**, 74312–74316.

Pronko, P. P., Dutta, S. K., Du, D. and Singh, R. K. (1995), 'Thermophysical effects in laser processing of materials with picosecond and femtosecond pulses', *Journal of Applied Physics*, **78**, 6233–6240.

Roberts, S. (1955), 'Interpretation of the optical properties of metal surfaces', *Physical Review*, **100**, 1667–1671.

Siegman, A. E. (1986), *Lasers*. Sausalito: University Science Books.

Sigel, R., Fytas, G., Vainos, N., Pispas, S. and Hadjichristides, G. (2002), 'Pattern formation in homogeneous polymer solutions induced by a continuous wave visible laser', *Science*, **297**, 67–69.

Steen, W. M. and Mazumder, J. (2010), *Laser material processing*. London: Springer.

Urech, L. and Lippert, T. (2010), 'Photoablation of polymer materials'. In Allen, N. S. (ed.), *Photochemistry and photophysics of polymer materials*. Hoboken: Wiley, pp. 541–568.

Vainos, N. A., Mailis, S., Pissadakis, S., Boutsikaris, L., Dainty, P., Parmitter, Ph. and Hall, T. J. (1996), 'Excimer laser use for microetching computer-generated holographic structures', *Applied Optics*, **35**, 6304–6319.

Vainos, N. A., Grivas, Ch., Fotakis, C., Eason, R. W., Anderson, A. A., Gill, D. S., Shepherd, D. P., Jelinek, M., Lancock, J. and Sonsky, J. (1998), 'Planar waveguides of Ti:Sapphire, Nd:GGG and Nd:YAG grown by pulsed laser deposition', *Applied Surface Science*, **129**, 514–519.

Vainos, N. A., Klini, A. and Prassas M. (1999), 'PLD grown waveguide amplifiers for telecoms', *Private and Confidential Results for Corning Inc.*

Vainos, N. A., Tsigara, A., Manasis, J., Giannoudakos, A., Mousdis, G., Vakakis, N., Kompitsas, M., Klini, A. and Roubani-Kalantzopoulou, F. (2004), 'Metal/metal-oxide/metal etalon structures grown by pulsed laser deposition', *Applied Physics A: Materials Science & Processing*, **79**, 1395–1397.

Yariv, A. (1986), *Quantum electronics*. New York: Wiley.

Youden, K., Eason, R. W., Gower, M. C. and Vainos, N. A. (1991), 'Epitaxial growth of $Bi_{12}GeO_{20}$ thin-film optical waveguides using excimer laser ablation', *Applied Physics Letters*, **59**, 1929–1931.

Zergioti, I., Mailis, S., Vainos, N. A., Papakonstantinou, P., Kalpouzos, C., Grigoropoulos, C. P. and Fotakis, C. (1998), 'Microdeposition of metal and oxide structures using ultrashort laser pulses', *Applied Physics A: Materials Science & Processing*, **66**, 579–582.

Zergioti, I., Papazoglou, D. G., Karaiskou, A., Vainos, N. A. and Fotakis, C. (2002), 'Laser microprinting of InO_x active optical structures and time resolved imaging of the transfer process', *Applied Surface Science*, **197**, 868–872.

Part I
Laser-induced growth of materials and surface structures

2
Emerging pulsed laser deposition techniques

R. W. EASON, T. C. MAY-SMITH, K. SLOYAN, R. GAZIA, M. DARBY and A. SPOSITO, University of Southampton, UK

Abstract: This chapter describes our progress to date in the growth of thin film optical waveguides using both single-beam and multiple-beam/multiple-target geometries using single-crystal garnet targets. More than one target offers the possibility of tuning both the optical and physical properties of the films produced via control of the temporal sequencing of the laser plumes from separate targets. Mixed, layered and superlattice growth, which involve vertical engineering of the film structure, allow precise control of the refractive index profile, a fundamental parameter in the design of optical waveguides. Multi-plume geometries can also be used to structure the thin film in the horizontal plane, for fabricating complex buried structures, a valuable capability for designing thin disk lasing devices.

Key words: pulsed laser deposition, laser ablation, thin crystal films, optical waveguides, garnet.

2.1 Current state-of-the-art in pulsed laser deposition (PLD)

While the use of pulsed lasers for deposition of thin films from laser-produced plasmas dates back more than 40 years, it was not until the late 1980s that pulsed laser deposition (PLD) was demonstrated to be a viable technique for growing functional materials with useful end applications. These early publications addressed the growth of thin films of high temperature, superconducting complex quaternary oxides (Dijkkamp *et al.*, 1987), and it is these first reports of successful growth by PLD that continue to fuel activities within the thin film materials community using this rapid, flexible and comparatively simple method for growing a vast array of single- and multi-component materials (Eason, 2007).

As outlined in the previous chapter, in its simplest form PLD is a physical vapour deposition process that involves four basic components: a pulsed laser (typically a nanosecond pulse duration UV laser such as an excimer), a vacuum chamber, a target that provides the necessary source of atoms and ions for deposition and a suitable substrate located some few centimetres from the target, on which the desired film can grow. Even with this basic level of equipment, thin films can be grown at a rate of a few microns per

55

hour, using a typical 10 Hz pulsed UV laser with ~ 100–500 mJ per pulse, over a substrate area of ~ 1 cm². If substrate heating is required, which is a routine requirement for growth of single-crystal films, then a radiative or filament-based heater can be used, and temperatures of order 800°C are easily achievable. The choice of substrate will reflect the usual requirements for epitaxial growth, which necessitates appropriate lattice matching between the single-crystal substrate and the crystalline thin film.

In the area of photonic structures, PLD has enormous benefit in both rapid prototyping, and also the ability to grow dielectrics, single crystals, glasses and functional materials such as piezoelectrics, ferroelectrics and transparent conducting oxides. Where there are restrictions such as cross-contamination, which present severe problems in molecular-beam epitaxy (MBE) growth for example, where minute traces of Li can severely alter the desired mobilities for a particular semiconductor device structure, PLD is much more tolerant, and optical hosts such as different compound oxides can be grown sequentially in the same chamber. Current activity in optical waveguide growth, planar optical sensors, microwave devices and surface acoustic wave devices as described in Eason (2007) illustrates the intrinsic versatility of photonic material growth, and this is an increasingly viable technique for research and development activity.

A typical schematic of a PLD chamber is shown in Fig. 2.1, which also shows details of the vacuum pumps and gas handling services. Typically, for growth of oxide materials, a background gas pressure of ~ 10^{-2} mbar of oxygen is used, although other gases such as nitrogen or argon or even ammonia can also be useful, for both moderating the dynamics and kinetics of the laser plume and also as a source of nitrogen (from laser dissociation of the NH_3), if nitride growth is the intended goal, for example.

Also shown in Fig. 2.1, however, is a more sophisticated and controllable mechanism for heating of the substrate: a CO_2 laser operating at a wavelength of 10.6 μm. The beam can be either focused and raster-scanned across the substrate or spatially re-formatted into a square profile ideally matching the exact size of the substrate (May-Smith *et al.*, 2007a). Laser heating is an extremely efficient method, as almost all substrates that are routinely used for PLD growth absorb strongly at 10.6 μm, and a laser power of ~ 25 W is capable of heating 1 cm² substrates to temperatures of > 1000°C. Additionally the use of a non-contact heating method means that substrates can easily be rotated while being heated, potentially resulting in much more uniform film growth.

Shown in Fig. 2.2a is a raster-scanned pattern of the incident CO_2 laser beam (made visible by the collinear visible red diode laser witness beam) for a stationary substrate. In contrast, a circular ring pattern, like that in Fig. 2.2b, allows the substrate to be rotated while still experiencing a uniform heating profile. Such an arrangement is far from standard on

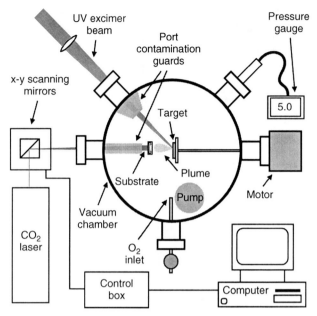

2.1 Schematic of a single-beam/single-target PLD chamber incorporating indirect laser heating of substrates.

most PLD set-ups, but the flexibility to heat while rotating the substrate is of tremendous value. The substrate is held during rotation by alumina supports, resulting in minimal heat-sinking of the substrate. Using a CO_2 laser power of ~ 100 W has allowed us to reach a temperature of greater than 2000°C, sufficient to melt a single-crystal sapphire substrate (melting temperature of 2040°C). Other advantages of non-contact laser heating are the faster substrate temperature ramping and benefits in terms of reduced chamber temperature from such a localised lower total power requirement.

Numerous laboratories use such PLD facilities to grow thin films for materials research applications, but there are few organisations that undertake commercial research, where wafer-scale films are the accepted standard. For substrates that are larger than a few square centimetres, the necessarily higher laser energy can become prohibitive – single pulse energies of more than 1 J are required. An alternative procedure of automated target tilting during growth may be used, so that the laser plume can be 'spray-painted' across the surface of the substrate. However, this also has drawbacks, as the elemental distribution within the plume can be angularly dependent, and hence the stoichiometry of the film may no longer mirror the stoichiometry of the original target, an end result that is not usually desirable.

(a)

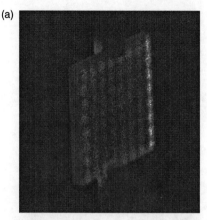

(b)

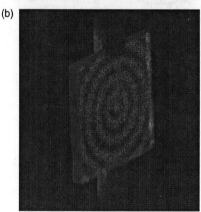

2.2 (a) Substrate under CO_2 laser heating via an 8 × 8 point raster scan. (b) Substrate under CO_2 laser heating via concentric circle profile patterning.

2.2   Problems for growth of thick films and designer refractive index profiles

An area of particular interest for PLD-grown thin films is the fabrication of optical waveguides, where both the substrate and final thin film are both in single-crystal format. Techniques such as evaporation, sputtering and other physical vapour deposition approaches are in general not suitable for growth of single-crystal optical materials. Amorphous or even polycrystalline films are often unsuitable for optical applications, as are films with unacceptably high propagation losses in-plane. If the material being deposited is isotropic (for example the cubic crystalline garnet hosts as discussed later) then poly-crystallinity is much less problematic, as the propagating light does not see any significant variation in refractive index, which would lead to optical scattering and hence unacceptably high values of optical loss. If, however, the thin film material is anisotropic (for example uniaxial or biaxial crystals) then polycrystallinity or disordered structures can produce extremely lossy films, which render them effectively useless for any final waveguide applications. Aligned epitaxial single-crystal films are therefore the prime end goal of our waveguide growth programme.

2.2.1 Growth of thick optical films

The next and most important aspect of optical waveguides concerns the thicknesses required to allow the propagation of fundamental mode optical beams: in almost all cases we wish to grow waveguides that will only guide the lowest order Gaussian spatial profile. The thicknesses needed in turn depend on the refractive index difference between the core that guides the mode, and the substrate, as illustrated schematically in Fig. 2.3. Optical waveguiding also requires that the core has a higher refractive index than the substrate.

For practical optical waveguides grown from targets doped with an active lasing ion, for example Nd in the crystal host $Y_3Al_5O_{12}$ (YAG), the inclusion of a typical dopant concentration of ~ 1% will only raise the refractive index by around 0.1%. If the substrate is also YAG, for example, relatively thick waveguides of ~ 10 μm are hence required for fundamental mode propagation. Such homoepitaxial growth is comparatively straightforward, but waveguide thicknesses of 10 μm may take a comparatively long time to grow. Heteroepitaxial growth of another garnet with much higher refractive index than the substrate, for example $Gd_3Ga_5O_{12}$ (GGG), whose refractive index is higher than YAG by around 7%, may be preferable, as single mode guiding can be achieved in a much thinner film.

Table 2.1 comprises a list of garnet crystals that we have investigated so far, which includes the previously mentioned YAG and GGG. Immediately

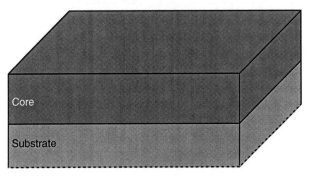

2.3 Basic planar waveguide structure of core and substrate.

Table 2.1 Physical properties of nine garnet crystals

Material	Refractive index (at 1.06 μm)	Thermal expansion (× 10⁻⁶K⁻¹)	Lattice constant (Å)
YAG	1.82	6.9	12.006
YbAG	1.83	8.6	11.939
YSAG	1.86	—	12.271
GSAG	1.89	7.7	12.389
GAG	—	—	12.113
YGG	1.91	—	12.273
YSGG	1.93	8.1	12.446
GSGG	1.94	8.0	12.544
GGG	1.95	8.3	12.383

Note: Where '—' appears, no reliable data have yet been found in the literature.
Source: After May-Smith and Eason (2007).

apparent are the differences in both thermal expansion coefficient and lattice constant for this range of garnets, so that there is an intrinsic mismatch between the substrate and core. Lattice mismatch is rarely a significant problem for thick film growth as the lattice of the film will fully relax to its natural value some small distance (of order ~ 1 μm) from the interface, and the presence of the interfacial defects does not usually pose much of a problem. A lattice mismatch of a few per cent is readily accommodated in single-crystal PLD growth. For thicknesses of a few microns, mismatches of < 5% are acceptable (Tomashpolskii and Sevostianov, 1980) while for PLD growth where bonding is weak, and epitaxy occurs via island growth, a lattice mismatch of > 13% has been reported (Wang *et al.*, 1991). More serious, however, is the mismatch in thermal expansion coefficients; a film grown at 800°C, for example, can exhibit severe distortion when cooled down to room temperature and removed from the chamber, to the extent that the substrate is bent from its original flatness into a shape as shown (with some exaggeration) in Fig. 2.4. Providing the deformation is within reasonable limits, most structures will survive this degree of bending, and do not show signs of cracking or delamination. However, the end-face polishing step that is a prerequisite for efficient launching of light into the waveguide will often cause the inbuilt tensile or compressive stress to be released. The core can shatter as a result, even to the extent of cracking the substrate into several pieces.

Film thicknesses of up to ~ 1 μm are generally not subject to such problems; however, it is experimentally difficult to efficiently launch light into so thin a waveguide. We have grown films of Nd-doped GGG on YAG at thicknesses of >100 μm, all of which have undergone cracking, particularly where two end faces meet and the tension/compression is a local maximum. Growth of films of several tens of microns thickness required for multi-mode

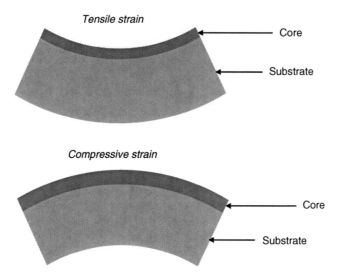

2.4 Schematic of effects of cooling from elevated temperature on thick PLD films with dissimilar thermal expansion coefficients between core and substrate.

planar amplifier structures also often result in cracking or poor yield. What is needed is a growth methodology that yields films that are tolerant to thermal coefficient mismatch (May-Smith *et al.*, 2011) and/or uses combinations of several targets to grow films whose properties, such as expansion coefficient, can be averaged to more closely match that of the substrate. We will return to this topic in subsequent sections where we discuss multiple target PLD geometries.

2.2.2 Choice of targets and minimisation of particulates

Since the very first demonstration of the PLD technique, one of its chief limitations has been the presence of 'particulates' in the film being deposited. These particulates are small (nanometre to micron-scale particles), and can originate both from the laser–target interaction, via the liquid phase and subsequent 'splashing' and exfoliation, and during the plasma transport to the substrate. Such particulates are a permanent distraction and their removal is a universal goal for all engaged in PLD research. Of particular annoyance for optical waveguide applications is the fact that particulate size is comparable to the optical wavelengths used, and hence scattering and waveguide propagation loss is a constant problem.

Particulate removal has been attempted by many groups using a wide variety of techniques, including velocity filtering (Yoshitake and Nagayama, 2004),

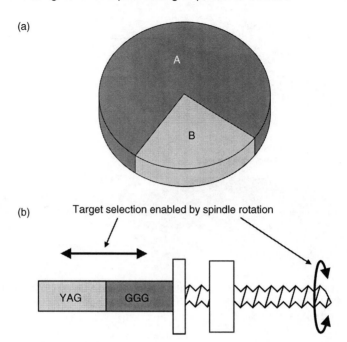

2.5 (a) Composite sector target of dissimilar materials A and B.
(b) Composite target formed by joining of two dissimilar rods.

shutters (Chen, 1994), synchronised pulsed gas jets (Barrington *et al.*, 2000) and also use of ultra-short pulse (picosecond and femtosecond) lasers (Gamaly *et al.*, 2007). It is now accepted, however, that all single-target geometries will suffer from particulate problems to a greater or lesser extent. One simple solution that at least mitigates the problems of particulates lies in the nature of the target itself. For high quality optical waveguide growth, we routinely use single-crystal (and hence fully dense) targets. If alternative sintered or compressed target materials are used, then the inevitable reduction of density from the ideal 100% of 'bulk' value leads to the presence of voids, or other inclusions, and these are a ready source of particulate generation.

Composite or sectored multi-component targets can also be used in single-beam PLD, in an attempt to grow films whose composition reflects the target geometry. In its most simple implementation, a sector target as shown in Fig. 2.5a can be fabricated, but this is usually made from sintered precursors so again suffers from the problems of particulates. An obvious alternative, and one that uses single-crystal targets, is shown in Fig. 2.5b; by controlling the point of incidence of the pulsed laser, as well as the rotation speed and/or pitch of the lead-screw, multi-component laser plumes can be generated.

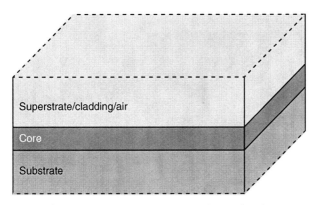

2.6 Basic planar waveguide structure incorporating a superstrate or upper cladding layer.

In practice, a more favourable approach can be to move to multi-target and multi-beam geometries, discussed in detail in Section 2.3. However, before we leave single-beam geometries behind us, one final point needs to be made concerning particulates: if you can't eliminate them, then one option is to deposit more material over them. In the case of cubic and hence optically isotropic garnet films, particles present within the volume of the film, that have since been buried, will contribute little to the final loss of the optical waveguide grown, provided the surface density of particles is not excessive. Admittedly, for a simple core/substrate waveguide, the optical mode does sample the top surface layer of the core region, and hence particulates are a problem on the surface. In this case, the solution is to grow a capping or superstrate layer, as shown in Fig. 2.6, so that the final waveguiding region itself is buried. Surface contamination of the top surface of the superstrate is then no longer a contributing factor to final waveguide propagation loss. To date, and as a benchmark of the quality of the garnet films grown by PLD so far, the lowest loss reported in a doped GGG on YAG waveguide was less than 0.1 dB cm^{-1} (Grivas *et al.*, 2004; May-Smith *et al.*, 2004).

2.3 Multi-beam PLD

With only a very few exceptions, such as the work reported in György *et al.* (2006) and Kompitsas *et al.* (2007), the use of more than one laser source to perform PLD is so far uncommon. If sequential deposition of materials is required to fabricate layered or mixed compounds, then target carousels can be used and specific targets selected as appropriate. However, if more sophisticated multilayers (addressed in a later section) are required and many thousands of sequential depositions are needed, then it becomes

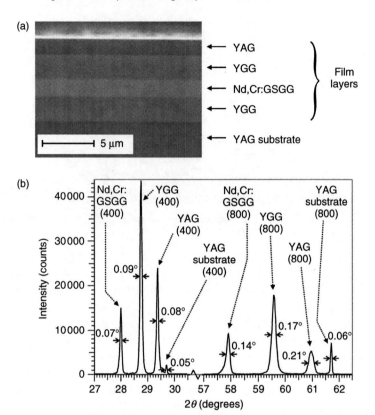

2.7 (a) SEM image of five layer symmetric multilayer structure (after May-Smith *et al.*, 2007b). (b) XRD spectrum of multilayer structure shown in (a) (after May-Smith *et al.*, 2007b).

increasingly impractical to alternate targets at timescales of potentially once per second. However, growth from two (or more) different targets using a single laser source can yield high-quality structures (see Fig. 2.7).

Figure 2.7a shows moderately thick (1–2 μm) single-crystal layered growth of YGG ($Y_3Ga_5O_{12}$), and the Nd and Cr doped quaternary garnet GSGG (Nd,Cr:$Gd_3Sc_2Ga_3O_{12}$), with YAG as both substrate and superstrate (May-Smith *et al.*, 2007b). The excellent crystallographic quality is shown in Fig. 2.7b, where all XRD peaks are clearly depicted for each single-crystal layer. Note, however, that due to non-optimised growth conditions for the YAG capping layer, the PLD-grown YAG shows an XRD peak position that is not coincident with the YAG substrate. This situation is not uncommon in all our PLD-grown garnets, as the basic garnet formula of $A_3B_5O_{12}$ is tolerant to variation between the A and B atomic ratios. This topic will be re-addressed in Section 2.3 when we consider Ga content in GGG growth via multi-beam deposition for stoichiometric correction.

(a)

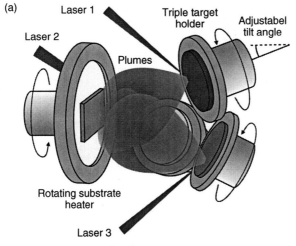

(b)

2.8 (a) Schematic of three laser/three target PLD set-up with independently addressable targets and laser-heated rotatable substrate (after Eason *et al.*, 2009). (b) Close-up view of three target assembly. Each target rotates and also has adjustable tilt angle.

2.3.1 Multi-target, multi-beam geometry

Shown in Fig. 2.8a is the multi-target PLD set-up, which has three separately controllable lasers that can be directed to three separate targets, $T_{1,2,3}$. This arrangement now brings an enormous further potential to PLD growth, as the three laser plumes generated can be synchronous, time-delayed or of different relative strengths or repetition rates, and films can be grown that have graded, layered, superlattice, or fully mixed character. Shown in Fig. 2.8b is a close-up picture of the three target assembly with three different single-crystal garnet targets in place. Each target rotates and has a

Table 2.2 Relevant physical properties for YSGG, GSGG and GGG extracted from Table 2.1

Material	Refractive index (at 1.06 µm)	Thermal expansion (× 10⁻⁶K⁻¹)	Lattice constant (Å)
YSGG	1.93	8.1	12.446
GSGG	1.94	8.0	12.544
GGG	1.95	8.3	12.383

programmable tilt capability for directional plume management. The rest of this chapter will therefore address a range of these possibilities, and show our progress to date in implementing such novel film geometries.

2.3.2 Layered growth: different targets with plumes temporally coincident

This arrangement is the next logical extension to the results presented in Section 2.3, but involves growth from more than one target, and mixed layers can therefore be produced. For optical waveguide growth this is exactly what is needed for producing specific refractive index profiles for the core region, or the interfacial layers that form the core–cladding boundaries. This mixed layer approach also presents the intriguing possibility of engineering not only refractive index but also thermal expansion coefficient, where the properties of each separate target $T_{1,2,3}$ can be averaged in the ratio of their intrinsic values and their relative abundances in the final film. As an example, we consider the three garnets GSGG, GGG and $Y_3S_2Ga_3O_{12}$ (YSGG), where a mixed film of GSGG and GGG is grown on a YSGG substrate. The aim is to grow a waveguiding layer with the requisite refractive index higher than that of the substrate, but with the thermal expansion coefficient matched as closely as possible, to eliminate the bending and/or cracking on cooling to room temperature discussed earlier. Using the data as in Table 2.2, it is easy to design the correct a priori recipe for growth, under the assumption that such mixing and averaging of materials properties holds true.

Figure 2.9 illustrates such a scenario whereby materials mixing can produce a waveguide structure whose end properties are closely matched using a film grown with a composition ratio of 66% GSGG to 33% GGG. Of particular relevance is the fact that matching of thermal expansion coefficient and lattice constant (to a difference of less than 0.4%) should be possible between film and substrate, while maintaining the required higher refractive index for the core. With this design concept in mind, a test structure was grown incorporating such hybrid garnet phases, using GGG and YSGG as

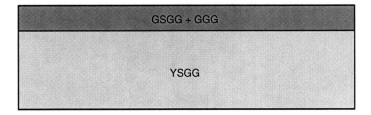

2 parts of GSGG + 1 part of GGG:

Film refractive index = ((2×1.94)+(1.95))/3 = 1.943

Film thermal expansion coefficient = ((2×8.0)+(8.3))/3 = 8.1×10^{-6}K^{-1}

Film lattice constant = ((2×12.544)+(12.383))/3 = 12.49 Å

2.9 Schematic of planar waveguide structure of GSGG + GGG grown on a YSGG substrate, and estimate of materials properties derived though simple averaging.

the film constituents; YSGG was chosen over GSGG for this test as a target was already available. The structure consisted of seven layers arranged symmetrically to achieve a symmetrical refractive index profile. The compositions and sequence of the layers are shown in Table 2.3. Layers 1 and 7 are made of pure YSGG, layers 2 and 6 are made of Nd:GGG and YSGG mixed in the ratio 1:2, layers 3 and 5 consist of the same precursors but mixed in a ratio 2:1, and finally layer 4 is the core of the waveguide and is made from Nd:GGG (Gazia *et al.*, 2008).

The growth time here is long, as the lasers used for ablation were low energy (100 mJ per pulse) Q-switched frequency quadrupled Nd:YAGs, whose wavelength of 266 nm is longer than the more conventional excimer wavelengths of 248 or 193 nm. However, the aim here was to determine whether such mixed layers could be grown, and to look at their crystal quality and final stoichiometry. Shown in Fig. 2.10a is a schematic of the layered film structure, while in Fig. 2.10b is an energy dispersive X-ray analysis (EDX) line scan of elemental composition acquired from a cross-sectional region of the film. It is clear (within the limited spatial resolution afforded by EDX) that the elemental composition follows the intended growth profile and that the final ~ 10 μm thick structure is indeed fully single crystal, as identified by XRD (not shown here) where peaks corresponding to each separate garnet phase as well as the hybrid mixed phases are seen. As a final example of the growth of an even more complicated layered structure, Fig. 2.11 shows the results for a thinner 11 layer composite, again using GGG and YSGG garnet targets (Eason *et al.*, 2009).

Table 2.3 Deposition parameters for each of the seven layers in the multilayer sample

Layer	Concentration of YSGG (%)	Concentration of Nd:GGG (%)	Laser repetition rate for YSGG target (Hz)	Laser repetition rate for Nd:GGG target (Hz)	Duration of deposition (min)
1	100	0	10	0	256
2	66	34	10	2	202
3	36	64	5	3.33	156
4	0	100	0	10	98
5	36	64	5	3.33	156
6	66	34	10	2	202
7	100	0	10	0	256

Source: After Gazia *et al.* (2008).

(a)

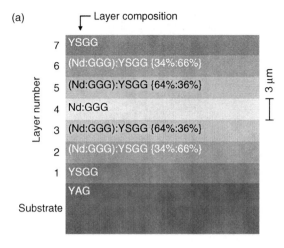

(b)

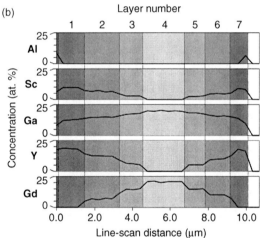

2.10 (a) Schematic of seven layer symmetric waveguide grown on a YAG substrate (after Gazia *et al.*, 2008). (b) EDX line scan results of elemental composition overlaid on a schematic of the structure in (a) (after Gazia *et al.*, 2008).

2.3.3 Mixed growth: different targets with plumes temporally coincident

As well as growing layers that are intended for optical waveguide end applications, there is an even more obvious need to grow materials as single layer films, where the stoichiometry can be tuned by growth from more than one target. As has been mentioned in Section 2.3, Ga is an element whose

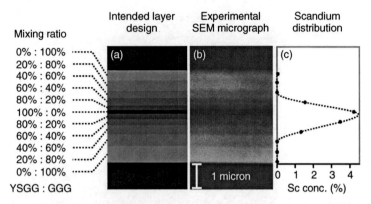

2.11 Composite picture of design schematic, SEM image and EDX line scan of scandium concentration in an eleven layer symmetric structure (after Eason *et al.*, 2009).

concentration in the final film can be less than that present in the target due to preferential re-sputtering from the growing film, as well as its intrinsic volatility and low melting point. Both can routinely lead to a deficiency of ~ 10% compared to the ideal 5:3 ratio between Ga and Gd in stoichiometric GGG, for example. If two targets are used, one being a standard GGG single-crystal target while the other is gallia (Ga_2O_3), then by varying either the number of pulses used for each target, or the laser fluence used to ablate either target, an empirical correction procedure can be adopted to alter the final Ga concentration in the resultant film.

As a demonstration of this dual beam ($T_1 = GGG, T_2 = Ga_2O_3$) technique, a series of eight depositions were performed under different relative values of fluence for T_1 and T_2 while all other parameters were kept the same. Figure 2.12 shows a composite XRD plot for five of the films that showed XRD peaks together with the values of Gd/Ga ratio as determined by EDX analysis (Darby *et al.*, 2008). It is seen that compositions that are both Ga deficient, (Gd/Ga > 0.6) and, more surprisingly, Ga rich (Gd/Ga < 0.6) can be grown; this situation is far from our normal experience when growing from single stoichiometric targets, which are almost always Ga deficient by ~10%. The introduction of an increasing amount of gallium from the Ga_2O_3 target is shown to decrease the lattice parameter of the GGG film (XRD peak appears at higher values of 2θ) and, for an optimum fluence used on the Ga_2O_3 target, make the film closer to stoichiometric GGG. The lattice shifting from the database stoichiometric value is believed to occur due to a varying amount of Gd^{3+} occupying lattice sites that would usually be occupied by Ga^{3+}, but are left open due to a Ga deficiency in the films. For Ga concentrations above that of stoichiometric GGG, the lattice parameter continued to reduce. This is thought to be due to the occupation of some

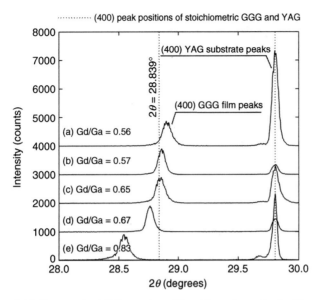

2.12 Composite XRD spectra of five films deposited with various fluences on the Ga_2O_3 and Nd:GGG targets of (a) 1.7 and 1.5 J/cm², (b) 1.7
and 2.3 J/cm², (c) 1.1 and 2.3 J/cm² and on the Nd:GGG target only of
(d) 2.3 J/cm², (e) 1.5 J/cm². The dotted line corresponds to the position
of the (400) GGG peak for stoichiometric composition ($2\theta = 28.839°$).
The intensities of the GGG (400) peaks have been normalised (after
Darby *et al.*, 2008).

lattice sites normally occupied by Gd^{3+} by the excess Ga^{3+} and/or to the
incorporation of the excess Ga in the lattice in the form of defects.

This mixed growth PLD approach brings increased flexibility and control,
specifically for materials that have previously proven difficult to deposit with
their ideal stoichiometric composition. A critically important application of
this technique is the minimisation of the lattice mismatch between the film
and the substrate by choosing a combination of materials, when mixed in
the correct ratio, that will grow with the appropriate lattice parameter. An
additional bonus to using multiple targets lies in adopting the 'cross-beam'
geometry that has been successfully adopted in Gaponov *et al.* (1982) and
Gorbunoff (2007) to minimise the concentration of particulates present in
the final film.

As seen in Fig. 2.13, a suitable blockage placed in front of the substrate
can provide a geometrical line-of-sight barrier to the incidence of particulates. These particulates, being massive in comparison to the atoms and ions
present in the plasma plumes, are ballistic in nature and travel in essentially
straight lines, thereby reaching the growing film in much fewer numbers
that would be present with a single-target geometry. The atomic and ionic

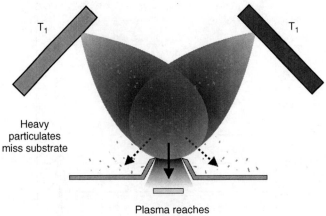

2.13 Schematic of cross-beam PLD technique for minimising particulates on the substrate (after Gorbunoff, 2007).

constituents of the laser plasma, however, will undergo numerous collisions during their transit to the substrate, and hence experience a deflection along the direction of their common intersection. An ideally particulate-free film should result, albeit with a relatively low growth rate. The expansion to three plasma sources (not yet attempted by us) allows out-of-plane plasma plume interactions and this has implications for even better particulate trajectory and removal strategies.

2.3.4 Layered growth: different targets and plumes temporally alternating

A further implementation involves two different targets and depositions that are no longer temporally coincident, with the intention of producing layered growth of materials whose thicknesses are controlled by the relative number of pulses on each target. For materials that have a compatible set of growth parameters, this is easy to achieve, and, as shown in Fig. 2.14, a simple mechanical shutter arrangement can be used, under computer-programmable control.

Targets T_1 or T_2 can be ablated simultaneously, to grow a mixed film as in Section 2.3.3, or sequentially, to grow multilayered structures. As a preliminary experiment, ten samples were grown and characterised via XRD in order to determine the number of pulses required to produce a film where individual layers of materials from these dissimilar targets can be resolved. The number of pulses on each target was varied between 5 and 5000 over the ten films. Each pulse produced a thickness of ~ 0.02 nm, and depositions

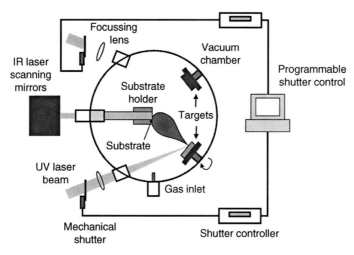

2.14 Schematic of dual-target PLD chamber with programmable shutter control (after Sloyan *et al.*, 2010).

were performed to yield periodic structures, where the number of pulses on T_1 equalled that on T_2, as well as aperiodic and chirped, where the number of pulses varied as a function of time. The growth of layered structures has particular relevance to the fabrication of Bragg structures where the refractive index contrast between repeated layers of these crystalline materials can produce single-crystal reflective stacks, for use in integrated thin film lasing disks and on the end faces of crystalline waveguides. Chirped structures allow a broader spectral tunability of reflection, albeit at reduced overall efficiency.

From the comparatively small thickness per pulse, and considering that the unit cell dimensions for GSGG and GGG are 1.270 and 1.247 nm respectively, it is reasonable to assume that more than ~60 pulses are required to grow a single unit cell. However, the garnet structure is not simple, containing eight planes within each unit cell. It is therefore feasible that some layering could be present even at thicknesses less than one unit cell, although this cannot be measured in the current set-up. Shown in Fig. 2.15 is a set of ten XRD spectra for depositions where the number of pulses was varied over three orders of magnitude, for the GSGG/GGG layered film growth (Sloyan *et al.*, 2010).

For the five shot case, the appearance of a peak at a value of 2θ approximately midway between the database values for GSGG and GGG is to be expected, suggesting a truly mixed film. However, when the number of pulses per layer exceeds ~100, satellite peaks can be observed either side of this main central peak. These peaks are indicative of a superlattice structure (Craven *et al.*, 2003; Ishibashi *et al.*, 2000) and their symmetrical 2θ displacement from the central mixed peak can be used to precisely calculate the periodicity of the superlattice grown. The intensity of the central peak

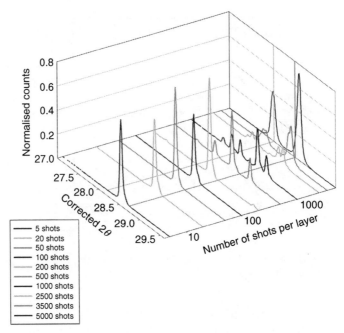

2.15 X-ray diffraction spectra for superlattices fabricated via alternating pulse bursts per target. The number of shots per burst on each target ranges from 5 to 5000 over the range of samples. The peak at $2\theta = 29.74°$ (not shown here) corresponds to the YAG substrate and is a convenient feature for normalisation of all the XRD spectra.

decreases as the number of shots per burst increases from 100 to 1000. From 2500 shots per layer onwards, the central peak is not observed, the satellite peaks become less significant and the pattern approaches that of the two individual component materials.

Observation of these satellite peaks, and more importantly their relative intensities compared to the central mixed peak are good indicators of the quality and uniformity of the layers grown. For PLD conditions that are stable, the thickness of a film deposited for 100 laser pulses, for example, can be extremely repeatable, and this is a prerequisite for the observation of such superlattice satellite peaks. Deposition runs have also been performed where the number of shots per layer has been varied from 700 to 100, in steps of 50, to fabricate a simple chirped stack, and also from 200 to 600, also in steps of 50, as shown in Fig. 2.16, but with varying numbers of total shots per section, to fabricate a compensated chirped structure so that each section has approximately the same thickness.

In fact, given the ease with which this shuttered PLD growth can be automated, growth of layers or stacks to any desired final design is now routine, and our next step is to fabricate designer Bragg stacks that we have

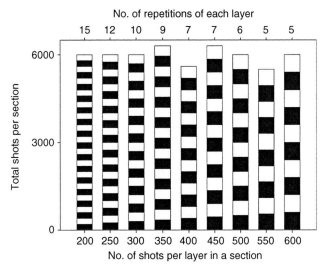

No. of repetitions of each layer

2.16 Diagram illustrating the number of repetitions of each layer in each component superlattice section. Layers of GGG and GSGG are represented by black and white sections respectively. The number of layer repetitions was chosen so that the overall thickness of each section is approximately equal (~ 6000 shots in total) (after Sloyan *et al.*, 2010).

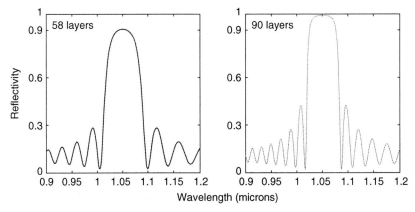

2.17 MATLAB simulations for Bragg multilayers for reflectivities of 90% (58 layers) and 99% (90 layers), using alternate YAG and GGG crystalline layers (after Sloyan *et al.*, 2010).

modelled theoretically and are shown in Fig. 2.17. A structure consisting of 58 alternating YAG and GGG layers on a YAG substrate would theoretically result in 90% reflectivity, while 99% reflectivity could be achieved with a structure consisting of 90 layers. Such structures would have total thicknesses of ~ 8 and ~ 12.5 μm respectively, excluding substrates, a value that is easily achievable with our current PLD set-up.

2.3.5 Mixed growth: different targets with variable relative plume time delay

As a final example of the flexibility afforded by a multi-target PLD system, we show that there is the potential to further tailor material properties by varying the time delay between successive laser pulses on two different targets. These properties include the lattice constant and hence by definition the refractive index of the deposited film, a parameter of fundamental importance in the design of optical waveguides.

At first sight, the relative time delay between successive pulses might not appear to have any major effect. When using more than one plasma plume the most logical choices would be to use either simultaneous plume arrival at the substrate, in which case a mixed film would be grown, or alternating plume arrival, where the film would grow from consecutive deposition from each target. Given our previous results as shown in Section 2.3.4, however, where no evidence of any layering occurred when layer thicknesses were less than 50 shots per target (~ 1 nm thickness), and XRD spectra indicated mixed growth in all cases, the use of alternating laser pulses even with variable time delays was not expected to prove important.

The results, however, showed that for two laser plumes, the time delay can be a crucially important factor in the final film properties (Sloyan et al., 2009). For time delays of order of the transit time for one plume to reach the substrate (some ~ 100 μs for example) the second plume is travelling in the first plume's wake, and hence does not experience the same background gas conditions such as pressure, collisional interaction and dynamic modification of the ion velocity, as would be the case if the two plumes were either synchronous or time delayed by much longer than this characteristic transit time.

In Fig. 2.18 we show a plot of the variation in film lattice constant normal to the film plane, derived from XRD spectral analysis, as a function of increasing relative delay between the two plumes that ranged from 3.2 μs to 50 ms. All delay values here took into account the relative difference in Q-switching times between the two lasers used and so are absolute delay values.

Two distinct regions of lattice constant (A and B) may be observed, with a transition region (C) occurring between ~ 150 μs and ~ 1 ms. The exact position of this transition region appeared to vary slightly dependent on plume order, however the same overall trend was observed both in the case of the GGG plume arriving at the substrate first (followed by the GSGG plume) and vice versa. The measured lattice constant for a delay of 100 ms (i.e., completely synchronous plumes for the 10 Hz laser repetition rate used) shows the same behaviour (within experimental error) as the lattice constant for a delay of 50 ms (i.e., completely asynchronous). An ion probe positioned

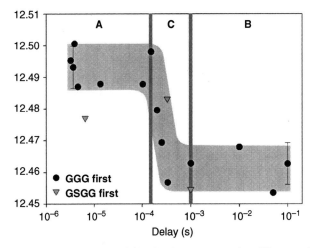

2.18 Plot of measured film lattice constant for different plume delay times. Data have been included for both plume arrival orders (GGG plume first, GSGG plume first) (after Sloyan *et al.*, 2009).

in place of the substrate was used to examine ion signals that occurred for single-target ablation and dual-target ablation under various relative time delays, to help elucidate the reasons behind this change in lattice constant that occurred within region C.

Analysis of these ion signals showed that for time delays of > 100 μs, the second plume was travelling in effectively a lower background gas pressure, as the first plume had swept away the gas in front of the target-probe region. For longer time delays around ~ 400 μs, the ion signal of the second plume was very similar to that of a single plume, and the partial vacuum created by the first plume had been refilled by the background gas. The time to refill this interaction region depends on several factors, but is predominantly a function of the size of the refilled region and the sound velocity in the gas. A rough estimate, however, of the time taken to refill this depleted region yields a value of 170 μs, which is consistent with the delay beyond which the ion signal of the second plume becomes gas-like again. The higher value of lattice constant for delays below ~ 200 μs may therefore be due to higher energy ion bombardment of the growing film, which can lead to in-plane compressive stress. For delays greater than 200 μs, the partial vacuum created by the first plume will be refilled and collisions with oxygen will slow down a significant fraction of the ions in the second plume.

It is interesting to note that the difference in values of lattice constant between regions A and C amounts to some 0.3%, a value that closely matches the difference between core and cladding in thin optical waveguides. The control parameter of relative time delay between plumes may therefore be a simple and controllable tool for growing a refractive index-tailored waveguide (under the assumption that refractive index is a linear function of lattice parameter).

2.4   Use of three different targets: combinatorial growth

In previous sections, we have discussed the use of PLD techniques that use multi-target geometries, but so far have only considered two targets at most. If three targets are used, then the parameter space for conducting depositions opens out still further. Shown in Fig. 2.19 is a schematic of a ternary phase diagram, where essentially any composition can be generated from three targets, A, B and C, under appropriate ablation conditions. The flexibility to grow films with specific or designer properties is therefore enormous, albeit at the experimental cost of running three lasers in place of the previous two for dual-target deposition.

While we have operated our PLD system with three independent lasers and targets, so far dual beam has afforded us a wealth of growth opportunities that we are still exploring. Three laser/three target geometries can also play a role, however, when materials are deposited that require horizontal patterning. In such cases the growth rate and the area of deposition can both benefit from multiple plume geometries.

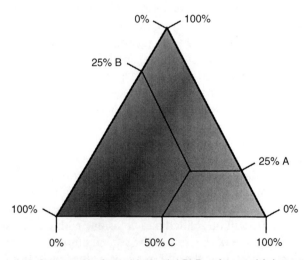

2.19 Schematic of combinatorial PLD using multiple targets, here labelled A, B and C. By control of either laser fluence or laser repetition rate on each target, any film composition is accessible within this ternary phase diagram.

2.4.1 Horizontally patterned structures: masking and local deposition

This final section addresses the use of multiple pulses and targets to grow single-crystal structures in which there is a horizontal (in-plane) patterning from the use of masks or cones. Shown schematically in Fig. 2.20 is the underlying principle behind such horizontal patterned growth.

Using either growth through a slit, as in Fig. 2.20a, or through a cone, as in Fig. 2.20b, there are the inevitable problems of diffusion or shadowing around the physical hard edge of the mask or cone, as the plasma plume will never be confined by such a physical geometry. For a mask, it can be impractical and inconvenient to fully contact to the substrate as removal

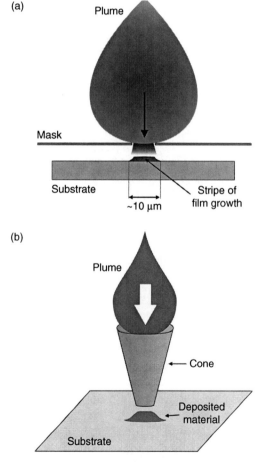

2.20 (a) PLD in restricted geometries: growth through a slit. (b) PLD in restricted geometries: growth through a cone.

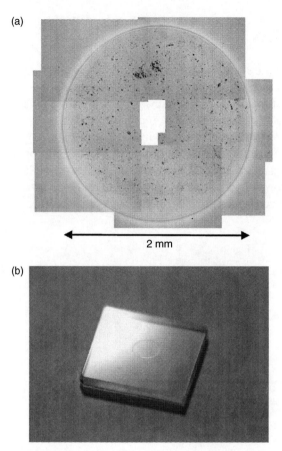

(a)

2 mm

(b)

2.21 (a) PLD growth through a 2 mm diameter laser-machined Si mask, placed in close proximity to the substrate. Notice the pronounced edge-effects around the perimeter, indicating lack of clear geometrical shadowing. (b) PLD growth through a cone of ~ 2 mm exit diameter (photograph by Sam Berry).

after substrate heating and deposition may be difficult and lead to inevitable degradation of edge quality. The alternative of proximity coupling, using mechanical attachment can work, and we have used a thin (25 μm) Si single-crystal mask with a 2.0 mm central laser-machined circular aperture, placed over the substrate and attached via solvent capillary action. Growth through a metal cone has also been tried and results are so far satisfactory.

However, as shown in Fig. 2.21a and 2.21b for physical masking and cone growth respectively, there is an inevitable edge-shadowing effect. For some applications where a secondary overgrowth of another deposited layer is needed, the edge slope, which has been measured to be ~ 0.5°, can prove restrictive. One of our main goals is to fabricate doped thin disk lasing structures (Giesen and Speiser, 2007) with a central doped core of a few millimetres in diameter, surrounded by an undoped cladding region. Even for such a small difference in refractive index of <0.1%, such a slope region will lead to unacceptably high reflectivity for the diode pumps that will be used to pump the central doped core in a waveguide geometry. Our attention has therefore turned to a three target arrangement as shown in Fig. 2.22, which has a central doped plume deposited via a cone and two undoped plumes simultaneously depositing the cladding region.

Success here would be a major achievement, and would result in an ideal thin disk structure that can be waveguide pumped, is all single-crystal format, and has the dopant located in the central lasing region. The final benefit of this three plume arrangement is that the transition region between the undoped cladding and the doped core would be graded, over perhaps a region of ~ 1 mm, an ideal scenario to assist in Gaussian profiling of the laser mode intensity.

2.5   Future work in complex PLD geometries

Figure 2.23a and 2.23b show two generic geometries that are still high on our priority list that are fundamental to all lasing waveguide geometries: graded layers and Gaussian doping. While a graded structure would only require a dual beam set-up, and practically we have already demonstrated this capability in the chirped growth reported in Section 2.3.4, the geometry of Gaussian doping in the plane of the core with additional growth of a

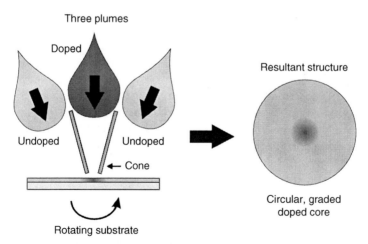

2.22 Three target/three plume geometry for horizontally structured growth for thin disk laser applications.

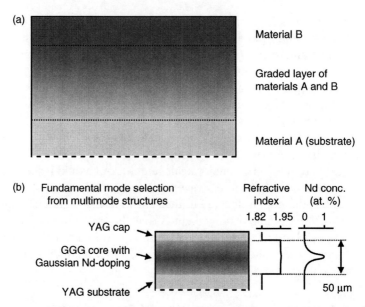

2.23 (a) Schematic of graded interface between materials A and B. (b) Schematic of Gaussian Nd concentration profile within a GGG planar waveguide structure: the final refractive index and Nd concentrations are shown for comparison.

suitable superstrate layer does require a three target set-up, and we intend to report our results on this technique soon.

What is certain, however, is that a multi-target, multi-beam PLD set-up is an extremely versatile and valuable piece of equipment, and will lead to many new techniques, geometries and, eventually, devices for application to a wide range of thin film materials fabrication.

2.6   Conclusions

The extension to multiple lasers and targets in PLD research has been shown to provide an important further degree of freedom. Through the judicious choice of targets and the sequence of ablation on each target as sequential, simultaneous or time-delayed, a range of complex material deposition can be achieved. Multiple laser plumes also allow for horizontally structured growth, in which millimetre scale features can be grown that are fully embedded within a dissimilar material matrix. Although few labs to date have invested in multiple beam PLD geometries, it seems that this may well become routine once the benefits and technological opportunities for thin film growth afforded by multiple plume interaction geometries are appreciated and progressively adopted.

2.7   Acknowledgements

The authors are pleased to acknowledge the contributions and collaborations that we have had in the past on PLD research and publications with the following colleagues: Steve Barrington, Sam Berry, Taj Bhutta, Christos Grivas, Steven Huband, James Lunney, Alistair Muir, Dave Shepherd, Pamela Thomas, David Walker and Michalis Zervas.

2.8   References

Barrington, S. J., Bhutta, T., Shepherd, D. P. and Eason, R. W. (2000), 'The effect of particulate density on performance of Nd:Gd$_3$Ga$_5$O$_{12}$ waveguide lasers grown by pulsed laser deposition', *Optics Communications*, **185**, 145–152.

Chen, L. C. (1994), 'Particulates generated by pulsed laser ablation'. In Chrisey, D. B. and Hubler, G. K. (eds.), *Pulsed laser deposition of thin films*. New York: John Wiley, pp. 167–198.

Craven, M. D., Waltereit, P., Wu, F., Speck, J. S. and DenBaars, S. P. (2003), 'Characterization of a-plane GaN/(Al,Ga)N multiple quantum wells grown via metalorganic chemical vapor deposition', *Japanese Journal of Applied Physics*, **42**, 235–238.

Darby, M. S. B., May-Smith, T. C. and Eason, R. W. (2008), 'Deposition and stoichiometry control of Nd-doped gadolinium gallium garnet thin films by combinatorial pulsed laser deposition using two targets of Nd:Gd$_3$Ga$_5$O$_{12}$ and Ga$_2$O$_3$', *Applied Physics A: Materials Science and Processing*, **93**, 477–481.

Dijkkamp, D., Venkatesan, T., Wu, X. D., Shaheen, S. A., Jisrawi, N., Min-Lee, Y. H., McLean, W. L. and Croft, M. (1987), 'Preparation of Y-Ba-Cu oxide superconductor thin films using pulsed laser evaporation from high T$_c$ bulk material', *Applied Physics Letters*, **51**, 619–621.

Eason, R. W. (ed.) (2007), *Pulsed laser deposition of thin films: Applications-led growth of functional materials*. Hoboken, NJ: John Wiley.

Eason, R. W., May-Smith, T. C., Grivas, C., Darby, M. S. B., Shepherd, D. P. and Gazia, R. (2009), 'Current state-of-the-art of pulsed laser deposition of optical waveguide structures: existing capabilities and future trends', *Applied Surface Science*, **255**, 5199–5205.

Gamaly, E. G., Rode, A. V. and Luther-Davies, B. (2007), 'Ultrafast laser ablation and film deposition'. In Eason, R. W. (ed.), *Pulsed laser deposition of thin films: Applications-led growth of functional materials*. Hoboken, NJ: John Wiley, pp. 99–130.

Gaponov, S. V., Gudkov, A. A. and Fraerman, A. A. (1982), 'Processes occurring in an erosion plasma during laser vacuum deposition of films. III. Condensation in gas flows during laser vapourization of materials', *Soviet Physics – Technical Physics*, **27**, 1130–1133.

Gazia, R., May-Smith, T. C. and Eason, R. W. (2008), 'Growth of a hybrid garnet crystal multilayer structure by combinatorial pulsed laser deposition', *Journal of Crystal Growth*, **310**, 3848–3853.

Giesen, A. and Speiser, L. (2007), 'Fifteen years of work on thin-disk lasers: Results and scaling laws', *IEEE Journal of Selected Topics in Quantum Electronics*, **13**, 598–609.

Gorbunoff, A. (2007), 'Cross-beam PLD: Metastable film structures from intersecting plumes'. In Eason, R. W. (ed.), *Pulsed laser deposition of thin films: Applications-led growth of functional materials*. Hoboken, NJ: John Wiley, pp. 131–160.

Grivas, C., May-Smith, T. C., Shepherd, D. P. and Eason, R. W. (2004), 'Laser operation of a low loss (0.1 dB/cm) $Nd:Gd_3Ga_5O_{12}$ thick (40 μm) planar waveguide grown by pulsed laser deposition', *Optics Communications*, **229**, 355–361.

György, E., Sauthier, G., Figueras, A., Giannoudakos, A., Kompitsas, M. and Mihailescu, I. N. (2006), 'Growth of $Au-TiO_2$ nanocomposite thin films by a dual-laser, dual target system', *Journal of Applied Physics*, **100**, 114302–114305.

Ishibashi, Y., Ohashi, N. and Tsurumi, T. (2000), 'Structural refinement of X-ray diffraction profile for artificial superlattices', *Japanese Journal of Applied Physics*, **39**, 186–191.

Kompitsas, M., Giannoudakos, A., György, E., Sauthier, G., Figueras, A. and Mihailescu, I. N. (2007), 'Growth of metal-oxide semiconductor nanocomposite thin films by dual-laser, dual target deposition system', *Thin Solid Films*, **515**, 8582–8585.

May-Smith, T. C. and Eason, R. W. (2007), 'Comparative growth study of garnet crystal films fabricated by pulsed laser deposition', *Journal of Crystal Growth*, **308**, 382–391.

May-Smith, T. C., Grivas, C., Shepherd, D. P., Eason, R. W, and Healy, M. J. F. (2004), 'Thick film growth of high optical quality low loss (0.1 dB cm^{-1}) $Nd:Gd_3Ga_5O_{12}$ on $Y_3Al_5O_{12}$ by pulsed laser deposition', *Applied Surface Science*, **223**, 361–371.

May-Smith, T. C., Muir, A. C., Darby, M. S. B. and Eason, R. W. (2007a), 'Design and performance of a ZnSe tetra-prism for homogeneous substrate heating using a CO_2 laser for pulsed laser deposition experiments', *Applied Optics*, **47**, 1767–1780.

May-Smith, T. C., Shepherd, D. P. and Eason, R. W. (2007b), 'Growth of a multilayer garnet crystal double-clad waveguide structure by pulsed laser deposition', *Thin Solid Films*, **515**, 7971–7975.

May-Smith, T. C., Sloyan, K. A., Gazia, R. and Eason, R. W. (2011), 'Stress engineering and optimization of thick garnet films grown by pulsed laser deposition', *Crystal Growth & Design*, **11**(4), 1098–1108.

Sloyan, K. A., May-Smith, T. C., Eason, R. W. and Lunney, J. G. (2009), 'The effect of relative plasma plume delay on the properties of complex oxide films grown by multi-laser, multi-target combinatorial pulsed laser deposition', *Applied Surface Science*, **255**, 9066–9070.

Sloyan, K. A., May-Smith, T. C., Zervas, M., Eason, R. W., Huband, S., Walker, D. and Thomas, P. (2010), 'Growth of crystalline garnet mixed films, superlattices and multilayers for optical applications via shuttered combinatorial pulsed laser deposition', *Optics Express*, **18**, 24679–24687.

Tomashpolskii, Y. Y. and Sevostianov, M. A. (1980), 'Epitaxial ferroelectric films', *Ferroelectrics*, **29**, 87–90.

Wang, S. Z., Xiong, G. C., He, Y. M., Luo, B., Su, W. and Yao, S. D. (1991), 'High misfit epitaxial growth: superconducting $YBa_2Cu_3O_{7-x}$ thin films on $(100)BaF_2$ substrates', *Applied Physics Letters*, **59**, 1509–1511.

Yoshitake, T. and Nagayama, K. (2004), 'The velocity distribution of droplets ejected from Fe and Si targets by pulsed laser ablation in a vacuum and their elimination using a vane-type velocity filter', *Vacuum*, **74**, 515–520.

3

The formation of nanocones on the surface of semiconductors by laser-induced self-assembly

A. MEDVID', Riga Technical University, Latvia and Institute
of Semiconductor Physics of NAS Ukraine, Ukraine

Abstract: A new laser method is elaborated for the formation of
nanocones on the surface of such semiconductors as elementary
semiconductors Si and Ge and semiconductor solid solutions – Si_{1-x}
Ge_x/Si and $Cd_{1-x}Zn_xTe$. Strong changes in the optical, mechanical and
electrical properties of the semiconductors after irradiation by Nd:YAG
laser are explained by the presence of the quantum confinement effect
(QCE) in nanocones where radius is equal to or smaller than Bohr's
radius of electron, hole or exciton. The phenomena of 'blue shift' of
photoluminescence spectra and 'red shift' of the phonon LO line in
the Raman spectrum are explained by exciton and phonon QCE in
nanocones, respectively. The asymmetry of the photoluminescence band
in the spectrum of Si nanocones is explained by the 1D graded band gap
structure.

Key words: quantum confinement effect, nanocones, laser, thermogradient
effect, photoluminescence, Raman back scattering.

3.1   Introduction

Semiconductor nanostructures, such as quantum wells – 2D (Fowler *et al.*,
1966), quantum wires (QWs) – 1D (Xia and Yang, 2003) and quantum dots
(QDs) – 0D (Alivisatos, 1996), have attracted extensive interest due to
their unique optical, electrical and mechanical properties and their potential applications in many fields. Such attention to nanostructures is due to
the presence of quantization of the particles moving in small crystals – the
so-called quantum confinement effect (QCE) (Brus, 1984; Furukawa and
Miyasato, 1988). The QCE takes place in a semiconductor or metal when
at least one dimension of the crystal is less than or equal Bohr's radius
of electron, hole or exciton. Today we have some well-developed methods for the formation of nanostructures (NSs). They are molecular beam
epitaxy (MBE), ion implantation and laser ablation (Morales and Lieber,
1998; Yoshida *et al.*, 1998). The above-mentioned methods need subsequent
thermal annealing of the structures in a furnace. A lot of time and a high
vacuum or a special environment, such as inert Ar gas, is needed for nanocrystal growth using these methods. As a result, nanocrystals grow with

85

uncontrollable parameters, broad size distribution and chaotically: so-called self-assembly (Zhao *et al.*, 2006). Therefore, one of the important tasks for nanoelectronics and optoelectronics is the elaboration of new methods for the formation of NSs in semiconductors with controlled parameters.

A review of the basic methods for the formation of nanostructures on the surface of semiconductors and their properties is given in Section 3.1. Experimental data on the formation of nanocones on a surface of Si, Ge and their solid solution and CdZnTe crystal and their optical properties are presented in Section 3.2. In Section 3.3 we propose a two-stage mechanism of nanocone formation on a surface of semiconductors. The formation of nanocones can be applied, as described in Section 3.4, for the design of third generation solar cells, Si white light emitting diodes, photodetectors with selective or 'bolometer' type spectral sensitivity and Si tips for field electron emitting devices with low work function.

We have shown a new way of forming nanostructures on the surface of elementary semiconductors by pulsed Nd:YAG laser radiation (LR) (Medvid' *et al.*, 2007a, 2008b). Conglomeration of the QDs in a line leads to formation of homogeneous QWs. But, if the diameter of the QWs changes monotonously, then a cone-like structure is formed. Cone-like nanostructures possess unique physical properties, for example Si and Ge nanocrystals have luminescence in the visible part of the spectra with a high intensity of radiation. The photoelectrical, mechanical and electrical properties of cone-like nanostructures have been used for designing new devices, such as solar cells, photon detectors and light emitting diodes (LEDs), based on the formation of a graded band gap 1D structure in elementary semiconductors (Medvid' *et al.*, 2008a), and Si tips for field electron emission (Evtukh *et al.*, 2010), and for the elaboration of a new method for the increase of radiation hardness of a semiconductor crystal, as shown in Section 3.4. We explain such LR-induced change of semiconductor properties by QCE in nanocones.

Another possibility for controlling the position of the band maximum in the photoluminescence (PL) spectra of Ge_xSi_{1-x} by changing concentration of Ge atoms – x was shown in papers by Mooney *et al.* (1995) and Sun *et al.* (2005). Unfortunately, the intensity of the PL band is very low because Si and Ge are semiconductors with indirect band gap structures and the control of the PL band range position is too limited by 0.67–1.1 eV, Ge and Si band gaps, respectively, in comparison with the laser method using QCE. The decrease of NSs' diameter to 4 nm leads to a 'blue shift' of the PL spectrum with a maximum position of up to 1.65 eV (Medvid' *et al.*, 2007a) in comparison with the PL spectrum of a bulk crystal.

New possibilities for the construction of devices are opening up using QCE resulting in, for example, an increase of efficiency for the third generation solar cells (Green *et al.*, 2004) and the construction of light sources with controllable wavelength radiation. Therefore, research into controlling the formation of NSs on a surface of the semiconductors by LR and studying their optical properties are the main tasks of this study. With this aim, it is necessary to understand the mechanism of interaction of high power LR with a semiconductor. The basic model used nowadays for a description of LR effects in a semiconductor is the thermal model (Beigelsen *et al.*, 1985), at least for nanosecond laser pulse duration. It implies that energy of light is transformed into thermal energy. But it is only the first step in the understanding of this process. Different models have been proposed for the explanation of the self-assembly of NSs on the surface of a semiconductor by LR. One of them is the photo-thermo-deformation (PTD) model (Emel'yanov and Panin, 1997). According to the PTD model, conversion of light into heat and deformation of the crystalline lattice of a semiconductor takes place due to inhomogeneous absorption of light. Formation of a periodical structure on the surface starts due to redistribution of interstitials and vacancies. Disadvantages of the PTD model are as follows: lack of explanation of the NSs formation in semiconductors, for example, Ge (Medvid' *et al.*, 2005) and 6H-SiC (Medvid' *et al.*, 2004) at high LR intensity when phase transition from solid state to liquid phase takes place, and accumulation and saturation effects (Medvid' *et al.*, 1999, 2002). It was shown that at high absorption of powerful LR in a semiconductor, high gradient of temperature occurs, which causes impurity atoms and intrinsic defects, interstitials and vacancies drift toward the temperature gradient, the so-called thermogradient effect (TGE) (Medvid', 2002). According to the TGE theory, the atoms which have a larger effective diameter than the atoms of the basic semiconductor material drift toward the maximum temperature, but the atoms with smaller effective diameter drift toward the minimum temperature, for example, Ge atoms in Ge_xSi_{1-x} solid solution drift toward irradiated surface. As a result, compressive mechanical stress arises in a semiconductor on the irradiated surface and tensile mechanical stress in the bulk of a semiconductor. This is only the first stage of nanocone formation on the irradiated surface of a semiconductor. Evidence of the TGE presence at these conditions is the formation of p-n junction on a surface of p-Si (Mada and Ione, 1986), p-InSb (Fujisawa, 1980; Medvid' and Fedorenko, 1999), p-CdTe (Medvid' *et al.*, 2001) and p-InAs (Kurbatov *et al.*, 1983).

Elementary semiconductors, such as Si and Ge crystals and their solid solutions, have been chosen for study because Si is the basic material in micro-electronics as it is cheap and Si technology is very well developed. CdTe crystal and its solid solution CdZnTe are very important materials for production of radiation detectors (Gupta *et al.*, 2006) and solar cells (Carini *et al.*, 2007; Gnatyuk *et al.*, 2006; Vigil-Galán *et al.*, 2005).

A new laser method designed for nanocone formation on the irradiated surface of semiconductors is reported. A two-stage model of nanocone formation is proposed. Studies of the nanocones' optical, mechanical and electrical properties which were used for construction of third generation solar cells, Si white LEDs, photon detectors with selective or 'bolometer' type spectral sensitivity, Si tip for field electron emitting with low work function of electron, are conducted.

3.2   Experiments and discussion

3.2.1   Experiments on Ge and Si crystals

All experiments for the formation of NSs were performed in an ambient atmosphere at a pressure of 1 atm, $T = 20°C$, and 60% humidity. Radiation from a pulsed Nd:YAG laser for Ge single crystals and Si_xGe_{1-x}/Si solid solution, basic frequency with the following parameters: pulse duration $\tau = 15$ ns, wavelength $\lambda = 1.06$ μm, pulse rate 12.5 Hz, power $P = 1.0$ MW and energy $W = 30.0$ mW was used. For Si single crystals with SiO_2 natural cover layer second harmonic with $\tau = 10$ ns and $\lambda = 532$ nm were used. The SiO_2 cover layer in experiments with CdZnTe for preventing evaporation of material was used. Usually, laser beams were directed to the irradiated surface of the sample. Ge(001) i-type single crystal samples with sizes $1.0 \times 0.5 \times 0.5$ cm^3 and resistivity $\rho = 45$ Ω cm were used in experiments. The samples were polished mechanically and etched in CP-4A solution to ensure the minimum surface recombination velocity $S_{min} = 100$ cm/s on all the surfaces. The spot of a laser beam of 3 mm diameter was scanned over the sample surface by a two coordinate manipulator in 20 μm steps. The surface morphology was studied by an atomic force microscope (AFM) and an electron scanning microscope (ESM). Optical properties of the irradiated and non-irradiated samples were studied by using PL and back scattering Raman methods. For PL the 442 nm line of a He-Cd laser and for micro-Raman back scattering an Ar+ laser with $\lambda = 514.5$ nm were used.

The AFM study of Ge surface morphology after irradiation by Nd:YAG laser basic frequency is shown in Fig. 3.1. The most interesting results were found by increasing the LR intensity up to 28.0 MW/cm^2, when nanocones arise on the irradiated surface of Ge, which are self-organized into a 2D lattice. The 2D picture of the irradiated surface of a Ge sample, as seen under ESM, is shown in Fig. 3.2. The 2D lattice is characterized by translation symmetry along and perpendicular to periodic lines with a pattern of C_{6i} point group symmetry and repetition period of 1 μm. C_{6i} patterns are marked by white circles. The patterns' orientation and their symmetry depending on orientation of the Ge surface were not observed. Unusual PL spectrum from the irradiated surfaces of Ge was found in the visible region of the spectrum with maximum at 1.65 eV, as shown in Fig. 3.3. The maximum band in a PL spectrum is usually situated at 0.67 eV, and the intensity of PL is very low due to the indirect band structure of Ge.

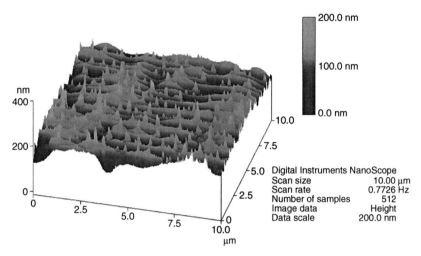

3.1 Three-dimensional AFM image of self-organized nanostructures formed under Nd:YAG laser radiation at intensity of 28 MW/cm².

3.2 SEM image of the Ge single crystal surface irradiated by Nd:YAG laser at intensity 28 MW/cm². C_{6i} patterns are marked by white circles.

Photoluminescence in the IR, red, and blue-violet region of the spectra of Si and Ge nanoparticles implanted into SiO_2 layer has been found by several scientific groups. Strong PL in the blue-violet region was achieved, with the values of excitation and emission energy being independent of the annealing temperature and, consequently, of the cluster mean size (Rebohle *et al.*, 2000; Werwa *et al.*, 1996). Different explanations of this PL mechanism have been proposed. One group of scientists explain the PL by QCE (Rebohle *et al.*, 2000) in Si and Ge nanoparticles, but another group (Fernandez *et al.*, 2002) by local Si-O and Ge-O vibration at Si-SiO_2 and Ge SiO_2 interface. A phonon confinement effect model was developed by Campbell and

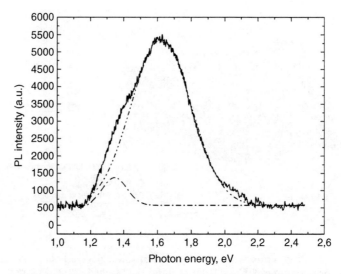

3.3 PL spectrum of the surface of Ge after irradiation by Nd:YAG laser at intensity up to 28 MW/cm².

Fauchet (1986) for Si nanocrystals. It showed that by simultaneously using micro-Raman back scattering spectrum and PL spectrum, it is possible to determine the origin of the PL. It means, if 'blue shift' in PL spectrum and 'red shift' of LO line in Raman back scattering spectrum are simultaneously present, then QCE takes place in NSs. Therefore, this 'blue shift' in PL spectrum of Ge we can explain by the presence of QCE on the top of nanocones where the radius of the balls is equal or less than Bohr's radii of electron and hole. Our calculation of the ball diameter on the top of nanocones using formula (1) from Efors and Efors (1982) and band gap shift from PL bands with maximums at 1.65 and 1.3 eV (Fowler *et al.*, 1966) at parameters of Ge: m_e = 0.12 m_0 and m_h = 0.379 m_0 for electron and hole effective masses, respectively, gives diameters of balls 4.0 and 6.0 nm.

$$E_g = E_g^0 + \frac{(\hbar)^2}{2d^2}\left(\frac{1}{m_e} + \frac{1}{m_h}\right) \tag{3.1}$$

An example of our suggestion is Raman back scattering spectra of the non-irradiated (curve 1) and of the irradiated (curve 2) surfaces of Ge crystal by the laser, as shown in Fig. 3.4. The line at 300 cm⁻¹ of the non-irradiated surface of Ge is attributed to bulk Ge (Ge–Ge vibration, LO phonon line). 'Red shift' of the LO phonon line in Raman back scattering spectra by 6 cm⁻¹ after irradiation of Ge surface takes place. The calculated position of line peak frequency of a Raman spectrum as a function of average crystal size

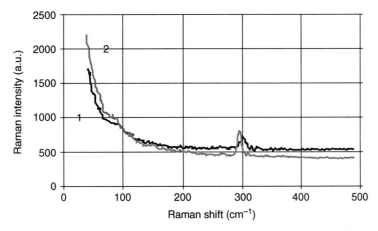

3.4 Micro Raman spectrum of the non-irradiated (curve 1) and of the irradiated surfaces of Ge single crystal by Nd:YAG laser (curve 2) at intensity up to 28 MW/cm².

(d_{ave}) for spherical Ge particles from Kartopu *et al.* (2004) corresponds to 4.0 nm diameter of Ge nanoball on the top of the nanocone. This value is in agreement with our calculation from formula [3.1] using the PL spectrum.

An AFM 3D image of the Si surface after irradiation by second harmonic of Nd:YAG laser at $I = 2.0$ MW/cm² of SiO_2/Si structure, and the same AFM 3D image of Si surface after subsequent wet etching by HF are shown in Figs. 3.5a and 3.5b respectively. Photoluminescence spectra of the irradiated (curves 1 and 2) and non-irradiated (curve 3) surface of SiO_2/Si at intensity of LR up to 2.0 MW/cm² are shown in Fig. 3.6. The surface morphology of SiO_2 layer is smooth and 'stone-block' like, but in reality, under the SiO_2 layer are very sharp Si nanocones, as shown in Fig. 3.5b, which arise on the SiO_2/Si interface after irradiation by the laser.

Photoluminescence of the SiO_2/Si structure in the visible range of a spectrum with maximum at 2.05 eV obtained after irradiation by the laser at intensity $I = 2.0$ MW/cm², as shown in Fig. 3.6, is unusual. The maximum band in PL spectrum is usually situated at 1.0 eV and the intensity of PL is very low due to the indirect band structure of Si crystals. PL of this structure after removal of the SiO_2 layer by wet etching in HF acid is similar and is obtained in the same region of the spectrum and having the same positions of maximums. It means that PL is not connected with local Si-O vibration at the Si–SiO_2 interface (Fernandez *et al.*, 2002). Therefore, we explain our results by the QCE in nanocones. Decrease of the PL intensity can be explained by increase of reflection index of the structure after removal of the SiO_2 layer. We can see that the visible PL spectrum of the SiO_2/Si structure is wide and asymmetric with gradual decrease of

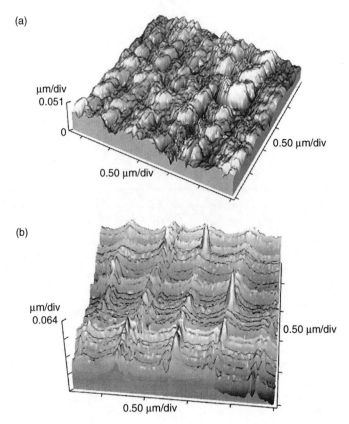

3.5 AFM 3D images of: (a) SiO$_2$ surface after irradiation of SiO$_2$/Si structure by Nd:YAG laser at $I = 2.0$ MW/cm^2 and (b) Si surface after subsequent removing of SiO$_2$ by HF acid.

intensity in IR range of the spectra. It is unique to PL spectrum typical
for graded band gap structure. These results present a dramatic rise of PL
intensity with a much higher band gap than for indirect Si. A schematic
image of a nanocone with a gradual decrease in diameter from p-Si substrate till top is shown in Fig. 3.7. An increase of energy of a radiation
quantum from substrate till top of the Si single crystal at PL of nanocone
takes place due to the QCE in nanowire, according to this formula (Li
and Wang, 2004):

$$\Delta E_g = \frac{2\hbar^2 \zeta^2}{md^2},$$

[3.2]

where $1/(m^*) = 1/(m_e^*) + 1/(m_h^*)$, ($m_e^*$ and m_h^* are electron and hole effective-
masses, respectively) and d is the diameter. For QWs, $\zeta = 2.4048$. In our

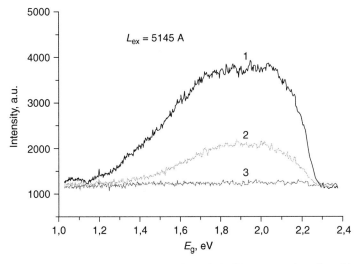

3.6 Photoluminescence spectra of the SiO$_2$/Si structure irradiated by the laser at intensity 2.0 MW/cm^2 (2. and 3. curves), after removing of SiO$_2$ layer by chemical etching in HF acid (3. curve). 1. curve corresponds to PL of the non-irradiated surface.

case, the diameter of nanocones/nanowires is a function of height $d(z)$, therefore it is a graded band gap semiconductor. Our calculation of the Si band gap as a function of nanowires d from PL spectrum using formula (2) from Li and Wang (2004) is shown in Fig. 3.7. We can see that the dependence is nonlinear and that the function of diameter is decreasing, especially rapidly at small diameters. In our case the maximum band gap is 2.05 eV which corresponds to the minimal diameter 2.3 nm on the top of nanocones/nanowires.

We have found for the first time a method of forming a 1D graded band gap semiconductor structure. The graded change of band gaps arises due to the Quantum Confinement Effect. Usually, a 2D graded band gap semiconductor structure is formed by conventional methods, for example, by MBE (Voitsekhovskii *et al.*, 2008), changing molecular components' concentration layer by layer.

3.2.2 Experiments on Si$_{1-x}$Ge$_x$/Si (x = 0.3; 0.4) and Cd$_{1-x}$Zn$_x$Te (x = 0.1) solid solutions

A 200 nm thick crystalline Si$_{1-x}$Ge$_x$ alloy layer was grown on Si(100) wafers with a thickness of 150 µm by MBE on top of Si. Alloys containing 30% and 40% Ge were used in the experiments. The surface of an Si Ge/Si structure was irradiated by basic frequency of the Nd:YAG laser.

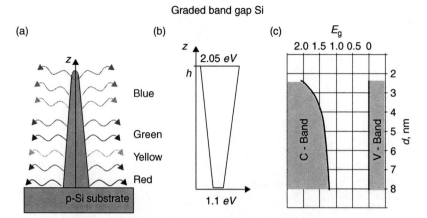

3.7 A schematic image of a nanocone with a gradually decreasing diameter from p-Si substrate till top, formed by laser radiation – (a) and (b). A calculated band gap structure of Si as a function of nanowires diameter using formula (2) from Li and Wang (2004) – (c).

The three-dimensional surface morphology of an $Si_{1-x}Ge_x/Si$ hetero-epitaxial structure recorded by AFM measurements after irradiation by the Nd:YAG laser at intensities of 7.0 MW/cm² (a) and 20.0 MW/cm² (b) is shown in Fig. 3.8. Figure 3.8a shows the nanocones with an average height of 11 nm formed by LR at the intensity of 7.0 MW/cm². Similar nanocones with an average height of 27 nm shown in Fig. 3.8b have been obtained by irradiation intensity of 20 MW/cm². Due to higher irradiation intensity, they are higher and more compact in diameter. After irradiation of the Si Ge/Si hetero-epitaxial structure by the laser at an intensity of 7.0 MW/cm² the surface structure begins to look like spots on unwetting material, for example, they look like water spots on a glass or 'quick silver' on a floor, as shown in Figs. 3.8c and 3.8d, respectively. It means that LR induces segregation of Ge phases at the irradiated surface of the sample due to drift and concentration of Ge atoms. This conclusion is in agreement with data from Kamenev *et al.* (2005) and Hartmann *et al.*, (2005). It shows that the Ge phase starts to form at 50% concentration of Ge atoms in an SiGe solid solution.

According to the TGE (Medvid', 2002), it is supposed that LR initiates the drift of Ge atoms toward the irradiated surface of the hetero-epitaxial structure (Medvid' *et al.*, 2009). PL spectra of the $Si_{1-x}Ge_x/Si$ hetero-epitaxial structures with the maxima at 1.60–1.72 eV obtained after laser irradiation at intensities of 2.0, 7.0 and 20.0 MW/cm² are shown in Fig. 3.9. The spectra are unique and unusual for the material because, depending on Ge concentration, the band gap of SiGe is situated between 0.67 and 1.12 eV (Sun *et al.*, 2005). As seen in Fig. 3.9, the $Si_{1-x}Ge_x$ structure emits light in the

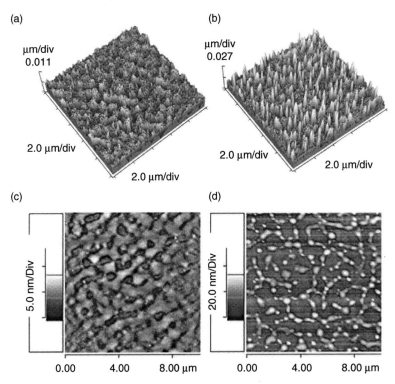

3.8 Three-dimensional AFM images of $Si_{1-x}Ge_x$/Si surfaces irradiated by the Nd:YAG laser at intensity (a) 7 MW/cm² and (b) 20 MW/cm² and two-dimensional surface morphology of the same spots of structure at intensities: (c) 7.0 MW/cm² and (d) 20.0 MW/cm².

visible range of spectrum and the intensity of PL increases with the intensity of irradiation. The maximum of the PL band at 1.70 eV is explained by the QCE (Efors and Efors, 1982). Position of the observed PL peak compared with the bulk material shows a significant 'blue shift'. The maxima of PL spectra of the $Si_{1-x}Ge_x$/Si hetero-epitaxial structure slightly shifts to higher energy when the laser intensity increases from 2.0 to 20.0 MW/cm², which is consistent with the QCE too. Our suggestions concerning Ge phase formation are supported by the back scattering Raman spectra as shown in Fig. 3.10. After laser irradiation at the intensity of 20.0 MW/cm², a Raman band at 300 cm⁻¹ appears in the spectrum. This band is attributed to the Ge–Ge vibration and is explained by formation of a new Ge phase (Kamenev *et al.*, 2005) in the $Si_{1-x}Ge_x$/Si hetero-epitaxial structure. It is proposed to use modified formula (3) from Efors and Efors (1982) for determination of concentration x in the nanocones of $Si_{1-x}Ge_x$/Si hetero-epitaxial structure formed by LR using PL spectra.

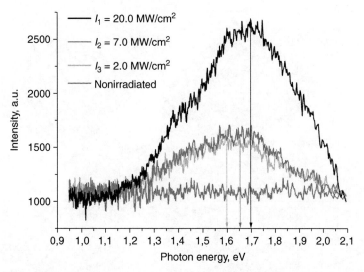

3.9 PL spectra of S$_{0.7}$Ge$_{0.3}$/Si hetero-epitaxial structures before and after irradiation by Nd:YAG laser.

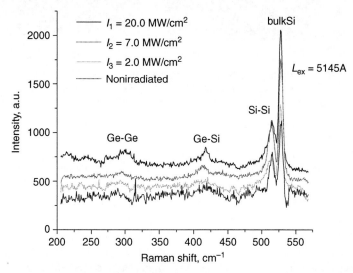

3.10 Back scattering Raman spectra of Si$_x$Ge$_{1-x}$/Si hetero-epitaxial structure before and after irradiation by the laser. Appearance of the 300 cm^{-1} Ge–Ge vibration band in Raman spectra is explained by the new phase formation in Si$_x$Ge$_{1-x}$/Si hetero-epitaxial structure.

Nanocone formation on a surface of Cd$_x$Zn$_{1-x}$Te solid solution with $x = 0.1$ after irradiation by strongly absorbed Nd:YAG LR with intensity $I =$ 12.0 MW/cm^2 was observed using AFM, as shown in Fig. 3.11. Studying the nanocones' optical properties, a new exciton band at energy up to 1.87 eV

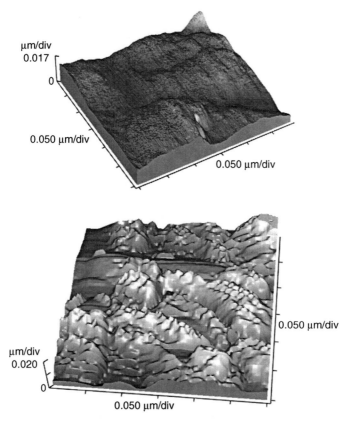

3.11 Atomic force microscope 3D images of the $Cd_{1-x}Zn_xTe$ ($x = 0.1$) surfaces: (a) before irradiation and (b) after irradiation by the laser at intensity of 12 MW/cm².

in the PL spectrum was seen for the first time. At the same time, a shift of A^0X and D^0X exciton lines on 3.1 and 2.5 meV correspondingly, toward the higher energy of quantum, that is the so-called 'blue shift', was observed, as shown in Fig. 3.12. Appearance of a new PL band is explained by Exciton Quantum Confinement (EQC) effect in nanocones and the 'blue shift' of the exciton bands – basically by mechanical compressive stress of the sample top layer formed on the irradiated surface of the samples. This process is described in the following way: irradiation of the $Cd_{1-x}Zn_xTe$ solid solution by the laser leads to the drift of Cd atoms toward irradiated surface and of Zn atoms – in the bulk of the semiconductor due to high gradient of temperature, the so-called TGE (Medvid', 2002). As a result, formation of a $CdTe/Cd_{1-x_1}Zn_{x_1}Te$ hetero-structure, where $x_1 > x$, takes place due to replacement of Zn atoms by Cd atoms at the irradiated surface. At the same time, the opposite process takes place under the top layer, in the buried layer of the semiconductor – Zn atoms replace Cd atoms. At least three factors determine A^0X and D^0X exciton lines position in PL spectrum. They are: concentration of Zn atoms in CdTe top layer and in CdZnTe buried layer (Reno and Jones, 1992), 2D EQC effect in the CdTe top layer when its thickness is comparable with Bohr's radius of exciton, and mechanical compressive stress of the CdTe top layer due to mismatch of CdTe and CdZnTe crystalline lattice. Decrease of the Zn atoms concentration in the top layer with an increase of LR intensity, according to the proposed model, leads to the 'red shift' of the exciton bands in PL spectra, as shown in Medvid' *et al.* (2007b), but increase of the Zn atoms concentration in buried CdZnTe layer manifests in 'blue shift' of the PL spectrum, as shown in Fig. 3.13. These effects do not compensate each other because they take place in different layers. This unusual situation can be explained by different input of these layers in the intensity of PL spectrum. If the top layer is excited by short wavelength light, then the 'red shift' of PL spectrum will be mostly observed, but if mainly the buried layer is excited, for example due to small thickness or transparency of top layer, then the 'blue shift' will be observed.

Of course, it is possible to observe both PL spectra simultaneously at intermediate situations. Exactly such a situation is observed in the PL spectrum shown in Fig. 3.12, after destruction of the CdTe top layer and formation of nanocones on the irradiated surface of the sample. Relaxation of the mechanical compressive stress in the CdTe layer comes to expression as self-assembly of nanocones on the irradiated surface of the structure like Stranski–Krastanov's mode, and simultaneous appearance of a new exciton band at 1.872 eV in PL spectrum at high intensity of LR takes place. Reconstruction of this band shows that it consists of three lines which look like A^0X, D^0X and A^0X–LO lines (the distance between the lines and their relation intensities are the same) in the non-irradiated PL spectrum of the semiconductor. Therefore, we connect the appearance of both the new band in PL spectrum and the nanocones' formation on the irradiated surface of the semiconductor with EQC in nanocones and denote them as A^0XQC and D^0XQC lines. Evidence of the mechanical stress relaxation process in the CdTe layer is non-monotonic dependence of the 'blue shift' as function of LR intensity, as shown in Fig. 3.13. Such non-monotonic dependence has been observed in p- and n-type Si after studying mechanical microhardness of Si after irradiation by Nd:YAG laser (Medvid' *et al.*, 2007b). Calculation of the mechanical compressive stress in the CdTe top layer using the maximum of the 'blue shift' of A^0X exciton line from Fig. 3.13 and $dE_g/dP = 10$ eV/Pa (Thomas, 1961), where E_g and P are the band gap of the CdTe crystal and mechanical stress respectively, gives $P = 4.62 \times 10^5$ Pa. This value corresponds to the ultimate strength limit of CdTe (Yonenaga, 2005). Calculation of the quantum dot diameter using the formula from

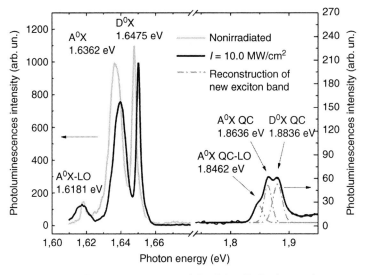

3.12 Photoluminescence spectra of the $Cd_{1-x}Zn_xTe$ ($x = 0.1$) measured at temperature 5 K: curve 1 before and curve 2 after irradiation by the laser at $I = 10.0$ MW/cm².

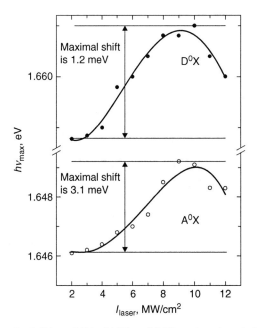

3.13 'Blue shift' of A^0X and D^0X exciton bands in PL spectra as function of the laser intensity.

Kayanuma (1988) and the 'blue shift' of A^0XQC line in PL spectrum on 0.27 eV gives the diameter of the quantum dots up to 10.0 nm. This data corresponds to the size of nanocones (height and diameter of the bottom of the cones are 10.0 nm) measured using 3D image of AFM. Evidence of the presence of the EQC effect in nanocones is the decrease of LO phonon energy by 0.7 meV in the PL spectrum, that is the so-called phonon QCE (Campbell and Fauchet, 1986). Our calculation of the Zn atom's distribution depending on intensity of LR using the thermo-diffusion equation has shown that the process of $CdTe/Cd_{1-x_1}Zn_{x_1}Te$ heterostructure formation is characterized by gradual increase of Zn atom's concentration in the buried layer with intensity of LR up to 8% ($x_1 = 0.18$). The thickness of the CdTe layer after irradiation by a laser with intensity of $I = 12.0$ MW/cm² becomes 10 nm. Therefore, we explain that the A^0X and D^0X lines do not return to their initial position in the PL spectrum by increase of Zn atoms concentration in buried layer till $x = 0.18$.

We connect the appearance of both the new band in PL spectrum and the nanocones formation on the irradiated surface of the semiconductor with EQC effect in nanocones and denote them as A^0XQC and D^0XQC lines. The process of nanocone formation is characterized by LR threshold intensity up to $I = 9.0$ MW/cm², as we can see in Fig. 3.13, maximum of 'blue shift' position VS the laser intensity.

3.3   Two-stage mechanism of nanocones formation in semiconductors

We propose the following method of nanocone formation on the irradiated surface of Ge_xSi_{1-x} and $Cd_xZn_{1-x}Te$ solid solutions. The mechanism is characterized by two stages – laser redistribution of atoms (LRA) and selective laser annealing (SLA) (Medvid' et al., 1996, 1997; Yonenaga, 2005). The first stage of the process, LRA, involves the formation of heterostructures, such as Ge/Si due to the separation of Ge and Si atoms in a Ge_xSi_{1-x} sample and $CdTe/Cd_{x_1}Zn_{1-x_1}Te$ in $Cd_{x_0}Zn_{1-x_0}Te$ ($x_1 > x_0$) solutions due to the separation of Cd and Zn atoms in the gradient of temperature. The dominant phenomenon in this process is the thermogradient effect (Medvid', 2002). The second stage, SLA, involves the formation of nanocones on the irradiated surface of a semiconductor. The heating of the structure, followed by the mechanical plastic deformation of the strained top layer leads to the formation of nanocones. The mechanical plastic deformation is caused by the relaxation of the mechanical compressive stress arising between these layers, owing to a mismatch of their crystal lattices and selective laser heating of the top layer. SLA occurs due to the higher absorption of the LR by the top layer than the buried layer. LRA is a nonlinear optical process: the concentration of the redistributed atoms

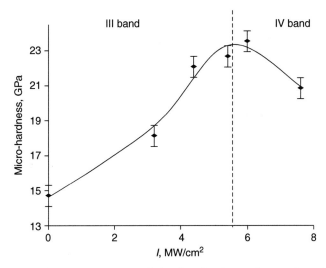

3.14 Micro-hardness of n-Si (111) wafer depending on laser intensity at load on indenter 20 g.

(Ge in the case of Ge_xSi_{1-x} and Cd in the case of $Cd_xZn_{1-x}Te$) increases with the number of laser pulses and at the same time, its absorption coefficient increases. As a result, the LRA stage gradually transits to the SLA stage. In the case of the elementary semiconductors, at the first stage of the process a thin top layer with a high concentration of interstitial atoms and mechanically compressive stress is formed on the irradiated surface of the semiconductor. For example, as shown in Fig. 3.14 for n-Si crystal, mechanical compressive stress increases with laser intensity at the first stage of the process (III band). This process takes place due to separation of interstitials and vacancies in the gradient of temperature field. Mechanical compressive stress of the top layer arises due to the presence of additional interstitials in the top layer and vacancies in the buried layer. At the second stage of the process, nanocones are formed on the irradiated surface of the semiconductor by plastic deformation of the top layer, mostly due to absorption of light by the top layer and therefore heating up. The cause of high absorption of light by the top is mechanical compressive stress which leads to a decrease of the semiconductor band gap (Kayanuma, 1988) and, of course, the increase of the absorption coefficient. Examples of the two-stage mechanism of nanocone formation on a surface of the semiconductors by LR are the following:

1. Non-monotonic dependence of Si crystal micro-hardness as a function of the laser intensity. There is an increase of the crystal micro-hardness at the LRA stage of laser irradiation and decrease of it at the LSA stage, as shown in Fig. 3.14, III and IV bands respectively.

2. An example of the presence of the first stage is the appearance and growth of LO phonon line intensity with a frequency of 300 cm^{-1} in the Raman back scattering spectrum of Ge$_x$Si$_{1-x}$ solid solution after irradiation by the laser. It means formation of a new Ge phase on the irradiated surface of Ge$_x$Si$_{1-x}$ (Medvid' et al., 2010a), as shown in Fig. 3.10.

3. The shift of the PL spectrum of a Cd$_x$Zn$_{1-x}$Te solid solution at a low intensity of LR toward lower energy of quantum – the 'red shift' (Medvid' et al., 2007b) – is the next evidence of the first stage of thin Cd$_x$Zn$_{1-x}$Te layer formation.

4. Appearance of nanocones on the irradiated surface of Ge$_x$Si$_{1-x}$ solid solution and their growth with the laser intensity has been found by measurements of the irradiated surface morphology by AFM, as shown in Fig. 3.8.

5. The shift of bands in the PL spectrum of a Cd$_x$Zn$_{1-x}$Te solid solution at a high intensity of LR toward higher energy of quantum – the 'blue shift' – and appearance of a new PL band at higher energy of quantum – EQC Effect, as shown in Fig. 3.12, are evidence of the second stage of the mechanism.

6. The 'blue shift' of the PL spectra and increase of the PL bands intensity of Si, Ge, Ge$_x$Si$_{1-x}$ and Cd$_x$Zn$_{1-x}$Te crystals with increasing intensity of the laser due to QCE, for example as shown in Fig. 3.9 for Ge$_x$Si$_{1-x}$ crystal, is the next evidence of the LSA stage.

7. Evidence of the first stage presence at nanocones formation process is the redistribution of intensity of LO-ZnTe, TO-CdTe and LO-CdTe phonon lines in Raman back scattering spectra, as shown in Fig. 3.15. It means that before irradiation of the sample by the laser intensity of LO-ZnTe, the phonon line was 3–4 times higher than the intensity of TO and LO-CdTe phonon lines but after irradiation, the opposite is observed in Raman spectra.

3.4   Applications in nanoelectronics and optoelectronics

3.4.1   High intensity Si source of white light and photon detector with selective or 'bolometer' type of spectral sensitivity on the base of nanocones

A result of investigation into PL spectra of Si and Ge$_x$Si$_{1-x}$ crystals has shown, in Figs. 3.2 and 3.9, that nanocones radiate a wide spectrum of light quantum like the sun or a glow lamp. At the same time, this structure can be used as a photon detector with a 'bolometer' type of spectral photo sensitivity (Capasso, 1987), as shown in Fig. 3.16, curve 1. If irradiation of the structure takes place

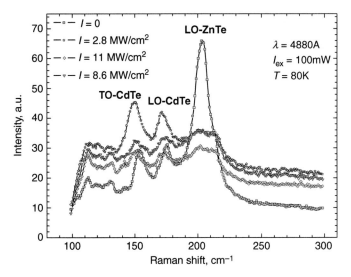

3.15 Raman back scattering spectra non-irradiated and irradiated CdZnTe sample at intensity of LR: $I = 0$ (curve 1), $I = 2.8$ MW/cm² (curve 2), $I = 8.6$ MW/cm² (curve 3) and $I = 11.0$ MW/cm² (curve 4).

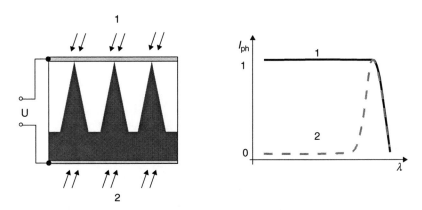

3.16 Scheme of photodetector with graded band gap structure and photoconductivity specters of photodetector with 'bolometric' type of photosensitivity – curve 1 (detector irradiate from nanocones side 1) and photodetector with selective type photosensitivity – curve 2 (detector irradiate from base side 2).

from the wide band gap part of a semiconductor with a gradient band gap structure, it means from the top of cones, or as a photon detector with selective type of spectral photo sensitivity, if irradiation of the structure takes place from narrow band gap part of graded band gap structure, as shown in Fig. 3.16, curve 2. Schematically graded band gap structure is shown in Fig. 3.16.

3.4.2 X- and γ-ray detector with increased radiation hardness

This part of the work deals with studying the possibility of increasing the radiation hardness of CdZnTe crystals using LR (Medvid' *et al.*, 2010b). An Nd:YAG laser with the following parameters – second harmonic and intensity range I = 0.5–2.0 MW/cm^{2-} – was used as a source of light. This method of PL was used in the experiments for estimation of crystalline lattice damage after irradiation by gamma rays. $Cd_{0.9}Zn_{0.1}Te$ crystal was the object of investigation. It is known that γ-radiation causes intrinsic defect generation in a semiconductor. The effect consists of an increase of the A^0X exciton line in the PL spectrum of CdZnTe by ten times after γ-irradiation by ^{60}Co source (E = 1.2 MeV) at room temperature with a dose of 5×10^5 Rad = 5 kGy, as shown in Fig. 3.17, curves 1 and 3. Experimental results showed that this effect is suppressed by five times in CdZnTe crystals after preliminary irradiation by the laser at an intensity of more than 2.0 MW/cm^2, as shown in Fig. 3.17, curve 4. The decrease of the A^0X exciton line intensity in the PL spectrum indicates the decrease of generation of Cd vacancies in CdZnTe after preliminary irradiation by the laser. The mechanism of this effect is explained in the following way: γ radiation leads to generation of additional Cd vacancies near the surface layer (Xia and Yang, 2003), which causes an increase of the A^0X exciton line in the PL spectrum. Laser radiation has

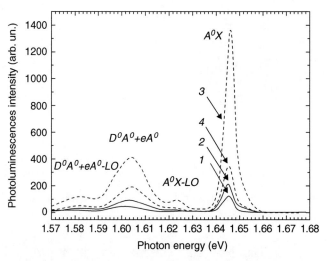

3.17 The PL spectra of $Cd_{1-x}Zn_xTe$ crystal before (curve 1) and after irradiation by Nd:YAG laser: curve 2 – irradiated by Nd:YAG laser with intensity 1.20 MW/cm^2; curve 3 – irradiated by γ-ray with a dose rate of 5.0 KGy; curve 4 – previously irradiated by Nd:YAG laser with intensity 1.20 MW/cm^2 and subsequently by γ-ray with a dose rate of 5.0 KGy.

the opposite effect on $Cd_{0.9}Zn_{0.1}Te$ crystals: interstitial Cd atoms are concentrated near the irradiated surface layer, but vacancies in the bulk of the semiconductor (Medvid', 2002). This leads to an A^0X line decrease and increase of the D^0X exciton line in the PL spectrum, as shown in Fig. 3.18. An increase in Cd_i atom concentration near the surface layer leads to an increase of the materials radiation hardness due to two facts: that Cd atomic weight is larger compared to other atoms – Zn and Te, and 'healing' of the crystal lattice by Cd_i atoms – recombination V_{cd} and Cd_i.

3.4.3 Electron field emitter on the base of Si nanocones

Nanocones have been formed on the irradiated surface of p-Si crystals by the second harmonic of Nd:YAG laser with an intensity of $2.0 \ MW/cm^2$. The measurements of electron field emissions were performed with a flat diode configuration with a glass spacer (Evtukh *et al.*, 2010). Distance between the cathode (silicon wafer) and the anode was 0.8 mm. The applied field emission setup allows achieving a vacuum of 1×10^{-5} Pa. The applied voltage varied between 1500 and 3600 V. The current–voltage characteristics of field emission current and corresponding curves in Fowler–Nordheim plots ($lgI/E^2 - 1/E$) are shown in Fig. 3.19. The electron field emission from such nanocones have some peculiarities, namely: (i) decrease of the threshold field from $E_{th} = 4 \times 10^4$ V/cm at the first measurement to $E_{th} = 3.5 \times 10^4$ V/cm

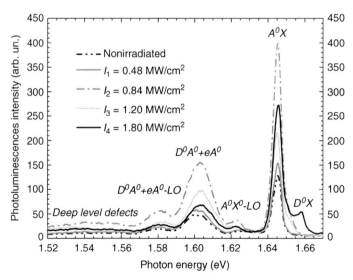

3.18 The PL intensity as a function of photon energy for $Cd_{1-x}Zn_xTe$ before and after irradiation by Nd:YAG laser.

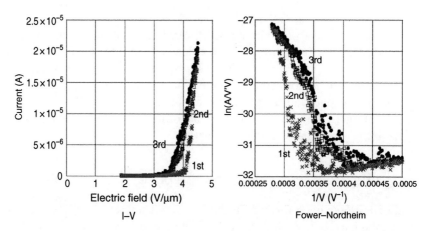

3.19 Current-voltage characteristics (a) and corresponding Fowler–Nordheim plots (b) of Si nanocones electron field emission. Nanocones have been formed by second harmonic of Nd:YAG laser with intensity 2.0 MW/cm².

in subsequent measurements, (ii) two slopes of Fowler–Nordheim curves (higher slope at low fields and lower slope at high fields) (Fig. 3.19). Analysis of the SEM micrographs and electron field emission curves allows us to estimate (i) the electron field enhancement coefficient, $\beta \approx 100$, (ii) work functions ($\Phi_1 = 6.8$ eV at the first measurement and $\Phi_2 = 3.9$ eV, $\Phi_3 = 2.38$ eV from the two slopes in subsequent measurements), (iii) effective emission area, $\alpha = 3 \times 10^{-8} - 1.8 \times 10^{-5}$ cm². The obtained experimental results were explained in the frame of the proposed model which takes into account formation of native oxide and/or positive dipoles (Si^+-O^-) due to oxygen adsorption on the Si surface. An improvement in the field emission after the first measurement may be due to the oxygen thermal field desorption or the transformation of the native oxide caused by a stress-induced leakage current process. As in metal-insulator-Si structures, the resistance of the thin oxide is reduced as a result of the current being passed through. The lower work function in relation to the known value for high doped n-type silicon $\Phi_0 = 4.15$ eV is explained by an increase of Si band gap on the top of nanocones, as shown in Fig. 3.7.

3.4.4 Third generation solar cells on the base of ITO/Si/Al structure

ITO/p-Si/Al structures, where the ITO top layer was 70 nm thick, an Si layer with a thickness of 500 μm and an Al back layer with a thickness of 100 nm in the experiments were studied. The structure was irradiated from the ITO side by an Nd:YAG LR second harmonic with a wavelength of $\lambda = 512$ nm and a pulse duration of $\tau = 10$ ns. The diameter of the laser beam spot was 0.9 mm. Irradiation of the samples in the scanning regime using a two coordinate manipulator with a 20 μm step was carried out. Experiments in ambient atmosphere at a pressure of 1 atm, $T = 20°C$, and 60% humidity were performed.

The irradiated structures were studied using AFM and PL spectroscopy methods. The measurements of current–voltage (I–V) characteristics to observe the changes of electrical parameters for ITO/p-Si/Al structure after irradiation by laser were carried out. ITO/p-Si/Al structure has been irradiated by an Nd:YAG laser with the aim of forming a p-n junction on the interface of ITO/Si and to grow nanocones on the interface of ITO/Si in which QCE takes place (Medvid' et al., 2011). The n-type Si layer was formed on p-type Si due to drift of interstitial Si atoms toward the irradiated surface, as a result of huge gradient of temperature induced by LR (Medvid', 2002).

Nanocones were formed on the irradiated Si surface with an average height of 30 nm by LR with an intensity of up to 2.83 MW/cm² as shown in Fig. 3.20. PL spectra of the ITO/p-Si structures with the maxima at 575 and 490 nm obtained after laser irradiation at intensities of 1.13 and 2.83 MW/cm² are shown in Fig. 3.21. Position of the observed PL peak compared with the bulk Si shows a significant 'blue shift'. The maxima of the PL band at 575 and 490 nm are explained by presence of the QCE (Medvid' et al., 2011) on the top of the nanocones. After irradiation of the ITO/p-Si/Al structure by the laser, photocurrent power increased twofold in comparison to the

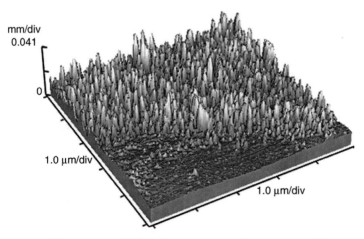

3.20 AFM images of ITO/p-Si structure irradiated by Nd:YAG laser: boarder of non-irradiated – smooth surface and irradiation by Nd:YAG laser at intensity 2.83 MW/cm² – surface with nanocones.

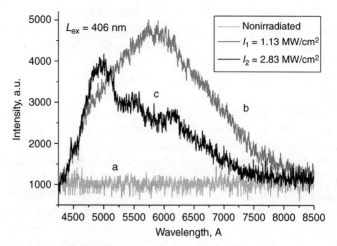

3.21 PL spectra of ITO/p-Si structure: before – (a) curve and after irradiation by the Nd:YAG laser, (b) and (c) curves.

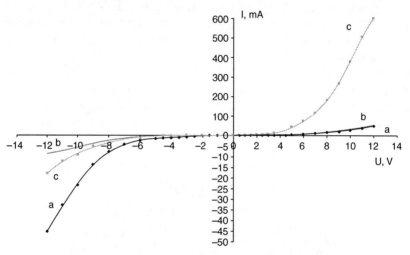

3.22 I–V characteristics of ITO/p-Si/Al structure: (a) non-irradiated; irradiated by Nd:YAG laser at intensities of (b) –1.13 MW/cm²; (c) –2.83 MW/cm².

non-irradiated structure. This effect is explained by QCE in nanocones. It means that the Si band gap is increased with the increase of the laser intensity. It is found that the dark I–V characteristic becomes diode-like with a rectification coefficient of $K = 105$ at 5 V caused by the laser irradiation with a intensity of $I_2 = 2.83$ MW/cm² as shown in Fig. 3.22. The improvements of the ITO/p-Si/Al structure as a solar cell can be explained by the increase of potential barriers between ITO and p-Si layers.

3.5   Conclusions

For the first time, a method for the formation of 1D graded band gap semiconductor structures has been found. Graded change of band gap arises due to the QCE in nanocones.

A new PL exciton band with energy of up to 1.87 eV was found in the PL spectrum of $Cd_{1-x}Zn_xTe$ solid solution after irradiation by a Nd:YAG laser second harmonic with an intensity of 12.0 MW/cm^2. Appearance of the PL exciton band is explained by exciton QCE in nanocones which were formed by LR. Two-stage mechanism of nanocones formation on a surface of the semiconductor by Nd:YAG LR consisting of laser redistribution of atoms and SLA is proposed. Possibility to control heterostructure parameters, for example, thickness, by LR intensity, which is the first stage of the process, is shown. The power of ITO/p-Si/Al solar cell structure has been increased twofold after irradiation by Nd:YAG laser second harmonic with intensity 2.83 MW/cm^2 due to formation of Si nanocones with p-n junction on their top. A twofold decrease of work function of electrons from Si nanocones formed by Nd:YAG laser second harmonic with intensity 2.0 MW/cm^2 is explained by increase of Si band gap on the top of nanocones due to QCE.

3.6   Acknowledgements

The author gratefully acknowledges financial support in part by Europe Project in the framework FR7-218000 'COCAE', ERAF Project 'Sol-gel and laser technologies for the development of nanostructures and barrier structures', No.2010/0221/2DP/2.1.1.0/10/AP IA/VIAA/145, the ESF Project No. 1DP/1.1.1.2.0/09/ APIA/VIAA/142, by the Latvian Council of Science according to the Grant 10.0032. 6.2 and the Riga Technical University project ZP-2010/19.

3.7   References

Alivisatos, A. P. (1996), 'Semiconductor clusters, nanocrystals, and quantum dots', *Science*, **271**, 933–937.

Beigelsen, D. K., Rozgonyi, G. A. and Shank, C. V. (1985), *Energy beam–solid interactions and transient thermal processing*. Pittsburgh: Material Research Society.

Brus, L. E. (1984), 'Electron–electron and electron–hole interactions in small semiconductor crystallites: The size dependence of the lowest excited electronic state', *Journal of Chemical Physics*, **80**, 4403–4409.

Campbell, H. and Fauchet, P. M. (1986), 'The effects of microcrystal size and shape on the one phonon Raman spectra of crystalline semiconductors', *Solid State Communications*, **58**, 739–774.

Capasso, F. (1987), 'Graded-gap and superlattice devices by bandgap engineering. In Dingle, R. (ed.), *Semiconductors and semimetals*, Volume 24. New Jersey: At&T Bell Laboratories Murray Hill, pp. 319–395.

Carini, G. A., Bolotnikov, A. E., Camarda, G. S. and James, R. B. (2007), 'High-resolution X-ray mapping of CdZnTe detectors', *Nuclear Instruments and Methods in Physics Research A*, **579**, 120–124.

Efors, A. L. and Efors, A. L. (1982), 'Band–band absorption of light in semiconductor ball', *Physics and Technics of Semiconductors*, **16**, 1209–1214 (in Russian).

Emel'yanov, V. I. and Panin, I. M. (1997), 'Defect-deformational self-organization and nanostructuring of solid surfaces', *Solid State Physics*, **39**, 2029–2035.

Evtukh, A., Medvid', A., Onufrijevs, P., Okada, M. and Mimura, H. (2010), 'Electron field emission from the Si nanostructures formed by laser irradiation', *Journal of Vacuum Science & Technology B*, **28**, C2B11–C2B13.

Fernandez, B. G., Lopez, M., Garcia, C., Perez-Rodriguez, A., Morante, J. R., Bonafos, C., Carrada, M. and Claverie, A. (2002), 'Influence of average size and interface passivation on the spectral emission of Si nanocrystals embedded in SiO_2', *Journal of Applied Physics*, **91**, 789–793.

Fowler, A. B., Fang, F. F., Howard, W. E. and Stiles, P. J. (1966), 'Magneto-oscillatory conductance in silicon surfaces', *Physical Review Letters*, **16**, 901–903.

Fujisawa, I. (1980), 'Type conversion of InSb from p to n by ion bombardment and laser irradiation', *Japanese Journal of Applied Physics*, **19**, 2137–2140.

Furukawa, S. and Miyasato, T. (1988), 'Quantum size effects on the optical band gap of microcrystalline Si:H', *Physical Review B*, **38**, 5726–5729.

Gnatyuk, V. A., Aoki, T., Hatanaka, Y. and Vlasenko, O. I. (2006), 'Defect formation in CdTe during laser-induced doping and application to the manufacturing nuclear radiation detectors', *Physica Status Solidi (c)*, **1**, 121–124.

Green, M. A. (2004), *Third generation photovoltaics: Advanced solar emerge conversion*. Berlin: Springer-Verlag.

Gupta, A., Viral, P., Compaan, L. and Alvin, D. (2006), 'High efficiency ultra-thin sputtered CdTe solar cells', *Solar Energy Materials & Solar Cells*, **90**, 2263–2271.

Hartmann, J. M., Bertin, F., Rolland, G., Semeria, M. N. and Bremond, G. (2005), 'Effects of the temperature and of the amount of Ge on the morphology of Ge islands grown by reduced pressure–chemical vapor deposition', *Thin Solid Films*, **479**, 113–120.

Kamenev, B. V., Baribeau, J.-M., Lockwood, D. J. and Tsybekov, A. (2005), 'Optical properties of Stranski-Krastanov grown three-dimensional Si/Ge nanostructures', *Physica E*, **26**, 174–178.

Kartopu, G., Bayliss, S. C., Hummel, R. E. and Ekinci, Y. (2004), 'Report on the origin of the orange PL emission band', *Journal of Applied Physics*, **957**, 2466–2472.

Kayanuma, Y. (1988), 'Quantum-size effects of interacting electrons and holes in semiconductor microcrystals with spherical shape', *Physical Review B*, **38**, 9797–9805.

Kurbatov, L., Stojanova, I., Trohimchuk, P. P. and Trohin, A. S. (1983), 'Laser annealing of $A^{III}B^{V}$ compound', *Report Academy of Science USSR*, **268**, 594–597.

Li, J. and Wang, L. W. (2004), 'Comparison between quantum confinement effects of quantum wires and quantum dots', *Chemistry of Materials*, **16**, 4012–4015.

Mada, Y. and Ione, N. (1986), 'p-n junction formation using laser induced donors in silicon', *Applied Physics Letters*, **48**, 1205–1207.

Medvid', A. (2002), 'Redistribution of the point defects in crystalline lattice of semiconductor in nonhomogeneous temperature field', *Defects and Diffusion Forum*, **210–212**, 89–101.

Medvid', A. and Fedorenko, L. (1999), 'Generation of donor centers in p-InSb by laser radiation', *Materials Science Forum*, **297–298**, 311–314.

Medvid', A. and Lytvyn, P. (2004), 'Dynamics of laser ablation in SiC', *Materials Science Forum*, **457–460**, 411–414.

Medvid', A., Dmytruk, I., Onufrijevs, P. and Pundyk, I. (2007a), 'Quantum confinement effect in nanohills formed on a surface of Ge by laser radiation', *Physica Status Solidi (c)*, **4**, 3066–3069.

Medvid', A., Fedorenko, L., Korbutjak, B., Kryluk, S., Yusupov, M. and Mychko, A. (2007b), 'Formation of graded band-gap in CdZnTe byYAG:Nd laser radiation', *Radiation Measurements*, **42**, 701–703.

Medvid', A., Fedorenko, L. and Snitka, V. (1999), 'The mechanism of generation of donor centres in p-InSb by laser radiation', *Applied Surface Science*, **142**(142), 280–285.

Medvid', A., Fukuda, Y., Michko, A., Onufrievs, P. and Anma, Y. (2005), '2D lattice formation on a surface of Ge single crystal by YAG:Nd laser', *Applied Surface Science*, **244**(1–4), 120–123.

Medvid', A., Hatanaka, Y., Korbutjak, D., Fedorenko, L., Krilyuk, S. and Snitka, V. (2002), 'Generation of the A-centres at the surface of CdTe(Cl) by YAG:Nd laser radiation', *Applied Surface Science*, **197–198**, 124–129.

Medvid', A., Kaupuzs, J., Madzulis, I. and Bl ms, J. (1996), 'Buried layer formation in silicon by laser-radiation', *Journal of Applied Physics*, **79**, 9118–9122.

Medvid', A., Knite, M., Kaupuzs, J. and Frishfelds, V. (1997), 'Mechanism of recording and erasing of optical information by laser radiation on SiO_2-(Co+Si)-SiO_2-Si multi-layer structure', *Applied Surface Science*, **115**, 393–398.

Medvid', A., Litovchenko, V. G., Korbutjak, D., Fedorenko, L. L. and Hatanaka, Y. (2001), 'Influence of laser radiation on photoluminescence of CdTe', *Radiation Measurements*, **33**, 725–729.

Medvid', A., Mychko, A., Strilchyuk, O., Litovchenko, N., Naseka, Y., Onufrijevs, P. and Pludonis, A. (2008), 'Exciton quantum confinement effect in nanostructures formed by laser radiation on the surface of CdZnTe ternary compound', *Physica Status Solidi (c)*, **6**, 209–212.

Medvid', A., Mychko, A., Gnatyuk, V. A., Levytskui, S. and Naseka, Y. (2009), 'Mechanism of nanostructure formation on a surface of CdZnTe crystal by laser irradiation', *Journal of Automation, Mobile Robotics & Intelligent Systems*, **3**, 127–129.

Medvid', A., Onufrijevs, P., Chiradze, G. and Muktupavela, F. (2010a), 'Impact of laser radiation on microhardness of a semiconductor', AIP Conference Proceedings, 30th International Conference on the Physics of Semiconductors, July 25–30, **338**, Coex, Seoul, Korea.

Medvid', A., Onufrievs, P., Dauksta, E., Barloti, J., Ulyashin, A., Dmytruk, I. and Pundyk, I. (2011), 'P-n junction formation in ITO/p-Si structure by powerful laser radiation for solar cells applications, *Advanced Materials Research*, **222**, 225–228.

Medvid', A., Onufrijevs, P., Dmitruk, I. and Pundyk, I. (2008), 'Properties of nanostructure formed on SiO_2/Si interface by laser radiation', *Solid State Phenomena*, **131–133**, 559–562.

Medvid', A., Onufrijevs, P., Lyutovich, K., Oehme, M., Kasper, E., Dmitruk, N., Kondratenko, O., Dmitruk, I. and Pundyk, I. (2010b), 'Self-assembly *of* nanohills in Si_xGe_{1-x}/Si by laser radiation', *Journal of Nanoscience and Nanotechnology*, **10**, 1094–1098.

Mooney, P. M., Jordan-Sweet, J. L., Ismail, K., Chu, J. O., Feenstra, R. M. and LeGoues, F. K. (1995), 'Relaxed Si0.7Ge0.3 buffer layers for high-mobility devices', *Applied Physics Letters*, **67**, 2373–2375.

Morales, A. M. and Lieber, C. M. (1998), 'A laser ablation method for the synthesis of crystalline semiconductor nanowires', *Science*, **279**, 208–211.

Rebohle, L., von Borany, F. J. H. and Skorupa, W. (2000), 'Blue photo- and electroluminescence of silicon dioxide layers ion-implanted with group IV elements', *Applied Physicas B: Lasers Optics,* **B70**, 131–134.

Reno, J. and Jones, E. (1992), 'Determination of the dependence of the band-gap energy on composition for $Cd_{1-x} Zn_x Te$', *Physical Review B*, **45**,1440–1442.

Sun, K. W., Sue, S. H. and Liu, C. W/ (2005), 'Visible photoluminescence from Ge quantum dots', *Physica E*, **28**, 525–530.

Talochkin, A. B., Teys, S. A. and Suprun, S. P. (2005), 'Resonance Raman scattering by optical phonons in unstrained germanium quantum dots', *Physical Review B*, **72**, 115416–115423.

Thomas, D. G. (1961), 'Excitons and band splitting produced by uniaxial stress in CdTe', *Journal of Applied Physics*, **32**, 2298–2304.

Vigil-Galán, O., Arias-Carbajal, A., Mendoza-Pérez, R., Santana-Rodríguez, G., Sastre-Hernández, J., Alonso, J. C., Moreno-García, E., Contreras-Puente, G. and Morales-Acevedo, A. (2005), 'Improving the efficiency of CdS/CdTe solar cells by varying the thiourea/$CdCl_2$ ratio in the CdS chemical bath', *Semiconductor Science and Technology*, **20**, 819–822.

Voitsekhovskii, A. V., Grygor'ev, D. V. and Smith, R. (2008), 'Radiation defect formation in graded-band-gap epitaxial structures $Hg_{1-x}Cd_xTe$ after boron ion implantation', *Semiconductor Science and Technology*, **23**, 1–7.

Vorobyev, L. E. (1996), 'Semiconductor parameters'. In Levinshtein, M., Rumyantsev, S. and Shur, M. (eds.), *Handbook Series on Semiconductor Parameters*, Vol. 1. London: World Scientific, 33–57.

Werwa, E., Seraphin, A. A., Chiu, L. A., Zhou, C. and Kolenbrander, K. D. (1996), 'Synthesis and processing of silicon nanocrystallites using a pulsed laser ablation supersonic expansion method', *Applied Physics Letters*, **64**, 1821–1825.

Xia, Y. and Yang, Y. (2003), 'Chemistry and physics of nanowires', *Advanced Materials*, **15**, 351–353.

Yonenaga, I. (2005), 'Hardness, yield strength, and dislocation velocity in elemental and compound semiconductors', *Materials Transactions*, **46**, 1979–1985.

Yoshida, T., Yamada, Y. and Orii, T. (1998), 'Electroluminescence of silicon nanocrystallites prepared by pulsed laser ablation in reduced pressure inert gas', *Journal of Applied Physics*, **83**(10), 5427–5432.

Zhao, Z. M., Yoon, T. S., Feng, W., Li, B. Y., Kim, J. H., Liu, J., Hulko, O., Xie, Y. H., Kim, H. M., Kim, K. B., Kim, H. J., Wang, K. L., Ratsch, C., Caflisch, R., Ryu, D. Y. and Russell, T. P. (2006), 'The challenges in guided self-assembly of Ge and InAs quantum dots on Si', *Thin Solid Films*, **508**, 195–199.

4

Fabrication of periodic photonic microstructures by the interference of ultrashort pulse laser beams

Q-Z. ZHAO, Shanghai Institute of Optics
and Fine Mechanics, CAS, China

Abstract: The fabrication of periodic photonic microstructures by the interference of ultrashort pulse laser beams has found applications in several fields. This interference can produce a periodically modulated light intensity distribution with a period of the same order as the laser wavelength. A periodically modulated microstructure can be obtained if the periodically modulated light intensity distribution is transferred to a photoreactive material. The interference of two beams creates a one-dimensional periodic pattern. In principle, by increasing the number of beams, 2D and 3D periodic patterns can be obtained. This technique (with its unique advantages of being maskless and photoresist-free, with inner encoding for transparent materials and single step processing) has opened up the possibility of fabrication of periodic functional microstructures in a large variety of materials. This chapter discusses the fabrication of periodic photonic microstructures by multi-beam interfered femtosecond laser pulses, and also looks at their applications for photonic devices.

Key words: femtosecond laser, interference, photonic device.

4.1 Review of periodic photonic devices induced by the interference of ultrashort pulse laser beams

Since the first demonstration of the mode-locking technique in 1964, ultra-short pulse lasers have made tremendous progress. In 1964, the laser pulse-width available was down to 100 picoseconds[1–3] whilst, by 1981, pulse-widths had been reduced to 100 femtoseconds using the colliding-pulse mode-locking technique.[4,5] With successive improvements in passive mode-locking and dispersion compensation, few-cycle pulses (of sub-6 fs) have been obtained.[6–8] Recently, laser intensities of 10^{22} W/cm^2 have been achieved.[9]

Parallel to these breathtaking achievements in ultrashort pulsed lasers, the applications of these lasers have flourished in numerous fields, as a result of their ultrashort and ultra-intense features. Of these, work with interference from ultrashort pulse laser beam-induced periodic photonic devices has made great progress and found applications in various fields.[10–25]

113

Femtosecond pulses are characterized by an ultrashort time domain with good coherence over the whole pulse duration. When two or more beams of femtosecond pulses split from one source beam and overlap in both space and time, a periodically modulated light intensity distribution with a period in the order of its wavelength can be produced. Furthermore, a periodically modulated microstructure can be obtained if the periodically modulated light intensity distribution is transferred to a photoreactive material.

Interference, diffraction and polarization are, of course, the three main research topics in wave optics. Since the first demonstration of double-slit interference by Thomas Young, optical interference has been extensively used in a wide range of scientific and technological areas. In recent years, laser interference lithography has also received great attention as it can produce periodic patterns at a micro/nano scale without using a mask.[26–29] For conventional interference lithography, interfered continuous wave (CW) or nanosecond laser beams are employed to irradiate a photoresist. After laser exposure, developing, fixing and etching procedures are needed to form the desired periodic microstructures. By comparison, the femtosecond laser interference technique can directly induce periodic microstructures without a photoresist. In addition, with the nonlinear interaction of a femtosecond laser with transparent materials, a periodic microstructure can be fabricated in the interior of these materials.

Section 4.2, below, presents the theoretical aspects of the interference of ultrashort pulse laser beams, including how to make femtosecond pulses overlap. The formation of microstructures by the interference of two femtosecond laser beams and their applications are reported in Section 4.3, whilst Section 4.4 looks at the formation of microstructures by the interference of multi-femtosecond laser beams and their applications. In Section 4.5, the transfer of periodic microstructures by the interference of multi-femtosecond laser beams is reported (combining the femtosecond laser interference technique and the laser-induced forward transfer technique, the transfer of periodic microstructures from a metal film on a supporting substrate to a receiving substrate has been successfully achieved). Finally, Section 4.6 presents a summary and outlook.

4.2   Theoretical aspects of the interference of ultrashort pulse laser beams

4.2.1 How to make femtosecond pulse interference

It is well known that the shorter the pulses, the smaller the area over which the pulses overlap at relatively large angles and with little coherence. For two beams crossed at an angle θ, the size of the overlap area is given by $L = c\tau(\sin\theta/2)^{-1}$, where c is the speed of light in the medium and τ is the

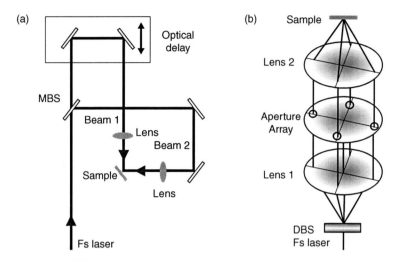

4.1 Beam delivery approaches for femtosecond pulses interference.
(a) MBS beam delivery; (b) DBS beam delivery.

pulse duration.[30] For example, for a 120 femtosecond duration and a 30°
crossed angle, the beams overlap only within a strip approximately 140 μm
wide. The number of interference fringes produced by two beams is independent of the angle and, for transform-limited pulses, is roughly $2c\tau/\lambda$,
where λ is the optical wavelength.[30] With 120-femtosecond pulses at 800
nm, only approximately 90 interference fringes can be produced.

Currently, there are two beam delivery approaches to achieve interference of femtosecond pulses. The first approach is to split the source beam by
a mirror beam splitter (MBS), with an optical delay in order to achieve the
spatial and temporal interference of femtosecond laser pulses. The second
is to split the source beam by a diffractive beam splitter (DBS), which can
divide one laser beam into several beams with equal intensity (illustrated in
Fig. 4.1), and then to use a confocal imaging system in which no optical delay
is needed. For the first approach, precise adjustments of the optical delay for
each optical path are needed by observing the sum-frequency generation, or
third-harmonic generations,[11] to obtain the temporal overlap of the femtosecond pulses; additionally, when the interference of more than two beams
is needed, a complicated optical set up is required, the precise adjustment
of which is very difficult. For the second approach, the optical set up is quite
simple and temporal overlap can be achieved without any adjustment.[30]

The intrinsic difference between the two beam delivery approaches is
that, in the case of diffraction off a grating (in the case of MBS), short
pulses become tilted, whilst pulses that correspond to different diffraction
orders (in the case of DBS) have parallel pulse fronts. Figure 4.2 illustrates
this difference.[10,30] When the source beam is split by an MBS (Fig. 4.2a),

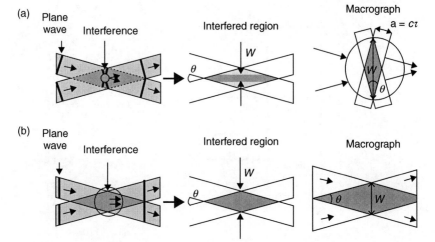

4.2 (a) Overlap of crossed femtosecond beams split by a mirror beam-splitter. (b) Overlap of crossed femtosecond beams split by a trans-mission grating with a confocal imaging system. (Reproduced from Ref. 10.)

the width of the overlapping area is $c\tau/\sin(\theta/2)$ because the pulse fronts of the two beams are not parallel. When the source beam is split by a DBS (Fig. 4.2b), the width of the overlapping area can reach the beam diameter of the incidence light because the wave fronts of the two beams are parallel. Thus, if a larger overlapping area is required, the second beam delivery approach is a better choice. The disadvantage of the second beam delivery approach, however, is that the power density in the interference plane is lower compared to the first beam delivery approach because the interference area in the second approach is larger than that of the first. It is hard to induce periodic structures on high damage threshold materials for the second approach.

For classification purposes, microstructures induced by using an MBS beam delivery are referred to as microstructures induced by two-beam femtosecond laser interference. Similarly, microstructures induced by using a DBS beam delivery are referred to as microstructures induced by multi-beam femtosecond laser interference.

4.2.2 Distribution of optical intensity in the interfered region

To investigate the interference patterns formed by multiple non-coplanar laser beams, we have taken a four-beam interference setup as an

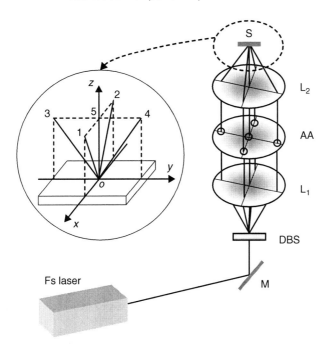

4.3 Optical setup for fabrication of periodic microstructures by interfering femtosecond laser beams. DBS: diffractive beam splitter, L_1 and L_2: lenses, and AA: aperture array. The inset shows the geometry of four-beam interference.

example, to deduce the optical intensity distribution within the region of interference.

Extending the different beam interferences is straightforward. Based on electromagnetic wave theory, the plane wave can be expressed as $E = A\exp[-i(\omega t - \mathbf{k} \cdot \mathbf{r})]$, which can be analyzed in coordinate as:

$$E = A \exp\left\{-i\left[\omega t - k\left(x \cos\alpha + y\cos\beta + z\cos\gamma\right)\right]\right\}. \qquad [4.1]$$

where A, k and ω are amplitude, wave vector and angular frequency of the plane wave, respectively, and α, β and γ are the azimuth angles of the plane wave. In Fig. 4.3 (inset), four laser beams, located in two orthogonal planes, irradiate a sample with incident angles of θ_1, θ_2, θ_3 and θ_4, respectively. The azimuth angles of the four beams can be calculated for this figure; see Table 4.1.

Table 4.1 Azimuth angle of four incident laser beams

Azimuth angle	α	β	γ
Beam 1	$-(90° + \theta_1)$	$90°$	$+(180° - \theta_1)$
Beam 2	$-(90° - \theta_2)$	$90°$	$-(180° - \theta_2)$
Beam 3	$90°$	$-(90° + \theta_3)$	$+(180° - \theta_3)$
Beam 4	$90°$	$-(90° - \theta_4)$	$-(180° - \theta_4)$

Substituting the azimuth angles of the four beams taken from Table 4.1 into Eq. [4.1], the complex amplitude of the four laser beams can be deduced as:

$$\begin{vmatrix} \tilde{E}_1 = A_1 \exp\left\{i\left[k(x\cos\alpha_1 + y\cos\beta_1 + z\cos\gamma_1)\right]\right\} = A_1 \exp\left\{ik\left[x(-\sin\theta_1) + z(-\cos\theta_1)\right]\right\} \\ \tilde{E}_2 = A_2 \exp\left\{i\left[k(x\cos\alpha_2 + y\cos\beta_2 + z\cos\gamma_2)\right]\right\} = A_2 \exp\left\{ik\left[x(-\sin\theta_2) + z(-\cos\theta_2)\right]\right\} \\ \tilde{E}_3 = A_3 \exp\left\{i\left[k(x\cos\alpha_3 + y\cos\beta_3 + z\cos\gamma_3)\right]\right\} = A_3 \exp\left\{ik\left[x(-\sin\theta_3) + z(-\cos\theta_3)\right]\right\} \\ \tilde{E}_4 = A_4 \exp\left\{i\left[k(x\cos\alpha_4 + y\cos\beta_4 + z\cos\gamma_4)\right]\right\} = A_4 \exp\left\{ik\left[x(-\sin\theta_4) + z(-\cos\theta_4)\right]\right\} \end{vmatrix}$$

[4.2]

The complex amplitude of the sum oscillation for the four beams in the interfering field is then:

$$\tilde{E} = \sum_{n=1}^{4} \tilde{E}_n, \quad (n = 1,2,3,4)$$

[4.3]

If we assume that the four beams have equal light intensity and approach the sample with the same angle of incidence (i.e., $A_1 = A_2 = A_3 = A_4 = A$ and $\theta_1 = \theta_2 = \theta_3 = \theta_4 = \theta$), then, by combining Eqs. [4.2] and [4.3], we obtain:

$$\tilde{E} = A\left\{\exp\left[ikx(-\sin\theta)\right] + \exp(ikx\sin\theta) + \exp\left[iky(-\sin\theta)\right] + \exp(iky\sin\theta)\right\}$$
$$\exp\left[ikz(-\cos\theta)\right]$$

[4.4]

and

$$\tilde{E}^* = A\left\{\exp(ikx\sin\theta) + \exp(-ikx\sin\theta) + \exp(iky\sin\theta) + \exp(-iky\sin\theta)\right\}$$
$$\exp(ikz\cos\theta)$$

[4.5]

where, '*' represents the complex conjugate.

Finally, we can get the distribution of light intensity within the interfering field:

$$I = \tilde{E}\tilde{E}^*$$

$$= A^2 \left\{ \exp\left[ikx(-\sin\theta)\right] + \exp(ikx\sin\theta) + \exp\left[iky(-\sin\theta)\right] + \exp(iky\sin\theta) \right\}$$

$$\exp\left[ikz(-\cos\theta)\right] \left\{ \begin{array}{l} \exp(ikx\sin\theta) + \exp(-ikx\sin\theta) \\ + \exp(iky\sin\theta) + \exp(-iky\sin\theta) \end{array} \right\} \exp(ikz\cos\theta)$$

[4.6]

Using the relation $\cos\alpha = \left(e^{i\alpha} + e^{-i\alpha}\right)/2$, Eq. [4.6] can be simplified to:

$$I_{4beams} = 2A^2 \left\{ \begin{array}{l} 2 + \cos(2kx\sin\theta) + \cos(2ky\sin\theta) \\ +2\cos\left[k(x-y)\sin\theta\right] + 2\cos\left[k(x+y)\sin\theta\right] \end{array} \right\} \quad [4.7]$$

Using a similar procedure, the distribution of light intensity for two- and five-beam interference can also be deduced as:

$$I_{2beams} = 2A^2\left[1 + \cos(2kx\sin\theta)\right] \qquad [4.8]$$

$$I_{5beams} = A^2 \left\{ \begin{array}{l} 5 + 2\cos(2kx\sin\theta) + 2\cos(2ky\sin\theta) + 8\cos(kx\sin\theta)\cos(ky\sin\theta) \\ +4\left[\cos(kx\sin\theta) + \cos(ky\sin\theta)\right]\cos\left[kz(\cos\theta - 1)\right] \end{array} \right\}$$

[4.9]

From Eqs. [4.7], [4.8] and [4.9], we can see that for two- and four-beam interference, the intensity distribution is independent of z, while for five-beam interference, it is not. Thus, the structures are basically independent of depth, that is, the same structure is observed in any xy plane in the case of the interference of four beams or less. However, with five-beam interference, the central beam has a different wave vector in the z axis. As a result, the structures obtained by five-beam interference is periodic in the z axis as well as in the x and y axes.

The interference fringe interval can be derived from the distribution of light intensity. For example, for two-beam interference, the period of the fringes, Λ, is given by:

$$\Lambda = \frac{2\pi}{2k\sin\theta} = \frac{\lambda}{2\sin\theta} \qquad [4.10]$$

When θ equals 90°, the interference fringes have a minimum period, which is half the wavelength.

4.3 Microstructures induced by the interference of two femtosecond laser beams

4.3.1 Experimental setup

For their experimental setup (schematically shown in Fig. 4.1a), most studies have used a regeneratively amplified Ti:sapphire laser with a wavelength of 800 nm, a pulse duration of 30–200 fs, and a pulse repetition of 1–1000 Hz. A single laser beam is split into two beams that are then redirected at approximately equal incident angles onto a sample surface, focused by two lenses. After the optical paths are adjusted to realize perfect overlap of the two beams both spatially and temporally, the sample surface is adjusted to be approximately normal to the perpendicular bisector of the two incident beams, so that the sample surface becomes the laser interfering plane. A single shot of the femtosecond laser with controlled energy for each interfered beam completes the encoding process.

4.3.2 Formation of periodic microstructures

Fabrication of surface relief gratings is the most common application for the femtosecond laser interference technique. The fabrication of surface relief gratings on a variety of transparent dielectrics including silica glass, sapphire, etc. have been extensively investigated by the Hosono group.[11-14, 19, 20] Figure 4.4 shows surface relief gratings encoded on amorphous SiO_2 bulk glass by irradiation with crossed femtosecond-laser pulses with different incident angles (θ, 10°, 45° and 80°, respectively). The period of grating, d, varies with θ, that is, it is 2.5 µm for 10°, 0.58 µm for 45° and 0.43 µm for 80°. These correspondences confirm the relation for two-beam

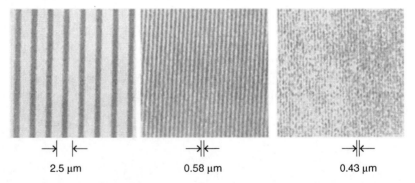

2.5 µm   0.58 µm   0.43 µm

4.4 Surface relief gratings encoded on SiO_2 glasses by holographic irradiation of femtosecond-laser pulses. The colliding angle between the two pulses is θ. (Reproduced from Ref. 11.)

4.5 Influence of the beam cross-angle on the shape of resulting double gratings. (a) Two-dimensional array of dual-structured hole is seen for $\theta = 45°$. (b) Double grating encoded under the crossing angle of 90° for both exposures. (Reproduced from Ref. 13.)

interference at the surface, according to Eq. [4.10]. In addition, by double exposure,[13] dual-structured gratings can be achieved (as shown in Fig. 4.5).

Our group (at the Shanghai Institute of Optics and Fine Mechanics) has fabricated surface relief gratings containing noble metal nanoparticles using two-beam femtosecond laser interference.[15] Au_2O-doped glass samples were first irradiated by two 800 nm interfered femtosecond laser pulses at room temperature and then heat-treated at 550°C to allow Au nanoparticle precipitation in the laser irradiation areas. One-dimensional periodic arrays of the Au nanoparticles were controlled by changing the pulse energy and the incident angle between the interfered laser pulses.

Kaneko *et al.* demonstrated fabrication of a submicron grating in a metal ion-doped polymer by two-beam femtosecond laser interference with two-photon photoreduction.[16]

As well as being used for the fabrication of surface relief gratings, two-beam femtosecond laser interference can also induce volume gratings inside transparent materials. Si *et al.* have fabricated volume gratings inside bulk azodye-doped polymers and perylene-orange-doped hybrid inorganic–organic materials.[17,18] The diffraction efficiency of the first-order Bragg gratings can be up to 90%. Figure 4.6 shows a volume and surface relief gratings fabricated in azodye-doped PMMA and the diffraction patterns.

It is much more difficult to fabricate volume-type gratings on the surface of non-photosensitive materials than it is to fabricate surface relief gratings. Kawamura *et al.*[19] attribute this to two reasons: first, the nonlinear interaction between a strong laser pulse and a material would deteriorate its coherency; second, free carriers cause enhanced optical absorption. It is

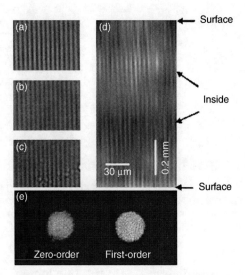

4.6 Optical microscopic photos (a–c) of the photoinduced gratings on the top surface (a), in the interior that is 0.4 mm from the top surface (b), and on the bottom surface of the sample (c). The photo on the right (d) is an orthogonal section image of the gratings observed using a confocal laser scanning microscope. (e) Beam patterns of the zero-order diffraction light (right) and the first-order diffraction light (left). The diffraction efficiency was estimated to be 76%. (Reproduced from Ref. 17.)

known that the peak frequency of the absorption band caused by free carriers, or plasma, which are created almost simultaneously by irradiation from a femtosecond laser pulse, is proportional to (carrier concentration)1/2. This implies that the optical absorption coefficient increases in the infrared (IR) region if the laser is powerful enough to generate carriers. The relaxation time of free carriers is nearly equivalent to the laser pulse duration so that, if the laser pulse duration is stretched, the volume gratings are likely to be encoded inside the material. Kawamura *et al.* used chirped (0.5–5 ps) IR laser pulse interference to induce volume gratings inside SiO_2 glass.[19] By moving the focal spot of the interfering laser beam, multi-layer gratings can be induced. Recently, Kawamura *et al.* have also demonstrated fabrication by a distributed-feedback color center laser in single lithium fluoride crystals, encoding gratings inside crystals using chirped femtosecond laser pulses interference.[20]

Li *et al.* have also fabricated multi-layer volume gratings inside non-photosensitive glass by using common femtosecond pulses,[21] though their results are not consistent with Kawamura's analysis (see above) and further investigation is required.

4.4   Microstructures induced by the interference of multiple femtosecond laser beams

Two-beam interference can induce one-dimensional structures and the double exposure of two-beam interference can induce two-dimensional structures. However, by multi-beam interference, two- or three-dimensional structures can be fabricated in one step. Cai *et al.* have demonstrated that all 14 Bravais lattices can be realized by the interference of four non-coplanar beams[31] and this technique will therefore be able to be used for the fabrication of metamaterials with special electromagnetic properties, such as photonic crystals and optical cloaks.[32]

4.4.1 Experimental setup

The setup for this experiment is shown schematically in Figs 4.1b and 4.3. A regeneratively amplified Ti:sapphire laser with a wavelength of 800 nm, a pulse duration of 30–200 fs and a pulse repetition of 1 kHz is used as the irradiation source. A DBS (the key element of this setup) splits the incident laser beam into several beams. The split beams are made parallel by a lens, L_1, and are then selected by an aperture array to obtain an aimed interference pattern. The selected beams are gathered by an objective, L_2, and create interference in a focused region, where the sample is placed. The temporal and spatial overlap of the five beams' pulses is perfectly achieved on the focal plane of L_2; no adjustment for the pulses' temporal overlap is needed. The distance between L_1 and L_2 is adjusted to make each beam's plane wave after lens L_2. The outer beams (of equal intensity) are placed symmetrically around the central beam, and make an angle with it. Irradiation power and duration can be adjusted to obtain clear periodic structures.

As examples, three types of samples were tested: a commercial silicon wafer (2 mm thickness), a commercial metallic (Ni) foil (20 μm thickness) and a thin metallic (Al) film deposited on a transparent quartz plate (200 nm thickness) were used. All the experiments were carried out at room temperature in ambient atmosphere.

The fabricated structures were observed by an optical microscope and a scanning electron microscope (SEM). The diffraction characteristics of the fabricated structures were also investigated using an He-Ne laser with a wavelength 633 nm, and with a diode-pumped all solid state laser with a wavelength of 532 nm.

4.4.2 Formation of periodic microstructures on silicon wafer

The formation of periodic microstructures on a silicon surface is important for its application in optoelectronic devices. Many types of micrometer-scale surface structures have been fabricated on the surface of silicon when irradiated by pulsed lasers.[33-36] Mazur's group, for example, created a silicon surface covered with a semi-ordered pattern of sharp conical microspikes by irradiating the surface with a femtosecond laser in the presence of halogen-containing gases.[35,36]

The fabrication of periodic microstructures on silicon wafers has been demonstrated using the interference of five femtosecond laser beams.[37] The typical input laser power (incident DBS) was set at 100 mW, whilst the exposure time was ~5 s. Figure 4.7 shows optical and SEM images of the microstructures fabricated by such five-beam femtosecond laser pulses. As can be seen, spots with a diameter of about 300 μm were fabricated in a one-step exposure process (~ 5 s) (Fig. 4.7a). By scanning the silicon wafer, the fabricated spots can be aligned. The detailed structures in the spot are shown in Fig. 4.7b, c and d. A four-fold symmetrical structure was induced on the silicon surface. The period of the array structures was about 2.5 μm. As the exposure time was up to 5 s with a laser power of 300 mW and a 1 kHz repetition rate, the ablated surface deposited much oxidized debris (as shown in Fig. 4.7d).

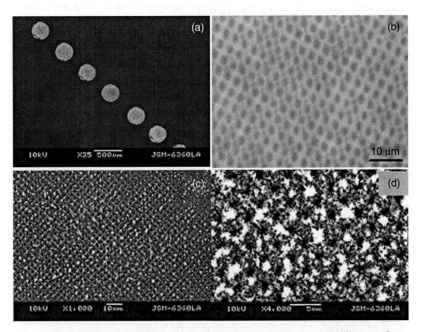

4.7 Optical and SEM images of the microstructures on silicon wafer fabricated by five-beam femtosecond laser pulses. (a) SEM images of fabricated spots. (b) and (c) Optical and SEM images for the details of the spots in (a). (d) Amplified SEM images for the details of the spots in (a).

The structures in Fig. 4.7 can be employed as a diffractive beam splitter, as shown in Fig. 4.8. To investigate the diffraction characteristics of the fabricated structures, a diode-pumped all solid state laser with a wavelength of 532 nm was coupled with a 10× optical objective into the structures and the diffraction patterns were recorded by a digital camera; see Fig. 4.8a. With a 532 nm laser wavelength, the induced structure works as a reflective beam splitter. Figure 4.8b shows the diffraction patterns of the periodic structures obtained. Two-order diffraction spots were observed for the periodic structures on a silicon surface. The first-order diffraction efficiency can be defined as the ratio of the averaged power of the first-order diffraction to the power of the zero-order diffraction. The measured diffraction efficiencies of the first-order diffraction spot were 4.84% (for a wavelength of 532 nm) for the periodic structures. The diffraction patterns can be changed by interval and by the shape of the structures, which are related to the laser energy, angle of the beams, focal length and focusing of the lens, irradiation time, laser wavelength, etc.

As already mentioned, there is much debris on the surface of the laser-structured area, which reduces its reflectivity. After washing the surface in an aqueous solution of HF acid to remove the debris, the first-order diffraction efficiency increased to 5.48%.

It should be noted that, with a wavelength of 532 nm, there is a high absorption level for silicon within the absorption spectra of the silicon sample (Fig. 4.9). But where there is 'strong' absorption, there is also 'strong' reflection, and the calculated reflectivity of Si from its optical constants is 37.34% with a wavelength of 532 nm. For IR light with a wavelength of more than 1200 nm, absorption is very low, so that the induced structures can be used as transmission beam splitters for that IR light. In addition, the

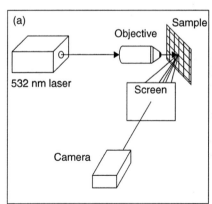
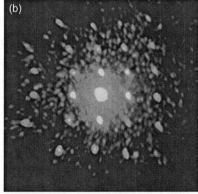

4.8 Diffraction patterns by the fabricated structures, tested by a 532 nm laser. Diffraction efficiency (averaged first order to zero order) is 4.84%. (a) Experimental scheme for testing the diffraction features. (b) Diffraction pattern.

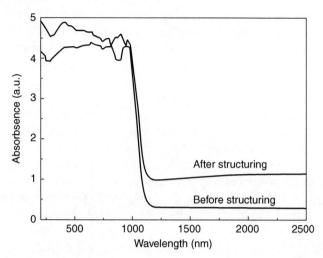

4.9 Absorption spectra before and after interfering femtosecond laser structuring for the silicon sample.

spectrum 'after structuring' shows a higher optical density than that before structuring, because the sample itself acts as a grating, and diffraction losses therefore appear as additional absorption.

4.4.3 Formation of periodic microstructures on metal foil and metal film

For some technological implementations, such as when micrometer-order array holes are necessary, metal materials are required for use in micro-fabrication. In the case of metal materials with high heat conductivities, the fabrication of micron-order structures by use of ultrashort pulse lasers have unique advantages.[38–40] By applying femtosecond laser pulses, the heat-affected zone is in the region of microns in size, allowing the creation of small lateral structures. However, short-pulse irradiation is not necessarily enough to such a small size. The theoretical limit for the smallest spot size achievable is defined by the numerical aperture of the imaging optics and the wavelength of the laser radiation. By decreasing the wavelength, Békési *et al.* employed a UV femtosecond laser to fabricate micron holes on a metal foil.[41] By employing an alternative method, an IR femtosecond laser interference technique, we were able in our experiments to fabricate micrometer-order array holes on both metal foil and metal film.[42] With the same experimental setup, we irradiated five-beam interfered femtosecond beams onto Ni-foil and onto Al-film deposited on a glass substrate, using a 150 mW laser (incident DBS), and with an exposure time of ~10 s; the structures fabricated on the Ni-foil are shown in the SEM images in Fig. 4.10. As with the structures fabricated on a silicon wafer, spots with a diameter of about 300 μm were fabricated on the Ni-foil in a one-step exposure process, and four-fold symmetric array holes were induced (with a period of about 2.5 μm). The depth of the ablated holes was about 10 μm (as measured by an optical microscope). As can be seen in Fig. 4.10b, the wall of holes is not very regular, which may be the reason for the longer ablation duration. Based on a two-temperature diffusion model, Chichkov et al. investigated the ablation behaviors of metal during femtosecond laser irradiation[37] and proposed that the ablation process could be considered as a direct solid–vapor transition, since very short time scales are involved in the ablation. In our case, the five-beam interference of femtosecond laser pulses produced a periodically modulated light field (the interference pattern) with a distribution of enhanced and weakened intensity. The enhanced optical intensity heated the lattice on a picosecond time scale,[38] resulting in the creation of vapor and plasma phases followed by a rapid expansion, and finally forming periodic hole structures on the metal foil.

Figure 4.11 shows the optical and SEM images of array holes fabricated by five-beam interfering femtosecond laser pulses on an Al-film. Similar four-fold symmetric structures were formed on the metal film. The period of the structures was also about 2.5 μm. In the more detailed SEM image in Fig. 4.11b, a particular 'bead in the hole' structure can be observed. Beads with a diameter of less than 900 nm are held in each hole. This structure is probably formed by surface tension: during ablation by interfering femtosecond laser pulses, the film is shrunk and torn, forming the 'bead in the hole' structure.

The structures in Fig. 4.11 work as diffractive beam splitters, as shown in Fig. 4.12. To investigate the diffraction characteristics of the structures, we coupled a diode-pumped all solid state laser (wavelength 532 nm) with an

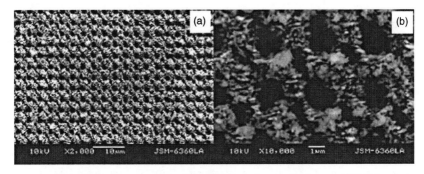

4.10 SEM photographs of the fabricated array holes on Ni-foil by the interference of five-beam femtosecond laser pulses. (a) SEM image of arrayed holes. (b) Magnified SEM holes in (a).

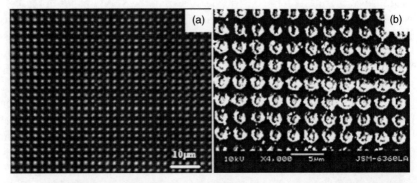

4.11 Optical microscope (a) and SEM (b) photographs of the fabricated array holes Al-film by the interference of five-beam femtosecond laser pulses.

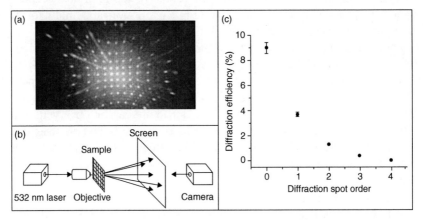

4.12 (a) Diffraction patterns by the structures. (b) Scheme for testing the diffraction. (c) Diffraction efficiency versus diffraction spot order.

optical objective lens onto the structures on the metal film, and the diffraction patterns were recorded by a digital camera (see Fig. 4.12a); the experimental setup is presented in Fig. 4.12b. Four-order diffraction spots were observed for the array holes. Here, the diffraction efficiency can be defined as the ratio of the averaged power of every order of diffraction spots to the power of the incident light. The measured diffraction efficiencies of all the order diffraction spots are shown in Fig. 4.12c.

Nakata *et al.* reported the generation of a nano-sized hollow bump array of gold thin film after uniformly spaced melting and inflation of the film induced by a single shot of four interfering femtosecond laser beams,[25] and they investigated the transition procedure of the induced periodic structures with increasing laser fluence.

4.5 Transfer of periodic microstructures by the interference of femtosecond laser beams

The laser-induced forward transfer (LIFT) technique has been extensively studied for microprinting of diffractive optical structures and computer-generated holograms,[43,44] writing active and passive mesoscopic circuit elements[45] and arranging pad arrays in microelectronic packaging,[46] etc. The LIFT technique utilizes pulsed lasers to remove a thin film material from a transparent support substrate and deposit it onto a suitable receiving substrate. The LIFT process was first shown by Bohandy *et al.* to produce direct writing of Cu lines by using single ns excimer laser pulses under a high vacuum.[47]

By combining the femtosecond laser interference technique and the LIFT technique, transferring of periodic microstructures from a metal Al film deposited on a supporting glass substrate to a receiving glass substrate has been demonstrated.[48] Using five-beam interference of near-IR 800 nm femtosecond laser pulses, micrometer-order periodic microstructures were successfully transferred to a receiving substrate, and the characteristics of the transferred structures (as diffractive optics) evaluated and the morphology of the structures investigated using an optical microscope and SEM (see Fig. 4.13).

The metal film is contacted with a receiving substrate. Two kinds of methods can be applied to focus the interfering laser beams onto the film surface. The first technique (called a front-side transfer), applicable for a transparent receiving substrate, is to focus the interfering beams through the receiving substrate onto the film surface. The second technique (called a rear-side transfer), and used in the case of a transparent supporting substrate, is to focus the interfering beams through the supporting substrate and onto the film surface. In our experiments, both front-side transfer and rear-side transfer could be employed, as the receiving and supporting substrates are both transparent to the laser wavelength. After interfering femtosecond laser irradiation, the supporting substrate and receiving substrate were separated. The induced structure on the supporting substrate and the transferred structure on the receiving substrate were then observed using a 10× optical objective lens, as shown in the inset of Fig. 4.13. As can be seen, the diameters of the induced spot on the supporting substrate and the transferred spot on the receiving substrate are about 300 μm. Concentric circular distributions can be seen in the spots, which were caused by the unequal intensity of the five-beam laser. We measured the intensity of the central beam to be about 20 times greater than that of the other beams. Using a filter can reduce the intensity of the central beam and make the distribution more even.

The details of the structures in the spots are shown in Fig. 4.14, which shows optical microscope and SEM images of the microstructures on the

(a) Contact metal film and
     receiving substrate

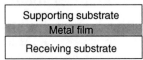

(b) Interfered fs laser
     irradiation

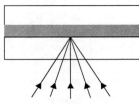

(c) Supporting substrate and
     receiving substrate separation

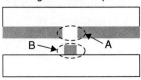

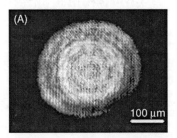

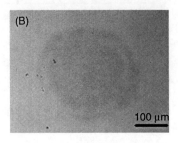

4.13 Schematic of transfer of periodic metal film microstructures by multi-beam interfering femtosecond laser pulses. (a) A receiving substrate is brought onto contact with the metal film deposited on the supporting substrate. (b) Five beams interfering femtosecond laser pulses irradiate the metal film by focusing them on the metal film through the receiving substrate. (c) The supporting substrate and receiving substrate are separated, transferring a periodic microstructure on the receiving substrate, meanwhile leaving a periodic microstructure on the metal film.

supporting substrate (a and c) and on the receiving substrate (b and d). A four-fold symmetric structure was transferred from the supporting substrate to the receiving substrate and a similar structure was also induced on the supporting substrate. The period of structures on both the receiving and supporting substrates was about 2.5 μm. The SEM image of the microstructures on the supporting substrate is clear and debris can be observed around every ablated hole. Although the SEM image of the structures on the receiving substrate illustrates the array structures, the structures look ill-defined when observed at the same amplified ratio as Fig. 4.14c. By adjusting the laser focusing parameters and the thickness of the film on the supporting substrate, the structures on the receiving substrate can be improved as desired.

The structures in Fig. 4.14 can work as a diffractive beam splitter (as shown in Fig. 4.15). To investigate the diffraction characteristics of the structures,

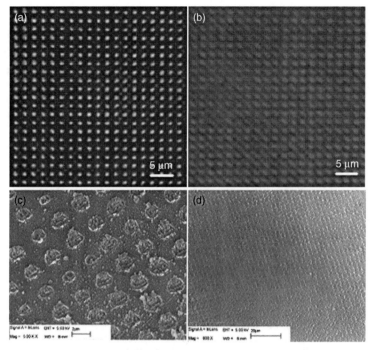

4.14 Optical microscope and SEM images of the structures on the
supporting substrate and on the receiving substrate fabricated by
five-beam interference of femtosecond laser pulses. (a) and (c) optical
microscope and SEM images of the microstructure on supporting substrate, (b) and (d) optical microscope and SEM images of the microstructure on receiving substrate.

we again coupled a diode-pumped all solid state laser with wavelength of
532 nm by a 10× optical objective lens into the structures on both the receiving and supporting substrates, the diffraction patterns being recorded by
a digital camera (see Fig. 4.15c). Figure 4.15a and 4.15b show the diffraction patterns of the periodic structures on both the receiving and supporting substrates. Four-order diffraction spots and two-order diffraction spots
were observed for the periodic structures on the supporting and receiving substrates, respectively. Here, the first-order diffraction efficiency can
be defined as the ratio of the averaged power of the first-order diffraction
to the power of the incident light. The measured diffraction efficiencies of
the first-order diffraction spot for the periodic structures on the supporting
and receiving substrates were, respectively, 3.73% and 1.93%. From the diffractive optics of the structures, we can conclude that the structures on the
supporting substrate should be periodic holes, whilst the structures on the
receiving substrate should be periodic pads. The diffraction patterns can be
changed by interval and shape of the holes and pads, which are related to

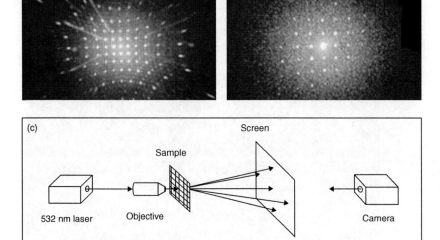

4.15 Diffraction patterns by the microstructures induced on supporting substrate (a) and the microstructures transferred on receiving substrate (b). The averaged diffraction efficiency of first order diffraction spot is 3.73% (a) and 1.93% (b), respectively. (c) Experimental scheme for testing the diffraction by the structures on the supporting substrate and on the receiving substrate.

the laser energy, angle of the beams, focal length and focusing of the lens, irradiation time, laser wavelength, etc.

Adrian *et al.* investigated the mechanism of metal deposition using the LIFT process.[49] They proposed that the LIFT process involves vapor-driven propulsion of metal from a film onto a target. In this case, the periodic holes generated in the supporting substrate are similar to those in a metal foil. The enhanced optical intensity drives the metal vapor onto the receiving substrate to form periodic pad structures, leaving periodic hole structures on the supporting substrate.

The interference transfer technique allows one-step, large-area, micrometer processing of materials for potential industrial applications, such as the fabrication of diffractive optical elements, arranging pad arrays in microelectronic packaging and for the fabrication of sensor elements.

4.6   Conclusions and future trends

This chapter has looked at the fabrication of periodic photonic microstructures by two-beam and multi-beam interfered femtosecond laser pulses.

The femtosecond laser interference technique with its unique advantages (being maskless, photoresist-free, with inner encoding and being single step, etc.) has opened up the possibility of fabrication of one-, two- and three-dimensional periodic functional photonic microstructures in a large variety of materials. Additionally, a new technique of transferring periodic microstructures on a metal film from a supporting substrate to a receiving substrate has also been introduced. By adjusting laser energy, beam angles, beam quantity, focal length, irradiation time and laser wavelength, etc., these fabricated structures can be changed in both period and shape.

The possibilities of using femtosecond laser interference for the fabrication of periodic structures with special functions are very promising. For example, fabrication of periodic dots with magnetic properties could be important for current data storage technologies in order to meet demands for increasing storage capacity. Fabrication of periodic dots with fluorescence will find applications in active photonic devices, whilst fabrication of periodic structures with quantum dots inside them will find numerous applications in emerging nano-devices. Fabrication of periodic structures containing DNA could have exciting applications in biomedical fields. Thus, interference fabrication could become a universal tool to prepare designed functional structures. Furthermore, fabrication of special architectures will also find applications in metamaterials and special lattice structured materials.

4.7   Sources of further information and advice

Books

Claude Rullière. *Femtosecond laser pulses: Principles and experiments*, 2nd edn. Springer Press, 2003 (available at: http://www.google.com/books).

Journal articles

1. P. G. Kryukov, 'Ultrashort-pulse lasers', *Quantum Electronics*, **31**, 95–119 (2001).
2. N. Bloembergen, 'From nanosecond to femtosecond science', *Reviews of Modern Physics*, **71**, S283–S287 (1999).
3. D. von der Linde and K. Sokolowski-Tinten, 'The physical mechanisms of short-pulse laser ablation', *Applied Surface Science*, **154–155**, 1–10 (2000).

Useful websites

Hosono-Kamiya Laboratory, http://www.khlab.msl.titech.ac.jp
Mazur group, https://mazur-www.harvard.edu
Prof. Kawata's Cyber Lab, www.skawata.com

4.8   References

1. DiDomenica, M., 'Small-signal analysis of internal (coupling-type) modulation of lasers', *Journal of Applied Physics*, 1964, **35**, 2870–2876.
2. Hargrove, L. E., Fork, R. L. and Pollack, M. A., 'Locking of He-Ne laser modes induced by synchronous intracavity modulation (diffraction by phonons in crystals E)', *Applied Physics Letters*, 1964, **5**, 4–5.
3. Yariv, A., 'Internal modulation in multimode laser oscillators', *Journal of Applied Physics*, 1965, **36**, 388–391.
4. Fork, R. L., Green, B. I. and Shank, C. V., 'Generation of optical pulses shorter than 0.1 psec by colliding pulse mode locking', *Applied Physics Letters*, 1981, **38**, 671–672.
5. Fork, R. L., Shank, C. V., Yen, R. and Hirliman, C. A., 'Femtosecond optical pulses', *IEEE Journal of Quantum Electronics*, 1983, **QE-19**, 500–506.
6. Morgner, U., Kärtner, F. X., Cho, S. H., Chen, Y., Haus, H. A., Fujimoto, J. G. and Ippen, E. P., 'Sub-two-cycle pulses from a Kerr-lens mode-locked Ti:sapphire laser', *Optics Letters*, 1999, **24**, 411–413.
7. Morgner, U., Ell, R., Metzler, G., Schibli, T. R., Kärtner, F. X., Fujimoto, J. G., Haus, H. A. and Ippen, E. P., 'Nonlinear optics with phase-controlled pulses in the sub-two-cycle regime', *Physical Review Letters*, 2001, **86**, 5462–5465.
8. Gallman, L., Sutter, D. H., Matuschek, N., Steinmeyer, G., Keller, U., Iaconis, C. and Walmsley, I. A., 'Characterization of sub-6-fs optical pulses with spectral phase interferometry for direct electric-field reconstruction', *Optics Letters*, 1999, **24**, 1314–1316.
9. Bahk, S.-W., Rousseau, P., Planchon, T. A., Chvykov, V., Kalintchenko, G., Maksimchuk, A., Mourou, G. A. and Yanovsky, V., 'Generation and characterization of the highest laser intensities (10^{22} W/cm²)', *Optics Letters*, 2004, **29**, 2837–2839.
10. Nakata, Y., Okada, T. and Maeda, M., 'Lines of periodic hole structures produced by laser ablation using interfering femtosecond lasers split by a transmission grating', *Applied Physics A*, 2003, **77**, 399–401.
11. Kawamura, K., Ito, N., Sarukura, N., Hirano, M. and Hosono, H., 'New adjustment technique for time coincidence of femtosecond laser pulses using third harmonic generation in air and its application to holograph encoding system', *Review of Scientific Instruments*, 2002, **73**, 1711–1714.
12. Kawamura, K., Ogawa, T., Sarukura, N., Hirano, M. and Hosono, H., 'Fabrication of surface relief gratings on transparent dielectric materials by two-beam holographic method using infrared femtosecond laser pulses', *Applied Physics B*, 2000, **71**, 119–121.
13. Kawamura, K., Sarukura, N. and Hirano, M., 'Periodic nanostructure array in crossed holographic gratings on silica glass by two interfered infrared-femtosecond laser pulses', *Applied Physics Letters*, 2001, **79**, 1228–1230.
14. Kawamura, K., Sarukura, N., Hirano, M. and Hosono, H., 'Holographic encoding of fine-pitched micrograting structures in amorphous SiO_2 thin films on silicon by a single femtosecond laser pulse', *Applied Physics Letters*, 2001, **78**, 1038–1040.
15. Qu, S., Qiu, J., Zhao, C., Jiang, X., Zeng, H., Zhu, C. and Hirao, K., 'Metal nanoparticle precipitation in periodic arrays in Au_2O-doped glass by two interfered femtosecond laser pulses', *Applied Physics Letters*, 2004, **84**, 2046–2048.
16. Kaneko, K., Sun, H., Duan, X. and Kawata, S., 'Two-photon photoreduction of metallic nanoparticle gratings in a polymer matrix', *Applied Physics Letters*, 2003, **83**, 1426–1428.

17. Si, J., Qiu, J., Zhai, J., Shen, Y. and Hirao, K., 'Photoinduced permanent gratings inside bulk azodye-doped polymers by the coherent field of a femtosecond laser', *Applied Physics Letters*, 2002, **80**, 359–361.
18. Qian, G., Guo, J., Wang, M., Si, J., Qiu, J. and Hirao, K., 'Holographic volume gratings in bulk perylene-orange-doped hybrid inorganic–organic materials by the coherent field of a femtosecond laser', *Applied Physics Letters*, 2003, **83**, 2327–2329.
19. Kawamura, K., Hirano, M., Kamiya, T. and Hosono, H., 'Holographic writing of volume-type micrograting in silica glass by a single chirped laser pulse', *Applied Physics Letters*, 2002, **81**, 1137–1139.
20. Kawamura, K., Hirano, M., Kurobori, T., Takamizu, D., Kamiya, T. and Hosono, H., 'Femtosecond-laser-encoded distributed-feedback color center laser in lithium fluoride single crystals', *Applied Physics Letters*, 2004, **84**, 311–313.
21. Li, Y., Watanabe, W., Yamada, K., Shinagawa, T., Itoh, K., Nishii, J. and Jiang, Y., 'Holographic fabrication of multiple layers of grating inside soda–lime glass with femtosecond laser pulses', *Applied Physics Letters*, 2002, **80**, 1508–1510.
22. Kondo, T., Matsuo, S., Juodkazis, S. and Misawa, H., 'Femtosecond laser interference technique with diffractive beam splitter for fabrication of three-dimensional photonic crystals', *Applied Physics Letters*, 2001, **79**, 725–727.
23. Kondo, T., Matsuo, S., Juodkazis, S., Mizeikis, V. and Misawa, H., 'Multiphoton fabrication of periodic structures by multibeam interference of femtosecond pulses', *Applied Physics Letters*, 2003, **82**, 2758–2760.
24. Nakata, Y., Okada, T. and Maeda, M., 'Fabrication of dot matrix, comb, and nano-wire structures using laser ablation by interfered femtosecond laser beams', *Applied Physics Letters*, 2002, **81**, 4239–4241.
25. Nakata, Y., Okada, T. and Maeda, M., 'Nano-sized hollow bump array generated by single femtosecond laser pulse', *Japanese Journal of Applied Physics*, 2003, **42**, L1452–L1454.
26. Zaidi, S. H. and Brueck, S. R. J., 'Multiple-exposure interferometric lithography', *Journal of Vacuum Science & Technology B*, 1993, **11**, 658–666.
27. Choi, J. O., Akinwande, A. I. and Smith, H. I., '100 nm gate hole openings for low voltage driving field emission display applications', *Journal of Vacuum Science & Technology B*, 2001, **19**, 900–903.
28. Pati, G. S., Heilmann, R. K., Konkola, P. T., Joo, C., Chen, C. G., Murphy, E. and Schattenburg, M. L., 'Generalized scanning beam interference lithography system for patterning gratings with variable period progressions', *Journal of Vacuum Science & Technology B*, 2002, **20**, 2617–2621.
29. Solak, H. H. and David, C., 'Patterning of circular structure arrays with interference lithography', *Journal of Vacuum Science & Technology B*, 2003, **21**, 2883–2887.
30. Maznev, A. A., Crimmins, T. F. and Nelson, K. A., 'How to make femtosecond pulses overlap', *Optics Letters*, 1998, **23**, 1378–1380.
31. Cai, L. Z., Yang, X. L. and Wang, Y. R., 'All fourteen Bravais lattices can be formed by interference of four noncoplanar beams', *Optics Letters*, 2002, **27**, 900–902.
32. Valentine, J., Li, J., Zentgraf, T., Bartal, G. and Zhang, X., 'An optical cloak made of dielectrics', *Nature Materials*, 2009, **8**, 568–571.
33. Fauchet, P. M. and Siegman, A. E., 'Surface ripples on silicon and gallium arsenide under picosecond laser illumination', *Applied Physics Letters*, 1982, **40**, 824–826.
34. Lu, Y. F., Choi, W. K., Aoyagi, Y., Kinomura, A. and Fujii, K., 'Controllable laser-induced periodic structures at silicon–dioxide/silicon interface by excimer laser irradiation', *Journal of Applied Physics*, 1996, **80**, 7052–7056.

35. Her, T. H., Finlay, R. J., Wu, C., Deliwala, S. and Mazur, E., 'Microstructuring of silicon with femtosecond laser pulses', *Applied Physics Letters*, 1998, **73**, 1673–1675.
36. Wu, C., Crouch, C. H., Zhao, L., Carey, J. E., Younkin, R., Levinson, J. A., Mazur, E.,
Farrell, R. M., Gothoskar, P. and Karger, A., 'Near-unity below-band-gap absorption
by microstructured silicon', *Applied Physics Letters*, 2001, **78**, 1850–1852.
37. Zhao, Q. Z., Qiu, J. R., Jiang, X. W., Zhao, C. J. and Zhu, C. S., 'Formation of array
microstructures on silicon by multibeam interfering femtosecond laser pulses', *Applied
Surface Science*, 2005, **241**, 416–419.
38. Chichkov, B. N., Momma, C., Nolte, S., von Alvensleben, F. and Tuenermann, A.,
'Femtosecond, picosecond and nanosecond laser ablation of solids', *Applied Physics
A*, 1996, **63**, 109–115.
39. Simon, P. and Ihlemann, J., 'Ablation of submicron structures on metals and semiconductors by femtosecond UV-laser pulses', *Applied Surface Science*, 1997, **110**,
25–29.
40. Pronko, P. P., Dutta, S. K., Squier, J., Rudd, J. V., Du, D. and Mourou, G., 'Machining
of sub-micron holes using a femtosecond laser at 800 nm', *Optics Communications*,
1995, **114**, 106–110.
41. Békési, J., Klein-Wiele, J.-H. and Simon, P., 'Efficient submicron processing of metals
with femtosecond UV pulses', *Applied Physics A*, 2003, **76**, 355–357.
42. Zhao, Q. Z., Qiu, J. R., Zhao, C. J., Jiang, X. W. and Zhu, C. S., 'Formation of array
holes on metal-foil and metal-film by multibeam interfering femtosecond laser beams',
Chinese Physics, 2005, **14**, 1181–1184.
43. Mailis, S., Zergioti, I., Koundourakis, G., Ikiades, A., Patentalaki, A., Papakonstantinou,
P., Vainos, N. A. and Fotakis, C., 'Etching and printing of diffractive optical microstructures by a femtosecond excimer laser', *Applied Optics*, 1999, **38**, 2301–2308.
44. Zergioti, I., Mailis, S., Vainos, N. A., Papakonstantinou, P., Kalpouzos, C.,
Grigoropoulos, C. P. and Fotakis, C., 'Microdeposition of metal and oxide structures
using ultrashort laser pulses', *Applied Physics A*, 1998, **66**, 579–582.
45. Chrisey, D. B., Pique, A., Fitz-Gerald, J., Auyeung, R. C. Y., McGill, R. A., Wu, H. D.
and Duignan, M., 'New approach to laser direct writing active and passive mesoscopic
circuit elements', *Applied Surface Science*, 2000, **154–155**, 593–600.
46. Bähnisch, R., Groß, W. and Menschig, A., 'Single-shot, high repetition rate metallic
pattern transfer', *Microelectronic Engineering*, 2000, **50**, 541–546.
47. Bohandy, J., Kim, B. F. and Adrian, F. J., 'Metal deposition from a supported metal
film using an excimer laser', *Journal of Applied Physics*, 1986, **60**, 1538–1539.
48. Zhao, Q. Z., Qiu, J. R., Zhao, C. J., Jiang, X. W. and Zhu, C. S., 'Optical transfer
of periodic microstructures by interfering femtosecond laser beams', *Optics Express*,
2005, **13**, 3104–3109.
49. Adrian, F. J., Bohandy, J., Kim, B. F. and Jette, A. N., 'A study of the mechanism of
metal deposition by the laser-induced forward transfer process', *Journal of Vacuum
Science & Technology B*, 1987, **5**, 1490–1494.

Part II
Laser-induced three-dimensional micro- and nano-structuring

5

Multiphoton lithography, processing and fabrication of photonic structures

J. FOURKAS, University of Maryland, USA

Abstract: This chapter discusses the use of multiphoton fabrication techniques to create two- and three-dimensional photonic devices. After reviewing the basic principles of multiphoton lithography, the materials that can be patterned with multiphoton techniques are introduced. Examples of photonic devices created with multiphoton lithography are then presented, followed by a discussion of future prospects for multiphoton fabrication techniques in photonics.

Key words: nonlinear optics, multiphoton absorption, photonic crystals, waveguides.

5.1   Introduction to multiphoton lithography

New applications in photonics are increasingly demanding the ability to create devices with intricate, 3D structures that potentially involve multiple materials. The fabrication of such structures is often challenging, if not impossible, using conventional lithographic techniques. Multiphoton lithographic techniques are gaining considerable prominence for their ability to create complex, 3D structures for applications in photonics and beyond (Farsari *et al.*, 2010; LaFratta *et al.*, 2007; Maruo and Saeki, 2008; Yang *et al.*, 2005).

The key to multiphoton lithography is that multiphoton absorption can be used to drive chemical or physical changes in extremely small (sub-femtolitre) volumes at precise locations in three dimensions. This localization arises through a combination of optical nonlinearity (the probability for multiphoton absorption depends nonlinearly on the intensity of the laser irradiation) and chemical nonlinearity (most photodeposition processes do not occur below a certain threshold exposure). The localized 3D patterning capability of multiphoton lithography allows for the deposition or removal of a wide range of different materials in 3D patterns with resolution that can be below 100 nm. The materials that can be deposited directly include polymers, inorganics and metals. In most cases, this deposition can be accomplished with just a few milliwatts of power from an ultrafast laser operating in the near-infrared region of the electromagnetic spectrum. Furthermore, with additional processing it is possible to coat or convert structures created

139

with multiphoton lithography with a broad range of other materials for specific applications. Due to these unique capabilities, multiphoton lithography is being used for an increasing number of applications in photonics, such as in the creation of photonic crystals, metamaterials, waveguide-based devices, microoptics and optical data storage media.

In this chapter we will introduce the basic principles of multiphoton absorption and how it can be used for the high-precision, localized deposition of materials in three dimensions. We will then review the types of materials that can be patterned with multiphoton lithography, after which we will discuss applications of multiphoton lithographic techniques in photonics. The chapter finishes by discussing future prospects for 3D photonic devices created with multiphoton lithography.

5.2   Principles of multiphoton absorption and lithography

In this section we review some of the essential principles of multiphoton absorption and its application to 3D fabrication. We also specifically consider the optical and chemical effects that determine the overall resolution of multiphoton fabrication techniques.

5.2.1 Multiphoton absorption

In conventional (linear) optical absorption, the energy of an individual photon matches the energy gap between the ground state and an excited state of an atom or molecule (Fig. 5.1a). As a result of this energy match, the probability for absorption scales linearly with the light intensity. In multiphoton absorption, an individual photon does not have enough energy to drive an electronic transition. Instead, the transition is driven by two or more photons acting collectively (Fig. 5.1b). Because there is no intermediate resonant state reachable by linear absorption, multiphoton absorption proceeds through one or more virtual states. Thus, all of the photons needed for a multiphoton absorption event must be present in the same place at the same time. The absorption probability therefore scales as the light intensity to the power of the number of photons that are needed for the absorption event (typically two, but in some cases three or more).

Consider a laser beam that is focused tightly, as through a microscope objective (Fig. 5.1c). The total number of photons per unit time passing through a plane that is transverse to the direction of propagation of the beam is the same regardless of the position of the plane along the beam axis (z). However, because the beam is focused, the area of the plane of the beam (A) is a function of z. The intensity in a plane at height z (which we

(a)

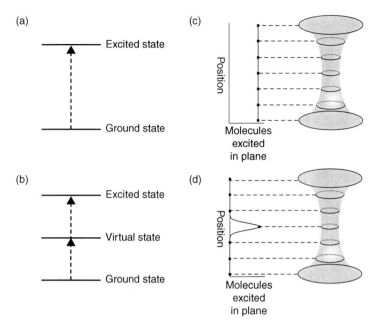

5.1 (a) Energy-level scheme for linear absorption. (b) Energy-level scheme for two-photon absorption. (c) Total number of molecules excited in transverse planes for linear absorption with a focused laser beam, ignoring depletion of the beam intensity. (d) Total number of molecules excited in transverse planes for two-photon absorption with a focused laser beam.

denote $I(z)$) is then inversely proportional to $A(z)$. However, the number of molecules within the volume $A(z)\,\mathrm{d}z$, which we will call $N(z)$, is proportional to $A(z)$. The total number of molecules excited in a plane at height z is proportional to the product of $I(z)$ and $N(z)$; this product is independent of $A(z)$. Thus, with linear excitation, the total number of molecules excited in a plane at height z (which we denote $N_e(z)$) is independent of z (ignoring any decrease of the intensity along the beam path due to absorption).

Now consider the case of two-photon absorption (Fig. 5.1d). The square of the intensity in any transverse plane is inversely proportional to $A^2(z)$. $N(z)$ is again proportional to $A(z)$, so $N_e(z)$ in this case is inversely proportional to $A(z)$. Thus, the probability of exciting molecules with two-photon absorption is much greater at the laser focus than in any other region of the beam.

If a continuous-wave laser is used for two-photon absorption, its intensity generally must be very high in order to have a significant probability of having two photons in the same place at the same time in the focal region. Thus, two-photon absorption is typically driven with a laser that produces pulses with a duration in the range of 50 fs to 1 ps. For instance, a typical Ti:sapphire oscillator produces pulses at a wavelength near 800 nm with a duration on the order of 100 fs at a repetition rate on the order of 80 MHz. The duty cycle of such a laser is on the order of 10^{-5}, so even when the peak intensity is high enough to drive two-photon absorption efficiently, the average power can still be quite low. Multiphoton fabrication is generally performed with an average laser power of a few milliwatts at the sample.

For an objective with a numerical aperture NA and a laser with wavelength λ, average power $\bar{P}$, pulse length τ and repetition rate R, the number of photons absorbed per molecule per unit time is given approximately by (Denk $et~al.$, 1990)

$$r_p \propto \frac{\bar{P}^2 \delta}{R^2 \tau} \frac{(NA)^4}{(2\hbar c\lambda)^2} \qquad \text{[5.1]}$$

Here, $\hbar$ is Planck's constant, c is the speed of light and δ is the two-photon-absorption cross-section. The units of δ are Göppert-Mayer (GM), in honor of Nobel laureate Maria Göppert-Mayer, who first predicted that two-photon absorption could occur. One GM unit is equal to 10–58 m^4 s photon^{-1}, and a typical dye molecule has a maximum δ value on the order of 1–100 GM. Molecules that are designed to have especially efficient two-photon absorption can have δ values that are as much as several orders of magnitude larger.

5.2.2 Resolution

Because multiphoton absorption has a nonlinear dependence on the light intensity, the effective point-spread function (PSF) for fabrication is smaller than the PSF for linear absorption at the same wavelength. Thus, multiphoton fabrication offers finer resolution than would otherwise be possible at the same wavelength. However, the resolution of the PSF is not superior to that available at the effective fabrication wavelength (e.g., half of the incident wavelength for two-photon absorption). Nevertheless, multiphoton absorption has a significant advantage over linear absorption in that molecules are excited only in the focal region. It is this effect that makes high-precision, 3D fabrication possible with multiphoton absorption.

In addition to optical nonlinearity, multiphoton lithography often also takes advantage of chemical nonlinearity. In many photochemical processes, there is a threshold exposure below which material is not deposited. As a result, the lithographic resolution of a multiphoton process can be considerably smaller than the diffraction-limited spot size of the laser beam, even when taking the optical nonlinearity into account.

As an example of the resolution obtainable with multiphoton fabrication, we will consider the most common multiphoton fabrication technique, multiphoton absorption polymerization (MAP, Fig. 5.2). If the laser beam

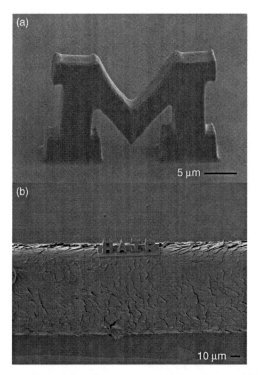

5.2 Representative structures created using MAP. (a) Letter M. (b) The word 'hair' written on a human hair (reprinted with permission from Baldacchini *et al.*, 2004).

is held in a fixed location while exposing a photoresist, a single volume element ('voxel') of polymer is created. In typical multiphoton photoresists, the best transverse resolution attainable with 800 nm light is on the order of 80 nm (Li *et al.*, 2009; Xing *et al.*, 2007). The ability to attain $\lambda/10$ resolution is attributable directly to the chemical nonlinearity of the photoresist. Resolution as fine as 20 nm using 800 nm excitation has been reported in special cases (Tan *et al.*, 2007), but not for free-form fabrication. In addition, the use of a shorter excitation wavelength has been shown to be able to improve the resolution of MAP to 60 nm (Haske *et al.*, 2007).

Due to the geometry of the PSF, the axial dimension of voxels created with MAP is typically a factor of three to five larger than the transverse dimension (Fig. 5.3a). Thus, voxels are generally shaped like cucumbers. This voxel asymmetry can be a limitation in the fabrication of detailed structures that require fine resolution, such as photonic crystals.

To address the issue of voxel asymmetry and to improve the resolution of MAP, a technique called Resolution Augmentation through Photo-Induced Deactivation (RAPID) lithography has been developed (Li *et al.*, 2009).

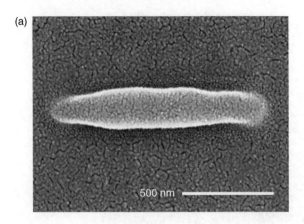

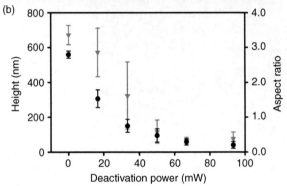

5.3 (a) Typical voxel created with MAP. (b) Resolution improvement attained using RAPID lithography as a function of deactivation power (both reprinted with permission from Li *et al.*, 2009).

RAPID lithography uses one laser to create a cross-linked polymer and a second laser to inhibit cross-linking. Through spatial phase shaping of the second laser beam to inhibit polymerization only in chosen regions, considerable improvements in resolution can be realized. For instance, voxels with a 40-nm axial dimension have been reported using 800-nm excitation, representing $\lambda/20$ resolution. In this case, the aspect ratio of voxels was reduced from the typical value of three to five down to one half (Fig. 5.3b). Further improvements in materials and optics for RAPID lithography promise to push the resolution down further, perhaps to the range of 20 nm or even smaller (Fourkas, 2010).

5.3   Materials for multiphoton lithography

In this section we review the types of materials that can be used for multiphoton lithography. We begin with a discussion of materials from which structures can be created directly. We then discuss the use of structures fabricated with multiphoton lithography as templates for the deposition of other materials.

5.3.1 Direct deposition

Direct deposition schemes use precursors that can be converted directly into the desired material using multiphoton absorption. The advantage of direct deposition is that the creation of devices typically requires a single fabrication step. However, the range of materials that can be deposited directly to form high-quality structures is somewhat limited. Nevertheless, direct deposition has found many uses in photonics, and here we will review some of the most important materials used in this type of fabrication.

Polymers

By far the most common multiphoton fabrication technique is MAP (LaFratta *et al.*, 2007; Maruo and Fourkas, 2008; Yang *et al.*, 2005). MAP typically employs a negative-tone photoresist that is hardened locally upon multiphoton excitation. MAP photoresists contain two key components: a cross-linkable monomer and a photoinitiator. Optical excitation of the photoinitiator starts a chain reaction through which monomers cross-link with one another, creating an insoluble polymer. Competing chemical processes limit the cross-linking reaction to regions with a sufficient concentration of excited initiator. After the desired 3D pattern has been scanned, the unexposed photoresist is removed by solvent development.

Two different types of negative-tone photoresist material have been employed widely for MAP. The first type is polymerized through a radical mechanism. Radical photoresists are typically composed of acrylic and/or methacrylic monomers (Baldacchini *et al.*, 2004; Cumpston *et al.*, 1999). Such monomers are widely available, and the properties of the final structures can be tuned in a straightforward manner by adjusting the composition of monomers used. Because radical polymerization takes place immediately upon excitation of a photoinitiator, structures created with these materials can be visualized during fabrication based on the change in refractive index caused by the cross-linking. This refractive index change can also lead to optical distortions that affect the fabrication of subsequent features, however, and so fabrication paths must be designed with care with these materials. Typical radical photoresists used for MAP are viscous liquids, which means that development can be accomplished with a few solvent washes. However, for photoresists that are liquids, flow during fabrication can also be an issue. Fillers that cause the resist to gel can circumvent this problem (Kuebler *et al.*, 2001), although the solvent development of such materials is somewhat more difficult.

The other major type of negative-tone photoresist used for MAP is polymerized by a cationic mechanism. Cationic photoresists are typified by epoxies, such as SU-8 (Kumi *et al.*, 2010; Witzgall *et al.*, 1998). Cationic polymerization in MAP begins with the excitation of photoacid generator (PAG). Rather than leading to an immediate polymerization chain reaction, the photoresist typically must be brought to an elevated temperature following exposure to facilitate cross-linking. After this 'postbake' step, solvent development is used to remove the unpolymerized material. Because the majority of polymerization occurs during the postbake, cationic photoresists such as SU-8 do not generally exhibit significant changes in refractive index during fabrication. This facet of cationic photoresists helps to minimize any optical distortions, but also makes it difficult to ascertain the location of structures relative to the substrate during fabrication.

Numerous photoinitiators and PAGs have been developed over the years for use with linear excitation. These materials tend to be optimized for high initiation velocities and efficient generation of radicals or acid. Many of these materials have been shown to be suitable for MAP as well. However, these initiators are intended for linear excitation. In this case, having too high an absorption cross-section can be a disadvantage because it is difficult to expose thick films of photoresist. Thus, commercially available initiators do not tend to have large multiphoton absorption cross-sections at appropriate wavelengths (Yanez *et al.*, 2009a). For this reason, some groups have focused on creating new photoinitiators or PAGs that have large δ values (Cumpston *et al.*, 1999; Yanez *et al.*, 2009a). These molecules typically allow MAP to be achieved at quite low intensities. However, due to the complexity of synthesizing these materials, as well as to solubility issues, the majority of researchers in the field still use commercially available components.

RAPID lithography requires a photoinitiator that can be excited by multiphoton absorption and deactivated by a second beam of light. To date, RAPID lithography has only been demonstrated for radical polymerization (Li *et al.*, 2009; Stocker *et al.*, 2011). Although the deactivation of one conventional photoinitiator has been reported using stimulated emission (Fischer *et al.*, 2010), other common dye molecules that are not typically viewed as photoinitiators appear to be promising candidates for RAPID initiators that can be excited and deactivated (by another mechanism) with high efficiency (Stocker *et al.*, 2011).

MAP can also be used to create patterns in positive-tone photoresists, that is, resists in which the exposed region can be removed by development, leaving the rest of the photoresist behind (Gansel *et al.*, 2009; Yu *et al.*, 2003; Zhou *et al.*, 2002). The challenge in using positive-tone resists for creating 3D structures is that it is difficult to remove material from small regions that are deep inside of large structures. Because of this mass-transfer issue, positive-tone photoresists have received far less attention than have negative-tone photoresists for MAP. Nevertheless, for some applications in photonics, positive-tone resists are advantageous and have been used.

Metals

Many applications in photonics, such as reflectors, gratings, photonic crystals and metamaterials, can benefit from the ability to deposit metals in three dimensions with nanoscale precision. As such, a considerable amount of effort has been put into the development of methods for the direct multiphoton deposition of metals. The creation of 3D metallic structures with multiphoton lithography presents significant materials challenges in comparison to the deposition of polymers. In the case of polymer fabrication, essentially 100% of the precursor (the photoresist) is converted to the final material (the cross-linked polymer). However, the deposition of metals relies on the photoreduction of metal ions, the dissociation of organometallic precursors, or the aggregation of nanoparticles by removing stabilizers. All of these processes leave organic byproducts. Furthermore, these precursors must generally be dispersed in another material. Thus, the fraction of the material that is converted in multiphoton metal deposition is typically considerably less than 100%.

One approach to multiphoton deposition of metallic structures is to begin with a liquid solution containing the precursor. A number of metals have been photodeposited from solution, including silver (Ishikawa *et al.*, 2006; Tanaka *et al.*, 2006), gold (LaFratta *et al.*, 2007; Tanaka *et al.*, 2006) and iron (LaFratta *et al.*, 2007). Solution deposition with multiphoton absorption is generally best suited for the creation of 2D patterns. Because a relatively small fraction of any volume of the solution can be converted to metal, it is more difficult to create 3D structures. Often one must scan over areas repeatedly, leaving enough time between scans for additional precursor to diffuse into the region of interest. Deposition solutions are also prone to the formation of gas bubbles, either through boiling or decomposition of the precursor, which can lead to distortions in the shape of the final structure. Additionally, metallic structures with submicron feature sizes often do not have sufficient mechanical integrity for the creation of free-standing structures. However, the fabrication of 3D nanostructures by deposition of silver nanoparticles from solution has been demonstrated (Cao *et al.*, 2009), although the fidelity of these structures does not yet approach what can be achieved for polymers with MAP.

Metal precursors can also be dispersed in polymer matrices for multiphoton deposition. This approach offers considerably greater latitude for the creation of 3D structures, especially if the polymer host need not be removed after fabrication. A number of groups have reported the creation of both 2D and 3D structures composed of metals such as silver and gold using polymer matrices (Baldacchini *et al.*, 2005; LaFratta *et al.*, 2006b; Maruo and Saeki, 2008; Stellacci *et al.*, 2002; Tosa *et al.*, 2007, 2008, Vurth *et al.*, 2008). In many cases the polymer matrix can be removed after fabrication to leave the metal structure behind.

Photodeposition of metals is an electroless process, and therefore tends to create structures that are somewhat porous. These structures therefore have electrical properties that are not as favorable as those of the bulk metal. However, the properties of these structures are usually more than sufficient for optical applications.

Inorganics

Inorganics are attractive for many applications in photonics that require high-index materials. However, in general the deposition of inorganics suffers the same difficulty as in the deposition of metals, that is, that there are not precursors available that offer nearly complete material conversion. One notable exception to this situation is chalcogenides. For instance, As_4S_6 can be 'photopolymerized' to create an insoluble, As_2S_3 glass, after which the unpolymerized starting material can be removed (Wong *et al.*, 2006). Because this process relies only on a phase change, the material conversion is 100%.

'Spin-on' silica glasses, which are composed of hybrid organic/inorganic precursors, offer another route to the creation of inorganic structures using multiphoton absorption. These materials were originally developed as electron-beam resists, but can also serve as photoresists that offer a relatively high degree of material conversion (Jun *et al.*, 2005; Nagpal *et al.*, 2008). Most of the organic material is removed in development, leaving a predominantly inorganic structure behind.

Most other approaches to inorganic fabrication using multiphoton absorption have relied on hybrid materials that also have a significant organic fraction in the final structure. For instance, alkoxy silanes that contain polymerizable organic groups can be condensed to form sol-gels (Farsari *et al.*, 2010; Houbertz *et al.*, 2003; Serbin *et al.*, 2003). Multiphoton absorption can then be used to cross-link the organic groups in desired regions, and the remainder of the sol-gel can be washed away. Structures formed in this way have mechanical properties that can be considerably superior to those of highly cross-linked organic polymers. Similar chemistry can be performed with other oxides, such as zirconia (Bhuian *et al.*, 2006) and titania (Sakellari *et al.*, 2010).

5.3.2 Indirect deposition

Given that it is difficult to use multiphoton fabrication to create high-quality structures from many materials that are desirable for applications in photonics, methods that use polymeric structures created with MAP as templates for final structures that include other materials provide an important alternative route for the creation of photonic devices. These methods fall into three general categories: surface coating, backfilling and molding.

Surface coating

For applications in which a thin layer of another material is sufficient to introduce the desired properties to a 3D structure, surface coating is an attractive route. For instance, a number of groups have demonstrated the use of electroless deposition to deposit metal coating on structures created with MAP (Chen *et al.*, 2006; Farrer *et al.*, 2006; Formanek *et al.*, 2006). Typically, after the 3D polymer structure is created, it is reacted with a chemical that can promote the subsequent deposition of metal from solution. This strategy has been employed (see, e.g., Fig. 5.4a) to create coatings of metals such as silver, gold and copper (Farrer *et al.*, 2006). Furthermore,

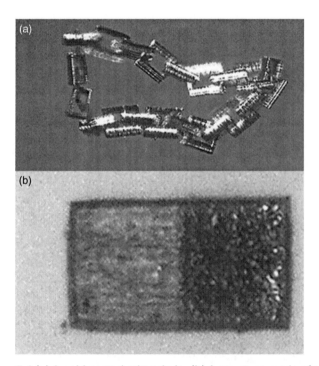

5.4 (a) A gold-coated microchain. (b) A structure made of a square of methacrylic polymer (left) and a square of acrylic polymer (right) that has been subjected to a copper metallization process. The acrylic polymer is coated selectively, while the methacrylic polymer remains pristine (both reprinted with permission from Farrer *et al.*, 2006).

by creating structures from multiple polymers with different chemical reactivity (Fig. 5.4b), the deposition of metals on selected areas of 3D structures has been demonstrated (Farrer *et al.*, 2006). Selective deposition is a powerful approach for the creation of complex, hybrid 3D structures for applications in photonics, electronics and other areas. Selective reactivity of different polymers can also be used to deposit or tether a wide range of other materials of interest for photonics, including oxides, dye molecules and biomolecules.

If conformal coatings are desired, vapor-phase methods are also suitable for the coating of 3D polymeric structures created with MAP. Beyond simple evaporation or sputtering for deposition of metals, structures can be coated with a wide variety of materials using techniques such as chemical vapor deposition (Rill *et al.*, 2008) and atomic layer deposition (Graugnard *et al.*, 2009; King *et al.*, 2006; Langner *et al.*, 2008).

Inversion

For many applications in photonics, coatings are not sufficient to achieve the desired optical properties, such as refractive index contrast. In this case, 3D polymeric structures can be backfilled with other materials. If the voids in the 3D structure are on the scale of hundreds of nanometers, solution-phase approaches are not suitable for backfilling. Instead, techniques such as atomic layer deposition must be used to accomplish uniform filling.

Once backfilling has been achieved, the polymer is often removed by calcination. In this case, the backfilled material occupies the negative space of the original structure, and so the process as a whole is known as 'inversion' (King *et al.*, 2006; Langner *et al.*, 2008). It is also possible to backfill the inverted structure with a third material, and then to etch away the original backfilling material. This 'double inversion' process creates a structure with essentially the same geometry as the original polymeric structure but composed of an entirely different material (Tetreault *et al.*, 2006).

Molding

Another approach to creating 3D structures from a wider range of materials is to use a polymeric structure fabricated with MAP to create a mold (LaFratta *et al.*, 2004, 2006a). Molds are generally created with an elastomeric material such as polydimethylsiloxane (PDMS). The 3D structure is immersed in the elastomer, which is then allowed to set. The elastomeric mold is then removed from the master structure, and can be filled with a molding material and pressed against a substrate to create a replica. This technique is known as microtransfer molding (μTM). Each master structure can be used to create many molds, and each mold can be used to create many replicas (Fig. 5.5).

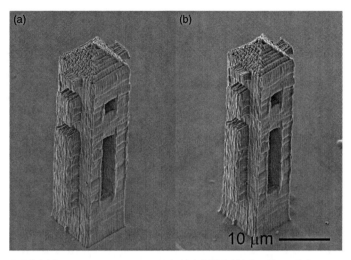

5.5 (a) Master structure created with MAP. (b) Replica of the structure created with microtransfer molding (both reprinted with permission from LaFratta *et al.*, 2004).

The geometries of the master structures that can be used for µTM are limited due to the constraint that the mold must be able to be removed from the master structure (and from each replica). However, due to the elastic nature of PDMS, the topological constraints are not nearly as stringent as might be imagined (LaFratta *et al.*, 2004). Additionally, a technique called membrane-assisted µTM (Fig. 5.6) even allows replicas with closed loops to be created (LaFratta *et al.*, 2006a).

5.4   Applications of multiphoton lithography in photonics

Due to its ability to create high-precision, 3D nano- and micro-structures, applications of multiphoton fabrication have grown exponentially over the past decade. Here we will discuss examples of some of the prominent applications of multiphoton fabrication in photonics.

5.4.1 Micro-optics

The fabrication of micro-optics, such as lenses, phase plates and Fresnel optics, is often challenging using conventional lithography. Because these structures are three-dimensional, a gray-scale technique must typically be used for their fabrication on microscopic scales. Multiphoton fabrication is suited ideally for the rapid prototyping and production of 3D micro-optics.

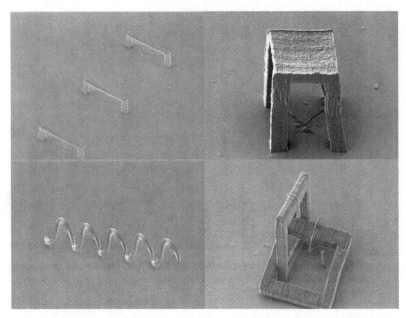

5.6 Examples of molded structures created with membrane-assisted microtransfer molding (reprinted with permission from LaFratta *et al.*, 2006a).

A number of groups have reported work in this area (Guo *et al.*, 2006; Malinauskas *et al.*, 2010a, 2010b; Yan *et al.*, 2010), and multiphoton fabrication is a highly promising technique for creating both individual micro-optics and arrays of micro-optics. Many of these devices are also amenable to molding after fabrication, making mass production feasible.

Multiphoton metal deposition has also been used to create micro-optics, such as diffraction gratings (Baldacchini *et al.*, 2005; Cao *et al.*, 2009; Kaneko *et al.*, 2003b). By using multiphoton absorption to pattern helical channels in a positive-tone photoresist and then back-filling these channels with gold via electrodeposition, the creation of broadband circular polarizers has also been demonstrated (Gansel *et al.*, 2009). This approach shows great promise for further applications in metamaterials.

It is also possible to incorporate dye molecules into structures created with MAP as a route to active micro-optics (Farsari *et al.*, 2008; Sun *et al.*, 2001; Yokoyama *et al.*, 2003). The challenge with this approach is that the multiphoton polymerization process destroys many of the different molecules that it would be desirable to be able to incorporate. However, devices such as micro-scale, distributed-feedback lasers (Yokoyama *et al.*, 2003) and dye-doped photonic crystals (Farsari *et al.*, 2008) have been reported.

5.4.2 Waveguide-based devices

Because it can be used to create extremely smooth structures with feature sizes of a micrometre or less, multiphoton fabrication is well suited for the creation of waveguide-based devices. Multiphoton fabrication offers some advantages over other techniques for creating polymeric and dielectric-loaded waveguides, particularly in its ability to take advantage of the third dimension. The ability to incorporate other materials in or on fabricated structures may also provide a path towards making active devices and sensors.

In some of the first work on the creation of waveguides with MAP, devices such as couplers and interferometers were demonstrated between the ends of optical fibers (Klein *et al.*, 2005, 2006). More recently, high-performance microring resonators (Fig. 5.7) have been demonstrated (Li *et al.*, 2008a), in addition to index-modulated gratings within waveguides (Dong *et al.*, 2007). Dielectric-loaded waveguides based on plasmonic metals are beginning to be explored as well (Reinhardt *et al.*, 2009, 2010).

One of the attractive features of the 3D capability of multiphoton fabrication for waveguide devices is that it can be used for coupling of fibers with one another or with waveguides and on-chip photonic devices. 3D couplers or on-fiber optics can reduce coupling losses in optical systems considerably. As a result, interest is growing in the multiphoton fabrication of structures directly on the ends of optical fibers (Cojoc *et al.*, 2010; Liberale *et al.*, 2010) or on side-polished fibers (Sherwood *et al.*, 2005).

5.4.3 Photonic crystals

Given the capability of multiphoton fabrication to create 3D structures with nanoscale resolution, its most popular application in optics has been in the creation of photonic crystals (Borisov *et al.*, 1998; Deubel *et al.*, 2004; Guo *et al.*, 2005; Jia *et al.*, 2009; Kaneko *et al.*, 2003a, Sakellari *et al.*, 2010; Serbin and Gu, 2006; Serbin *et al.*, 2003; Straub *et al.*, 2005; Sun *et al.*, 1999). MAP is well suited for the creation of photonic crystals with nearly arbitrarily complex geometries, with resolution and voxel asymmetry being the main limitations in determining the geometries and operational wavelengths that can be achieved. Furthermore, because MAP is a direct-write method, defects such as cavities and waveguides can be created wherever desired within a photonic crystal.

MAP does have some drawbacks for the creation of photonic crystals. First, because it is a point-by-point fabrication method, the ultimate size of the photonic crystals that can be fabricated is typically on the order of hundreds of micrometres on a side. Photonic crystal structures are complex and have large numbers of voids, which means that they are not compatible

(a)
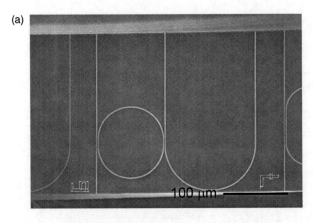

100 μm

(b)
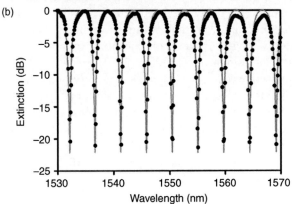

5.7 (a) Microring resonator created using MAP. (b) Typical data from the through port of a microring resonator (both reprinted with permission from Li *et al.*, 2008a).

with molding techniques. Second, material shrinkage upon exposure and development can distort photonic crystal structures, although this effect can generally be pre-compensated during fabrication. Finally, the index contrast between polymers and air is not sufficient to create a stop band. This last issue has been a major driver of research into the development of hybrid materials with higher refractive indices (Farsari *et al.*, 2010; Sakellari *et al.*, 2010; Serbin *et al.*, 2003) and of inversion schemes (King *et al.*, 2006; Langner *et al.*, 2008; Tetreault *et al.*, 2006).

5.4.4 Optical data storage

Another well-studied application of multiphoton absorption in photonics is in high-density optical data storage (Fig. 5.8) (Adam *et al.*, 2010;

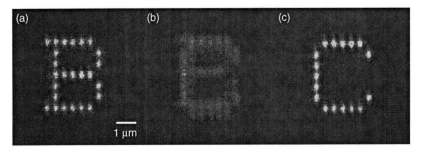

5.8 Data written in different planes of a molecular glass using multiphoton absorption (reprinted with permission from Olson *et al.*, 2002).

Cumpston *et al.*, 1999; Dvornikov *et al.*, 2009; Iliopoulos *et al.*, 2010; Li *et al.*, 2008b; Olson *et al.*, 2002; Wang and Stucky, 2004; Walker *et al.*, 2008; Yanez *et al.*, 2009b; Zhang *et al.*, 2009). Rather than depositing material, in these approaches to optical memory multiphoton absorption is typically used to create a photochemical or photophysical change within a small volume of a macroscopic substrate. This change must later be able to be read out optically, either with a multiphoton process or, often preferably, with a linear process. Data are typically written in a binary format, and multiphoton absorption allows many layers to be addressed individually. Storage of a terabyte of data in 200 layers of a disk the size of a DVD has been demonstrated recently (Dvornikov *et al.*, 2009; Walker *et al.*, 2008). The key to practical commercial success for such devices will be achieving inexpensive and reliable read-out.

5.5   Future prospects for multiphoton lithography in photonics

Multiphoton fabrication offers unique capabilities for the creation of high-resolution, 3D structures, and can be performed with a table-top instrument. This technique is being used increasingly for applications in photonics. While multiphoton lithography is now a relatively mature technique, significant enhancements continue to be made both in materials and instrumentation. We can expect the resolution of this technique and the range of materials that can be employed to continue to improve in the coming years, enabling further applications and enhanced performance in photonics. Multiphoton fabrication is suited ideally to the prototyping of photonic devices as well as to the creation of devices for research purposes. If methods can be developed to increase the scalability of this technique, it has the potential to become a mainstay in the mass production of micro- and nano-photonic devices.

5.6   References

Adam, V., Mizuno, H., Grichine, A., Hotta, J. I., Yamagata, Y., Moeyaert, B., Nienhaus, G. U., Miyawaki, A., Bourgeois, D. and Hofkens, J. (2010), 'Data storage based on photochromic and photoconvertible fluorescent proteins', *Journal of Biotechnology*, **149**, 289–298.

Baldacchini, T., Lafratta, C. N., Farrer, R. A., Teich, M. C., Saleh, B. E. A., Naughton, M. J. and Fourkas, J. T. (2004), 'Acrylic-based resin with favorable properties for three-dimensional two-photon polymerisation', *Journal of Applied Physics*, **95**, 6072–6076.

Baldacchini, T., Pons, A. C., Pons, J., Lafratta, C. N., Fourkas, J. T., Sun, Y. and Naughton, M. J. (2005), 'Multiphoton laser direct writing of two-dimensional silver structures', *Optics Express*, **13**, 1275–1280.

Bhuian, B., Winfield, R. J., O'Brien, S. and Crean, G. M. (2006), 'Investigation of the two-photon polymerisation of a Zr-based inorganic-organic hybrid material system', *Applied Surface Science*, **252**, 4845–4849.

Borisov, R. A., Dorojkina, G. N., Koroteev, N. I., Kozenkov, V. M., Magnitskii, S. A., Malakhov, D. V., Tarasishin, A. V. and Zheltikov, A. M. (1998), 'Fabrication of three-dimensional periodic microstructures by means of two-photon polymerisation', *Applied Physics B: Lasers and Optics*, **67**, 765–767.

Cao, Y. Y., Takeyasu, N., Tanaka, T., Duan, X. M. and Kawata, S. (2009), '3D metallic nanostructure fabrication by surfactant-assisted multiphoton-induced reduction', *Small*, **5**, 1144–1148.

Chen, Y. S., Tal, A., Torrance, D. B. and Kuebler, S. M. (2006), 'Fabrication and characterization of three-dimensional silver-coated polymeric microstructures', *Advanced Functional Materials*, **16**, 1739–1744.

Cojoc, G., Liberale, C., Candeloro, P., Gentile, F., Das, G., De Angelis, F. and Di Fabrizio, E. (2010), 'Optical micro-structures fabricated on top of optical fibers by means of two-photon photopolymerisation', *Microelectronic Engineering*, **87**, 876–879.

Cumpston, B. H., Ananthavel, S. P., Barlow, S., Dyer, D. L., Ehrlich, J. E., Erskine, L. L., Heikal, A. A., Kuebler, S. M., Lee, I. Y. S., McCord-Maughon, D., Qin, J., Rockel, H., Rumi, M., Wu, X.-L., Marder, S. R. and Perry, J. W. (1999), 'Two-photon polymerisation initiators for three-dimensional optical data storage and microfabrication', *Nature*, **398**, 51–54.

Denk, W., Strickler, J. H. and Webb, W. W. (1990), 'Two-photon laser scanning fluorescence microscopy', *Science*, **248**, 73–76.

Deubel, M., Von Freymann, G., Wegener, M., Pereira, S., Busch, K. and Soukoulis, C. M. (2004), 'Direct laser writing of three-dimensional photonic-crystal templates for telecommunications', *Nature Materials*, **3**, 444–447.

Dong, Y., Yu, X. Q., Sun, Y. M., Hou, X. Y., Li, Y. F. and Zhang, M. (2007), 'Refractive index-modulated grating in two-mode planar polymeric waveguide produced by two-photon polymerisation', *Polymers for Advanced Technologies*, **18**, 519–521.

Dvornikov, A. S., Walker, E. P. and Rentzepis, P. M. (2009), 'Two-photon three-dimensional optical storage memory', *Journal of Physical Chemistry A*, **113**, 13633–13644.

Farrer, R. A., Lafratta, C. N., Li, L., Praino, J., Naughton, M. J., Saleh, B. E. A., Teich, M. C. and Fourkas, J. T. (2006), 'Selective functionalization of 3-D polymer microstructures', *Journal of the American Chemical Society*, **128**, 1796–1797.

Farsari, M., Ovsianikov, A., Vamvakaki, M., Sakellari, I., Gray, D., Chichkov, B. N. and Fotakis, C. (2008), 'Fabrication of three-dimensional photonic crystal structures containing an active nonlinear optical chromophore', *Applied Physics A: Materials Science & Processing*, **93**, 11–15.

Farsari, M., Vamvakaki, M. and Chichkov, B. N. (2010), 'Multiphoton polymerization of hybrid materials', *Journal of Optics*, **12**, 124001.

Fischer, J., Von Freymann, G. and Wegener, M. (2010), 'The materials challenge in diffraction-unlimited direct-laser-writing optical lithography', *Advanced Materials*, **22**, 3578–3582.

Formanek, F., Takeyasu, N., Tanaka, T., Chiyoda, K., Ishikawa, A. and Kawata, S. (2006), 'Three-dimensional fabrication of metallic nanostructures over large areas by two-photon polymerisation', *Optics Express*, **14**, 800–809.

Fourkas, J. T. (2010), 'Nanoscale photolithography with visible light', *Journal of Physical Chemistry Letters*, **1**, 1221–1227.

Gansel, J. K., Thiel, M., Rill, M. S., Decker, M., Bade, K., Saile, V., Von Freymann, G., Linden, S. and Wegener, M. (2009), 'Gold helix photonic metamaterial as broadband circular polarizer', *Science*, **325**, 1513–1515.

Graugnard, E., Roche, O. M., Dunham, S. N., King, J. S., Sharp, D. N., Denning, R. G., Turberfield, A. J. and Summers, C. J. (2009), 'Replicated photonic crystals by atomic layer deposition within holographically defined polymer templates', *Applied Physics Letters*, **94**, 263109.

Guo, R., Li, Z., Jiang, Z., Yuan, D., Huang, W. and Xia, A. (2005), 'Log-pile photonic crystal fabricated by two-photon photopolymerisation', *Journal of Optics A: Pure and Applied Optics*, **7**, 396–399.

Guo, R., Xiao, S. Z., Zhai, X. M., Li, J. W., Xia, A. D. and Huang, W. H. (2006), 'Micro lens fabrication by means of femtosecond two photon photopolymerisation', *Optics Express*, **14**, 810–816.

Haske, W., Chen, V. W., Hales, J. M., Dong, W. T., Barlow, S., Marder, S. R. and Perry, J. W. (2007), '65 nm feature sizes using visible wavelength 3-D multiphoton lithography', *Optics Express*, **15**, 3426–3436.

Houbertz, R., Frohlich, L., Popall, M., Streppel, U., Dannberg, P., Brauer, A., Serbin, J. and Chichkov, B. N. (2003), 'Inorganic-organic hybrid polymers for information technology: From planar technology to 3D nanostructures', *Advanced Engineering Materials*, **5**, 551–555.

Iliopoulos, K., Krupka, O., Gindre, D. and Salle, M. (2010), 'Reversible two-photon optical data storage in coumarin-based copolymers', *Journal of the American Chemical Society*, **132**, 14343–14345.

Ishikawa, A., Tanakaa, T. and Kawata, S. (2006), 'Improvement in the reduction of silver Ions in aqueous solution using two-photon sensitive dye', *Applied Physics Letters*, **89**, 113102.

Jia, B. H., Kang, H., Li, J. F. and Gu, M. (2009), 'Use of radially polarized beams in three-dimensional photonic crystal fabrication with the two-photon polymerisation method', *Optics Letters*, **34**, 1918–1920.

Jun, Y., Leatherdale, C. A. and Norris, D. J. (2005), 'Tailoring air defects in self-assembled photonic bandgap crystals', *Advanced Materials*, **17**, 1908–1911.

Kaneko, K., Sun, H.-B., Duan, X.-M. and Kawata, S. (2003a), 'Submicron diamond-lattice photonic crystals produced by two-photon laser nanofabrication', *Applied Physics Letters*, **83**, 2091–2093.

Kaneko, K., Sun, H.-B., Duan, X.-M. and Kawata, S. (2003b), 'Two-photon photoreduction of metallic nanoparticle gratings in a polymer matrix', *Applied Physics Letters*, **83**, 1426–1428.

King, J. S., Graugnard, E., Roche, O. M., Sharp, D. N., Scrimgeour, J., Denning, R. G., Turberfield, A. J. and Summers, C. J. (2006), 'Infiltration and inversion of holographically defined polymer photonic crystal templates by atomic layer deposition', *Advanced Materials*, **18**, 1561–1565.

Klein, S., Barsella, A., Leblond, H., Bulou, H., Fort, A., Andraud, C., Lemercier, G., Mulatier, J. C. and Dorkenoo, K. (2005), 'One-step waveguide and optical circuit writing in photopolymerizable materials processed by two-photon absorption', *Applied Physics Letters*, **86**, 211118.

Klein, S., Barsella, A., Taupier, G., Stortz, V., Fort, A. and Dorkenoo, K. D. (2006), 'Optical components based on two-photon absorption process in functionalized polymers', *Applied Surface Science*, **252**, 4919–4922.

Kuebler, S. M., Rumi, M., Watanabe, T., Braun, K., Cumpston, B. H., Heikal, A. A., Erskine, L. L., Thayumanavan, S., Barlow, S., Marder, S. R. and Perry, J. W. (2001), 'Optimizing two-photon initiators and exposure conditions for three-dimensional lithographic microfabrication', *Journal of Photopolymer Science and Technology*, **14**, 657–668.

Kumi, G., Yanez, C. O., Belfield, K. D. and Fourkas, J. T. (2010), 'High-speed multiphoton absorption polymerisation: fabrication of microfluidic channels with arbitrary cross-sections and high aspect ratios', *Lab on a Chip*, **10**, 1057–1060.

Lafratta, C. N., Baldacchini, T., Farrer, R. A., Fourkas, J. T., Teich, M. C., Saleh, B. E. A. and Naughton, M. J. (2004), 'Replication of two-photon-polymerized structures with extremely high aspect ratios and large overhangs', *Journal of Physical Chemistry B*, **108**, 11256–11258.

Lafratta, C. N., Fourkas, J. T., Baldacchini, T. and Farrer, R. A. (2007), 'Multiphoton fabrication', *Angewandte Chemie International Edition*, **46**, 6238–6258.

Lafratta, C. N., Li, L. and Fourkas, J. T. (2006a), 'Soft-lithographic replication of 3D microstructures with closed loops', *Proceedings of the National Academy of Sciences USA*, **103**, 8589–8594.

Lafratta, C. N., Lim, D., O'Malley, K., Baldacchini, T. and Fourkas, J. T. (2006b), 'Direct laser patterning of conductive wires on three-dimensional polymeric microstructures', *Chemistry of Materials*, **18**, 2038–2042.

Langner, A., Knez, M., Müller, F. and Gösele, U. (2008), 'TiO$_2$ microstructures by inversion of macroporous silicon using atomic layer deposition', *Applied Physics A: Materials Science & Processing*, **93**, 399–403.

Li, L., Gattass, R. R., Gershgoren, E., Hwang, H. and Fourkas, J. T. (2009), 'Achieving λ/20 resolution by one-color initiation and deactivation of polymerisation', *Science*, **324**, 910–913.

Li, L., Gershgoren, E., Kumi, G., Chen, W. Y., Ho, P. T., Herman, W. N. and Fourkas, J. T. (2008a), 'High-performance microring resonators fabricated with multiphoton absorption polymerisation', *Advanced Materials*, **20**, 3668–3671.

Li, X. P., Chon, J. W. M. and Gu, M. (2008b), 'Confocal reflection readout thresholds in two-photon-induced optical recording', *Applied Optics*, **47**, 4707–4713.

Liberale, C., Cojoc, G., Candeloro, P., Das, G., Gentile, F., De Angelis, F. and Di Fabrizio, E. (2010), 'Micro-optics fabrication on top of optical fibers using two-photon lithography', *IEEE Photonics Technology Letters*, **22**, 474–476.

Malinauskas, M., Gilbergs, H., Žukauskas, A., Purlys, V., Paipulas, D. and Gadonas, R. (2010a), 'A femtosecond laser-induced two-photon photopolymerisation technique for structuring microlenses', *Journal of Optics*, **12**, 035204.

Malinauskas, M., Žukauskas, A., Purlys, V., Belazaras, K., Momot, A., Paipulas, D., Gadonas, R., Piskarskas, A., Gilbergs, H., Gaidukevi i t , A., Sakellari, I., Farsari, M. and Juodkazis, S. (2010b), 'Femtosecond laser polymerisation of hybrid/integrated micro-optical elements and their characterization', *Journal of Optics*, **12**, 124010.

Maruo, S. and Fourkas, J. T. (2008), 'Recent progress in multiphoton microfabrication', *Laser & Photonics Reviews*, **2**, 100–111.

Maruo, S. and Saeki, T. (2008), 'Femtosecond laser direct writing of metallic microstructures by photoreduction of silver nitrate in a polymer matrix', *Optics Express*, **16**, 1174–1179.

Nagpal, P., Han, S. E., Stein, A. and Norris, D. J. (2008), 'Efficient low-temperature thermophotovoltaic emitters from metallic photonic crystals', *Nano Letters*, **8**, 3238–3243.

Olson, C. E., Previte, M. J. R. and Fourkas, J. T. (2002), 'Efficient and robust multiphoton data storage in molecular glasses and highly crosslinked polymers', *Nature Materials*, **1**, 225–228.

Reinhardt, C., Seidel, A., Evlyukhin, A. B., Cheng, W. and Chichkov, B. N. (2009), 'Modeselective excitation of laser-written dielectric-loaded surface plasmon polariton waveguides', *Applied Physics B: Lasers and Optics*, **26**, B55–B60.

Reinhardt, C., Seidel, A., Evlyukhin, A., Cheng, W., Kiyan, R. and Chichkov, B. (2010), 'Direct laser-writing of dielectric-loaded surface plasmon-polariton waveguides for the visible and near infrared', *Applied Physics A: Materials Science & Processing*, **100**, 347–352.

Rill, M. S., Plet, C., Thiel, M., Staude, I., Von Freymann, G., Linden, S. and Wegener, M. (2008), 'Photonic metamaterials by direct laser writing and silver chemical vapour deposition', *Nature Materials*, **7**, 543–546.

Sakellari, I., Gaidukeviciute, A., Giakoumaki, A., Gray, D., Fotakis, C., Farsari, M., Vamvakaki, M., Reinhardt, C., Ovsianikov, A. and Chichkov, B. N. (2010), 'Two-photon polymerisation of titanium-containing sol-gel composites for three-dimensional structure fabrication', *Applied Physics A: Materials Science & Processing*, **100**, 359–364.

Serbin, J. and Gu, M. (2006), 'Experimental evidence for superprism effects in three-dimensional polymer photonic crystals', *Advanced Materials*, **18**, 221–224.

Serbin, J., Egbert, A., Ostendorf, A., Chichkov, B. N., Houbertz, R., Domann, G., Schulz, J., Cronauer, C., Frohlich, L. and Popall, M. (2003), 'Femtosecond laser-induced two-photon polymerisation of inorganic-organic hybrid materials for applications in photonics', *Optics Letters*, **28**, 301–303.

Sherwood, T., Young, A. C., Takayesu, J., Jen, A. K. Y., Dalton, L. R. and Antao, C. (2005), 'Microring resonators on side-polished optical fiber', *IEEE Photonics Technology Letters*, **17**, 2107–2109.

Stellacci, F., Bauer, C. A., Meyer-Friedrichsen, T., Wenseleers, W., Alain, V., Kuebler, S. M., Pond, S. J. K., Zhang, Y. D., Marder, S. R. and Perry, J. W. (2002), 'Laser and electron-beam induced growth of nanoparticles for 2D and 3D metal patterning', *Advanced Materials*, **14**, 194–198.

Stocker, M. P., Li, L., Gattass, R. R. and Fourkas, J. T. (2011), 'Multiphoton photoresists giving nanoscale resolution that is inversely dependent on exposure time', *Nature Chemistry*, **3**, 225–229.

Straub, M., Nguyen, L. H., Fazlic, A. and Gu, M. (2005), 'Complex-shaped three-dimensional microstructures and photonic crystals generated in a polysiloxane polymer by two-photon microstereolithography', *Optical Materials*, **27**, 359–364.

Sun, H.-B., Matsuo, S. and Misawa, H. (1999), 'Three-dimensional photonic crystal structures achieved with two-photon-absorption photopolymerisation of resin', *Applied Physics Letters*, **74**, 786–788.

Sun, H.-B., Tanaka, T., Takada, K. and Kawata, S. (2001), 'Two-photon photopolymerisation and diagnosis of three-dimensional microstructures containing fluorescent dyes', *Applied Physics Letters*, **79**, 1411–1413.

Tan, D., Li, Y., Qi, F., Yang, H., Gong, Q., Dong, X. and Duan, X. (2007), 'Reduction in feature size of two-photon polymerisation using SCR500', *Applied Physics Letters*, **90**, 071106.

Tanaka, T., Ishikawa, A. and Kawata, S. (2006), 'Two-photon-induced reduction of metal ions for fabricating three-dimensional electrically conductive metallic microstructure', *Applied Physics Letters*, **88**, 081107/1–081107/3.

Tetreault, N., Von Freymann, G., Deubel, M., Hermatschweiler, M., Perez-Willard, F., John, S., Wegener, M. and Ozin, G. A. (2006), 'New route to three-dimensional photonic bandgap materials: Silicon double inversion of polymer templates', *Advanced Materials*, **18**, 457–460.

Tosa, N., Vitrant, G., Baldeck, P. L., Stephan, O., Astilean, S. and Grosu, I. (2007), 'Two-photon laser deposition of gold nanowires', *Journal of Optoelectronics and Advanced Materials*, **9**, 641–645.

Tosa, N., Vitrant, G., Baldeck, P. L., Stephan, O. and Grosu, I. (2008), 'Fabrication of 3D metallic micro/nanostructures by two-photon absorption', *Journal of Optoelectronics and Advanced Materials*, **10**, 2199–2204.

Vurth, L., Baldeck, P., Stephan, O. and Vitrant, G. (2008), 'Two-photon induced fabrication of gold microstructures in polystyrene sulfonate thin films using a ruthenium(II) dye as photoinitiator', *Applied Physics Letters*, **92**, 171103.

Walker, E., Dvornikov, A., Coblentz, K. and Rentzepis, P. (2008), 'Terabyte recorded in two-photon 3D disk', *Applied Optics*, **47**, 4133–4139.

Wang, J. and Stucky, G. D. (2004), 'Mesostructured composite materials for multibit-per-site optical data storage', *Advanced Functional Materials*, **14**, 409–415.

Witzgall, G., Vrijen, R., Yablonovitch, E., Doan, V. and Schwartz, B. J. (1998), 'Single-shot two-photon exposure of commercial photoresist for the production of three-dimensional structures', *Optics Letters*, **23**, 1745–1747.

Wong, S., Deubel, M., Perez-Willard, F., John, S., Ozin, G. A., Wegener, M. and Von Freymann, G. (2006), 'Direct laser writing of three-dimensional photonic crystals with complete a photonic bandgap in chalcogenide glasses', *Advanced Materials*, **18**, 265–269.

Xing, J.-F., Dong, X.-Z., Chen, W.-Q., Duan, X.-M., Takeyasu, N., Tanaka, T. and Kawata, S. (2007), 'Improving spatial resolution of two-photon microfabrication by using photoinitiator with high initiating efficiency', *Applied Physics Letters*, **90**, 131106.

Yan, L., Yu, Y., Guo, L., Wu, S., Chen, C., Niu, L., Li, A. and Yang, H. (2010), 'High efficiency multilevel phase-type Fresnel zone plates produced by two-photon polymerisation of SU-8', *Journal of Optics*, **12**, 035203.

Yanez, C. O., Andrade, C. D. and Belfield, K. D. (2009a), 'Characterization of novel sulfonium photoacid generators and their microwave-assisted synthesis', *Chemical Communications*, **7**, 827–829.

Yanez, C. O., Andrade, C. D., Yao, S., Luchita, G., Bondar, M. V. and Belfield, K. D. (2009b), 'Photosensitive polymeric materials for two-photon 3D WORM optical data storage systems', *ACS Applied Materials & Interfaces*, **1**, 2219–2229.

Yang, D., Jhaveri, S. J. and Ober, C. K. (2005), 'Three-dimensional microfabrication by two-photon lithography', *MRS Bulletin*, **30**, 976–982.

Yokoyama, S., Nakahama, T., Miki, H. and Mashiko, S. (2003), 'Fabrication of three-dimensional microstructure in optical-gain medium using two-photon-induced photopolymerisation technique', *Thin Solid Films*, **438–439**, 452–456.

Yu, T., Ober, C. K., Kuebler, S. M., Zhou, W., Marder, S. R. and Perry, J. W. (2003), 'Chemically amplified positive resists for two-photon three-dimensional microfabrication', *Advanced Materials*, **15**, 517–521.

Zhang, Z. S., Hu, Y. L., Luo, Y. H., Zhang, Q. J., Huang, W. H. and Zou, G. (2009), 'Polarization storage by two-photon-induced anisotropy in bisazobenzene copolymer film', *Optics Communications*, **282**, 3282–3285.

Zhou, W., Kuebler, S. M., Braun, K. L., Yu, T., Cammack, J. K., Ober, C. K., Perry, J. W. and Marder, S. R. (2002), 'An efficient two-photon-generated photoacid applied to positive-tone 3D microfabrication', *Science*, **296**, 1106–1109.

6

Laser-based micro- and nano-fabrication of photonic structures

V. SCHMIDT, Joanneum Research, Austria

Abstract: The development of photonic micro-systems requires advanced and robust lithographic methods, which enable the reliable fabrication of intricate, high-quality, three-dimensional (3D) structures at high spatial resolution. In this context, the laser is presented as a versatile tool for fabrication and rapid prototyping beyond the limitations of earlier methods. Supported by the growing availability of tailored and efficient photo-sensitive materials, laser lithography builds specially designed multidimensional photonic micro- and nano-structures for emerging M(O)EMS/N(O)EMS applications. Laser-based fabrication methods for two-dimensional (2D) and 3D photonic structures and application examples are presented, whereas a special focus is put on two-photon absorption-based laser lithography, related material and technological aspects.

Key words: photonics, photonic structure, fabrication, laser manufacturing, rapid prototyping, two-photon absorption, laser lithography, laser structuring, photo-polymerization, photonic bandgap.

6.1 Introduction and motivation

Photonics – the management of light – is a key technology today for the benefit of mankind (Photonics21, 2010). The utilization of light comprises the generation, detection, amplification, processing and control over propagation and polarization of light. In photonic applications the concerted use of light is based on the interaction of light and matter, which is generally accompanied by a change of the properties of light (direction, flux, polarization, wavelength, etc.). While a simple medium influences light mainly by intrinsic material properties, periodic geometric arrangement of media with different refractive indices enables new possibilities. However, such structures require wavelength-sized features for a strong light–matter interaction. Thus, miniaturization plays an important role on the one hand improving integration, speed, cost and performance of multidimensional photonic systems, and on the other hand facilitating new physical phenomena such as light confinement and redirection by photonic bandgap structures (commonly known as photonic crystals), enhanced transmission in metal films and artificial materials (with both negative electric permittivity and negative magnetic permeability, so called left-handed materials).

162

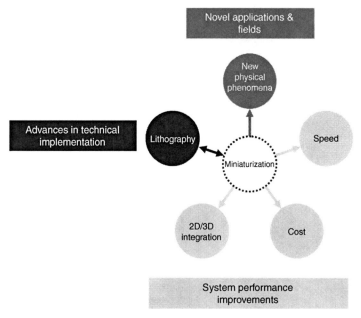

6.1 The role of miniaturization in the field of photonics.

With respect to the technical implementation of photonic systems, miniaturization motivates additionally novel and advanced lithographic methods (Fig. 6.1). Multidimensional photonic architectures and state-of-the-art fabrication methods are comprehensively reviewed in the literature (cf. Arpin *et al.*, 2010; Busch *et al.*, 2007; Jahns *et al.*, 2008).

With respect to micro- and nano-optics and the integration of optics into micro- and nano-optical and electrical systems (MOEMS/NOEMS), patterning by means of photolithography (and associated etching processes) has greatly and adapted and led the way to improved lithographic processes as exemplified in Section 6.2. The technical implementation of more sophisticated lithographic methods, which enable the direct 3D integration of optical structures, opened novel applications and systems not only in microelectronics, but also in micro- and nano-photonics and extended to other fields such as medicine, MEMS fabrication, micro-fluidics and biology.

6.2   Fabrication of 2D and 3D photonic micro-structures

6.2.1 Photolithography

The progress of miniaturization and development of photonic structures is closely related to advances in various lithographic methods (Fig. 6.2). Photolithography

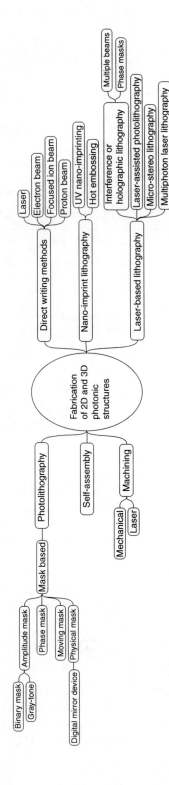

6.2 Fabrication methods for 2D and 3D photonic structures.

was a pioneering task for microscale patterning. It was originally developed for mass fabrication in micro-electronics and is progressively being used for the fabrication of optical and mechanical devices. The earliest methods for fabricating micro-optics were based on adapting the well-known integrated circuit wafer processes, which were developed for large-scale fabrication of micro-electronics, to produce wafer form micro-optics such as diffractive optical elements and lenses (Borek *et al.*, 2007). The use of alternative exposure tools with shorter wavelengths combined with technical resolution enhancements (e.g. immersion technique or phase shift masks) and improvements in the illumination system (e.g. micro-lens arrays based light homogenization) increased the achievable resolution of mask-based photolithography and hence the field of application towards further miniaturization (Voelkel *et al.*, 2010). The limitations of photolithography to expose only planar substrates have recently been circumvented by efforts to extend photolithography to non-planar surfaces using computer-generated holographic masks (Maiden *et al.*, 2005).

Conventional photolithography transfers a pattern from a mask in a single exposure step to a photosensitive material. The mask itself is either a binary or gray-tone mask used for the generation of either binary or locally variable exposure patterns in contact, proximity or projection mode. Other approaches are gray-tone or half-tone (mask) photolithography (Reimer *et al.*, 1997), moving mask lithography (Hirai *et al.*, 2007; Kim S.-K. *et al.*, 2011), and maskless photolithography based on digital mirror devices (Guo and Dong, 2010).

The use of dangerous or toxic agents during material processing in photolithography, special ambient conditions (e.g. yellow light, clean room, inert atmosphere conditions), pushed the adaptation of lithographic fabrication methods and triggered additionally the development of alternative structuring technologies and new materials.

A comparison of 3D micro-fabrication methods has been made (Jang *et al.*, 2007) distinguishing between self-assembly approaches, construction-based approaches and the interference lithography approach. Self-assembly techniques are based either on block copolymer self-assembly or colloidal self-assembly. The laser as exposure tool is found in the construction-based approaches such as layer-by-layer manufacturing (such as micro-stereo-lithography), direct laser writing methods including single and multiphoton techniques as well as in the interference lithography approach, which requires the overlay of multiple coherent beams either provided by sophisticated optical arrangements in the experimental setup or by phase masks.

6.2.2 Mechanical machining

In addition to photolithography, mechanical machining is today another approach to fabricate photonic components because of the technical improvements in the achievable precision. Single point grinding and polishing as well as ultra precision diamond machining is used especially for high quality aspheric or freeform optical surfaces and prototypes. Dimensions and surface quality (roughness) depend on tool, material and machine parameters. Diamond machining can be used for machining parts with dimensions in the range of a few microns with a roughness typically in the range of a few nanometers and a form deviation in the sub-micrometer range. The positioning accuracy of the machine is usually better than the achievable structure resolution in the micrometer range due to the brittle fracture mechanism associated with the material removal process (Borek *et al.*, 2007). The flexibility regarding complex 3D shapes of multi-axes machines is an advantage, though mechanical machining remains a time consuming and hence expensive technology. However, it is well suited for the fabrication of single parts or molds for replication. Examples of fabricated micro-optics are multi lens arrays, micro-mirror arrays and fiber-coupling lenses (Gläbe and Riemer, 2010).

6.2.3 Replication methods

Less expensive but precise polymer replicas of expensive micro-optical structure can be done by injection molding or by hot embossing. In contrast to injection molding, hot embossing creates micro-structures with lower

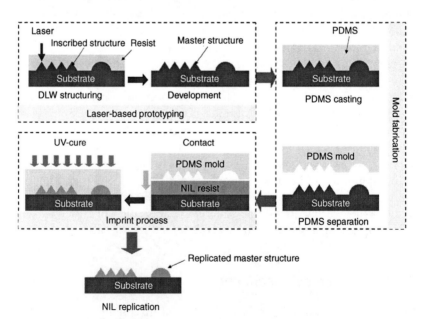

6.3 Process of DLW rapid prototyping and NIL replication.

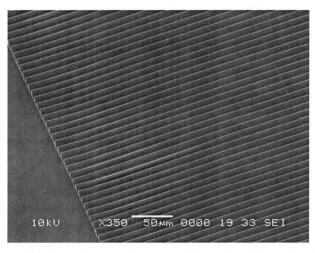

6.4 Replicated sawtooth structure. The structure is made with a poly-dimethylsiloxane (PDMS) mold in Ormostamp according to the process in Fig. 6.3.

internal stress and is applied to the fabrication of micro-lenses, Fresnel lenses, optical waveguides, optical benches, optical gratins, micro-spectrom-eters, and distributed feedback structures for organic semiconductor lasers (Worgull *et al.*, 2010).

Nano-imprint lithography (NIL) is a fast and cheap process for the repli-cation of structures with features down to a few nanometers, which makes this method very attractive for mass production. A crucial factor in NIL is the quality of the stamp. High quality stamps are processed by other litho-graphic methods such as direct writing methods or laser-based lithography. A typical work flow of the combination of direct writing laser rapid proto-typing for micro-optical elements and replication is shown in Figs 6.3 and 6.4. The replication of photonic structures in the visible spectral range requires stamps with nanometer scaled features. While direct writing methods are suitable for irregular and arbitrary patterned stamps, periodic patterns are processed defect-free over large areas by means of laser interference lithog-raphy. Subsequent replication via nano-imprint lithography yields photonic structures, for example LED applications (Kim *et al.*, 2007).

Two-photon-based laser lithography is used for the fast fabrication of 2D structures that are subsequently molded into PDMS in order to obtain a flexible inverse replica (Yang *et al.*, 2004; Yi *et al.*, 2004b). Beyond that, two-photon polymerized 3D structures can be replicated rapidly and with high quality by means of micro-transfer molding (LaFratta *et al.*, 2004). This method facilitates the replication of structures with high aspect ratios and re-entrant features with large overhangs. This kind of soft lithographic replication of 3D structures can be extended to the fabrication of freely moving components by means of so-called membrane assisted micro-transfer molding (LaFratta *et al.*, 2007).

6.2.4 Laser-based lithography

Laser-based lithography takes advantage of unique properties of laser radiation, such as high monochromaticity, coherence, directed emission of radiation and excellent beam quality along with high focusability. In a photolithographic approach, the laser may replace a conventional light source and provide high power radiation at a short wavelength or it can be tightly focused and used for patterning without masks. The large coherence length facilitates the generation of complex intensity patterns at sub-wavelength resolution, which can be transferred into a suitable material. Thus, there are manifold approaches for laser-based lithography for structuring a variety of materials. The laser power is used either for removal (e.g. ablation, milling, drilling and cutting), joining (welding), marking or addition (e.g. laser cladding, sintering) of material. These types of laser machining generally work on a macroscopic scale in combination with robots and may not be called lithography, but on a microscopic scale, the laser is used the same way, probably at lower power and different laser parameters, but still the laser deposits energy in a target material, which is then patterned. This microscopic and macroscopic direct patterning or writing process can be called laser lithography in the classical meaning of lithography.

Different laser sources (Table 6.1) have been widely used for laser precision macro-, micro- and nano-machining (Chong *et al.*, 2010) and optical lithography for various transparent materials such as glasses or polymers. Although local melting of glass by means of CO_2 laser heating results in the formation of glass lenses (Veiko *et al.*, 1986), laser direct writing of microlenses in soda-lime glass is achieved with an Nd:YVO$_4$ laser (Nieto *et al.*, 2010) or intense UV laser exposure (355 nm) leads to glass swelling and the formation of superficial glass bumps (Logunov *et al.*, 2011); femtosecond lasers are especially versatile instruments for processing. Thus, fs lasers are widely used in micro- and nanotechnology for the precise ablation of metallic and transparent materials (Gattass and Mazur, 2008; Kazansky *et al.*, 2004; Kondo *et al.*, 2004) and laser engineering of biomaterials (Stratakis *et al.*, 2009). Intense ultrafast laser pulses can induce structural and/or optical changes, and nonlinear effects, such as multiphoton ablation or multiphoton absorption. Regarding microscale patterning, laser-based photolithography based on multiphoton polymerization is well suited for the fabrication of 3D polymeric photonic structures without the need of layer stacking as in other lithographic methods, which require time consuming sequential alignment and processing.

Table 6.1 Some important laser types, emission characteristics and field of application

Laser type	Gain medium	Emission characteristics (mode of operation, wavelengths, power/pulse energy)	Application
Ar-ion laser	Gas	cw(*) operation λ = 488 nm, 514.5 nm some Watt	Spectroscopy, holography, machining
He-Ne laser	Gas	cw operation λ = 633 nm some 0.1 Watt	Alignment, spectroscopy, holography, interferometry
He-Cd laser	Gas	cw operation λ = 325 nm some 10 mW	Lithography, interferometry
Excimer laser	Gas	pulsed operation (ns) λ = 157 nm (F_2), 193 nm (ArF), 248 nm (KrF), 308 nm (XeCl) some Joule	Lithography, ablation, machining, surgery
CO_2 laser	Gas	cw, pulsed operation (μs) λ = 10.6 μm 10–100 kW	Machining, cutting, welding, drilling
Nd:YAG	Solid state	cw, pulsed operation (ns, ps) λ = 1064 nm, 532 nm (SH**), 355 nm (TH), 266 nm (FH) some kW	Material processing, Laser Pumping, Research, Surgery
Ti:S	Solid state	pulsed operation (ps, fs) λ = 670–1080 nm some mJ	Spectroscopy, nonlinear material processing
Nd:Glass	Solid state	pulsed operation (ms, ns, ps) λ = 1062 nm some 100 J	High energy multiple beam systems, Laser fusion
AlGaAs	Semiconductor	cw, pulsed operation (μs) λ = 780–880 nm some kW (laser diode bars)	Machining, medical, optical discs, laser pumping
AlGaInP	Semiconductor	cw operation λ = 630–680 nm	Machining, medical, optical discs, laser pumping
InGaAsP	Semiconductor	cw, pulsed operation (ps) λ = 1150–1650 nm	Machining, medical, optical discs, laser pumping
GaN	Semiconductor	cw operation λ = 405 nm	Optical discs, lithography
Dye laser	Dye	cw, pulsed operation (ns) λ = 300–1200 nm, depends on used dye	Spectroscopy, research, medical

Notes: Cf. Bäuerle (2008).
(*) cw = continuous wave; (**) SH = second harmonic; TH = third harmonic; FH = fourth harmonic.

In micro-fabrication, the laser represents a powerful tool for patterning and is used as a flexible (in terms of wavelength, pulse width, etc.) light source in laser-assisted (photo-) lithography. In this context, excimer lasers have been widely used as described in Section 6.2.1 due to their high photon energy and high output power, which can be technically exploited by efficient laser-ablation processing of many different materials. But with the technical use of the intense light of a femtosecond laser, an alternative light source entered laser-based lithographic processes, which is capable of processing a large variety of materials, ranging from metals to transparent dielectrics for diverse applications (Fig. 6.5).

Femtosecond laser pulses are used for the ablation of metals and dielectrics (2D and 3D structures), structural changes of materials, optical changes or mechanical processing such as cutting, drilling or milling in order to replace mechanical machining tools. The power of ultrafast lasers is sufficient to efficiently process hard materials such as steel (e.g. NAK80), which is a common material for molding in the plastics industry. Setting the power close to the material-dependent ablation threshold, structures with sub-diffraction limited features can be generated, fine enough to show color effects

6.5 Micro-fabrication with femtosecond lasers.

(Noh *et al.*, 2007). Moreover, the direct writing of 2D and 3D conducting structures can be realized by laser-based decal transfer of Ag silver nano-paste on a substrate, which can replace electrical bonding of electro-optical chips in micro-systems (Wang *et al.*, 2010). Another way of generating a direct laser-written structure is the laser-assisted diffusion of interstitial ions in semiconductor heterostructures for the generation of sub-micrometer LEDs (Makarovsky *et al.*, 2010).

Furthermore, intense ultrafast laser pulses can trigger nonlinear effects such as multiphoton ablation or multiphoton polymerization in a photosensitive resin via photo-initiation and photo-polymerization. There are many different approaches to how a laser can be used to build a physical structure: the focusing of an intense laser pulse may result in evaporation of material from a target (laser ablation), which uses the laser in a machining way to replace diamond tools for drilling, cutting or milling. High power lasers are used in rapid prototyping of 3D structures to melt powder-like materials selectively (selective laser sintering – SLS) or induce a photochemical reaction, which alters material properties used in laser-based photolithography. In most cases, the laser focus acts like a laser-pencil, which moves in a defined way (i.e. according to a template) across the target and writes a 2D or 3D structure in the material. Generally, the structure is built by the energy transfer from the laser to the material, which induces a material modification at the exposed locations. The efficiency of this process depends on the material chemistry, laser intensity, wavelength, pulse width, spectral width, beam profile, repetition rate, material absorption, focusing optics, etc.

Other related methods use the laser in a number of ways for the deposition of material from gaseous, liquid and solid precursors. The techniques involve either laser-induced chemical/electrochemical or physical reactions, which subsequently deposit a material on a substrate. Material transfer methods use a laser for removal of the transfer material from a substrate to another receiving substrate in close vicinity. These methods such as laser chemical vapor deposition (LCVD), laser-enhanced electroless plating (LEEP), laser-enhanced or activated electroplating, laser consolidation of thin solid films, laser-induced forward transfer (LIFT), matrix-assisted pulsed laser direct write (MAPLE), laser-induced backward transfer or laser contact-free trapping and transferring of particles in solution for 3D direct writing are surveyed in Hon *et al.* (2008). Each method can be treated extensively, which is clearly beyond the scope of this chapter, but not all of them are used in the field of photonics. The next section is dedicated to the most relevant methods related to the laser fabrication of photonic structures.

Direct writing methods take advantage of a patterned energy deposition from scanning an energetic beam over a sensitive material. Common scanning beam lithographic methods are electron beam lithography (Steingrüber *et al.*, 2001), focused ion beam lithography (Callegari, 2009), proton beam lithography (Bettiol *et al.*, 2004; Debaes *et al.*, 2006; Sum *et al.*, 2004; van Erps *et al.*, 2010) and laser lithography (Rhee, 2010).

Two- or multiphoton or laser-based lithography has been applied as a versatile tool in micro-fabrication for optical, medical and biological applications (see below). Because of the high spatial resolution and the 3D structuring capability, the two-photon technology can be used for 3D data storage. Examples of photonic micro- and nanostructures are presented in below.

6.3   Laser lithography for the fabrication of photonic structures

Optical lithography benefits from the high throughput due to parallelism in fabrication and hence low device costs. It is the work-horse in modern semiconductor industries. But it suffers from the diffraction limit imposed by physical optics, which is proportional to the wavelength of the light source and the numerical aperture of the imaging optics. Decreasing the wavelength from ultraviolet to deep ultraviolet opened the path for excimer lasers to become useful light sources in lithographic work stations. A simultaneous increase of the numerical aperture based on immersion technology further pushed the achievable critical dimensions for patterning to lower values. State-of-the-art values for numerical aperture (1.35), wavelength (193 nm) and taking advantage of resolution enhancement technologies such as phase shift masks, off-axis illumination, optical proximity correction and improved resists decreased the value for the critical dimension from 500 to 45 nm. Multiple exposure techniques further decrease the wavelength and hybrid lithography, which combines, for example, 157 nm interference lithography and electron beam lithography, may ensure that optical lithography remains of crucial importance (Rothschild, 2010).

Some aspects related to the laser fabrication of photonic components are presented in the next section.

6.3.1   Ultraviolet, deep ultraviolet and laser photolithography

Laser light sources in the ultraviolet spectral range are either frequency multiplied solid-state lasers or gas lasers. Especially excimer lasers provide pulsed high power ultraviolet to deep ultraviolet emission at a typical repetition rate of a few hundred Hz up to some kHz. The main scientific and industrial applications of excimer lasers are material processing, lithography and medicine, involving methods such as laser ablation, engraving, marking, surface and sub-surface modifications and coatings made via pulsed laser deposition in either projection or direct exposure mode. In projection mode, the laser light is projected via a mask onto a target for UV exposure. In direct exposure mode, the focused laser light directly ablates the material by moving the focused laser across the target. In projection mode the effect of the laser results either in patterned material removal or in exposure of the material without removal, which depends mainly on the laser fluence and applied pulse number.

The active medium of an excimer laser is a gas of electrically excited dimers ('excimer', or more precisely excited complexes), where an excited noble gas atom and a halogen form a noble gas halide, which decays after a short time (typically some ns) into the dissociated state (e.g. Kr*F →Kr + F) under emission of UV light. The type of excimer determines the wavelength of the emission. The technically most relevant excimers are ArF, KrF, XeCl and XeF (Basting *et al.*, 2002) and there are manifold applications for excimer laser processes such as excimer-based optical lithography (Elliott and Ferranti, 1989; Partel *et al.*, 2010), excimer laser chemical vapor deposition (Wang *et al.*, 1996), μ-fluidics (Shin *et al.*, 2006), laser annealing of thin silicon oxide films (Richter *et al.*, 2011), photochemical welding of silica micro-spheres (Okoshi *et al.*, 2009), excimer laser micro-machining (Chiu and Lee, 2011; Ihlemann and Rubahn, 2000; Wu *et al.*, 2006a) and eye surgery (Vossmerbaeumer, 2010). The high photon energy of the excimer radiation is capable of directly breaking intramolecular bonds of the target material, with only negligible thermal impact on the surrounding material. In 1989 KrF and ArF lasers demonstrated the fabrication of 0.4–0.5 μm feature sizes via direct photo-ablative decomposition and conventional latent image exposure with subsequent wet chemical development of deep UV resists (Elliott and Ferranti, 1989). In 1996, KrF lasers operating at 248 nm were available for lithography at stable conditions up to 1 kHz. At this wavelength such lasers were accepted tools for the manufacturing of structures with features down to 250 nm. At the time ArF lasers operating at 193 nm had not reached the same level of technical maturity, especially at kHz repetition rate, because they suffered from energy instabilities. But they were promising for lithography due to the shorter wavelength (Pätzel *et al.*, 1996). Besides decreasing the illumination wavelength, further efforts reaching for enhanced resolution and high fidelity patterns are improvements on the illumination system, such as off-axis illumination, where the laser beam hits the mask at an angle with respect to the optical axis of the imaging system. The imaging system acts as a low pass spatial frequency filter, which introduces imaging errors such as corner rounding and line-end pullback. Different types of off-axis illumination are easily implemented by special apertures in the illumination path and often combined with optical proximity correction which improves the imaging of dense features (cf. Fig. 6.6 and 6.7). Recently, freeform lenses were developed

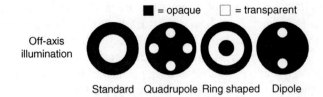

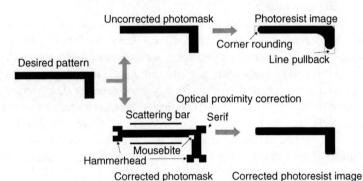

6.6 Principle of off-axis illumination and optical proximity correction (Fritze *et al.*, 2003).

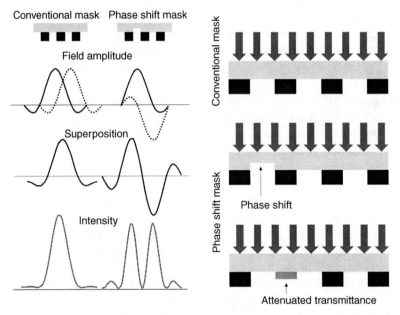

6.7 Working principle of phase shift masks.

for application in off-axis illumination (Wu *et al.*, 2011). The use of phase shift masks for sub-wavelength lithography (Fritze *et al.*, 2003) for contrast enhancement on the resist, and double exposure methods (Geisler *et al.*, 2008; Mosher *et al.*, 2009) combined with phase shift masks (Fay, 2002) or immersion technology (Yoshino *et al.*, 2010) are other approaches for resolution enhancement. A micro-optical beam homogenizer for the mask illumination improves the achievable spatial resolution due to reduced diffraction and aberration effects (Partel *et al.*, 2010).

KrF excimer lasers have been used for micro-structures in polymers for more than 20 years. The excimer laser machining directly from CAD files using ablation with mask projection was used for the rapid prototyping of micro-structures with optimized parameters for wall angle, ablation depth and stitching methods in polycarbonate samples (Mutapcic *et al.*, 2005). Originally for binary micro-structures such as holes, trenches, pores or gratings, excimer laser etching and replication was applied to the field of micro-fabrication of 3D microtopologies with a contour mask technique. Non-binary topological structures with a smooth or continuous height profile can be processed by either line scans with a specially designed contour mask or gray-scale mask projection lithography (Zimmer *et al.*, 1996). Both methods require a homogeneous beam profile, stable laser, and a good quality mask, because a controlled etch depth is achieved by the local transmittance of the mask and the number of applied pulses. Usually the homogeneous beam profile is achieved with micro-lens-based fly-eye integrators.

In the contour mask approach, the laser is projected via a stationary contour mask on a moving sample. Contour mask and sample motion are adapted to the desired topology and the specific ablation rate of the KrF laser pulses. Thus, point-like pulsed ablation with a specific ablation rate yields in intersecting scans the desired topology, which is shown for prismatic structures (Braun *et al.*, 1998). The time consuming excimer ablation process is followed by a replication process using acrylates, which facilitates several tens of replications at high accuracy. Hence, a combination of both processes increases throughput of the fabrication process. It is claimed that an *in situ* built debris layer and its interaction with the formed ablation plume above the processed region during material ablation causes an elevated temperature of the material, which supports the smoothness of the surface. For even more complex topologies the contour mask method is combined with a moving gray-scale mask, whereas the scanning speed of the contour mask and the sample have to be adjusted in a certain ratio according to the magnification of the projection lens (Fig. 6.8). 3D topologies can also be achieved with a two-step machining approach with a diagonal scanning method and two contour masks and a third aperture mask (Zimmer *et al.*, 2000).

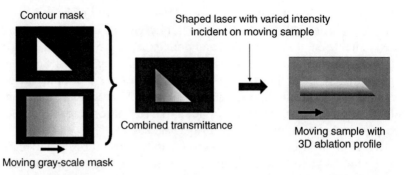

6.8 Contour and gray-scale mask projection lithography for 3D excimer laser ablation.

For optical/photonic applications, a high surface quality after laser treatment is essential. The application of UV or deep UV laser for material ablation is an efficient method, but also highly disruptive to the target material. Thus, residual significant surface roughness is a major issue for optical applications of laser-ablated materials. Nevertheless, laser-ablation methods in combination with suitable processing schemes are used for precise etching of fused silica for refractive and diffractive micro-optics. The suggested laser-induced backside wet etching process yields direct machined surface of almost optical quality with a surface roughness of less than 10 nm rms, whereas the machined surface of the transparent sample is in direct contact with a laser light absorbing liquid. Smooth surfaces can be achieved by control of the ablation rate, which was in the range of several nm per laser pulse (Zimmer and Böhme, 2005). Other approaches yield aspheric lenses with a 248 nm KrF laser applying a planetary contour scanning method (or respective motion of the mask). This method facilitates precise surface profile control (Chiu and Lee, 2011).

Recent applications of excimer lasers in nano-fabrication cover also excimer laser-assisted nano-imprint lithography, where the laser is used as heating source to selectively melt a polymer material, which is immediately imprinted. The laser pulse heating requires a transparent stamp (fused silica), but reduces the duration of the imprint process to a few hundred ns because the imprint resist is directly heated from the laser pulse. Additionally, selective melting of resist material is proposed to remove fabrication defects and reduce the roughness, which leads to an enhancement of the structure profile. A single pulse exposure melts only a very thin layer of the surface, whereas the surface tension in the liquid phase will smooth out rough edges and reshape fabricated nanostructures depending on the experimental boundary conditions. Metal nanoparticle monolayers and periodic arrays of nanoparticles as well as sub-10 nm nano-fluidic channels were demonstrated (Xia *et al.*, 2010). The laser-assisted imprint process is well suited for the fabrication of nm period gratings. The selective melting leads to a reshaping of fabricated squares to spherical structures, which could act as micro-lenses.

In addition to advanced optical lithography, excimer lasers are used in medical applications. The high UV power of these lasers is well suited for laser ablation of hard and soft tissue in dentistry and surgery (Murray and Dickinson, 2004). Especially ArF laser at 193 nm are used in ophthalmology for refractive corneal surgery. Due to the absorption maximum at 193 nm of the cornea, this wavelength is well suited for the removal of tissue from the cornea surface thus correcting the refractive power of the human eye. The photo-ablation leaves surrounding tissue unchanged and yields precise patterns, thus making the excimer laser a perfect tool for micro-surgery (Vossmerbaeumer, 2010).

Based upon laser ablation or surface modifications, excimer lasers are applied to fabrication of photonically relevant structures: spatially selective oxidation of silicon monoxide layers by ArF excimer laser annealing under oxygen atmosphere is reported in Richter *et al.* (2011). The conversion from SiO_x ($x \sim 1$) to SiO_2 is observed by increasing transmittance upon multipulse irradiation below the ablation threshold. Thus, patterned changes of transmittance and refractive index can be achieved, which is technically relevant for phase masks or patterned surface functionalization. Depending on the oxygen concentration, either smooth films for optical phase elements with a smooth surface and variations of the refractive index are obtained, or, in an oxygen-rich environment, submicron sized SiO_2 nanoparticles are fabricated. In contrast to the non-ablative annealing process, excimer laser ablation lithography is used for the fabrication of gratings on various glasses (Dyer *et al.*, 1996), binary reflective (Flury *et al.*, 2003) or transmitting multi-level (Winfield *et al.*, 2000) diffractive optical elements. The reflective binary diffractive optical elements are fabricated by laser-ablation lithography with excimer lasers and subsequent metal film coating. In a pixel by pixel scan across the target, a photoresist is ablated from the substrate by projecting the laser beam on the resist. In a following etching step, the pattern is transferred to a substrate and coated. This binary reflective kinoform can be used for high power laser beam shaping optics. The multilevel transmitting diffractive optics in glass were obtained by direct excimer laser ablation. In this process, a square aperture was imaged to a glass target and material was removed from the surface pixel by pixel until the desired depth was achieved.

It can be observed in glass targets that below the ablation threshold, the incident laser power leads to a local temperature increase (sub-threshold incubation), which can be measured by the deflection of a monitoring laser beam. The monitoring beam is guided parallel to the surface and focused above the laser irradiation zone. The duration of the deflection increases upon the start of ablation of material from the surface. This way, the material-dependent laser threshold for ablation can be measured. Depending on the material, the laser wavelength and a given laser fluence below the single pulse ablation threshold, incubation may occur and lead to ablation after a certain number of pulses, but incubation does not necessarily happen in all materials (Dyer *et al.*, 1996). Incubation is considered to play a role in various polymer blends, which influences the ablation results upon excimer laser irradiation (Kunz *et al.*, 1998). The interaction between the UV laser and the target depends on the spectral material absorption and thermal conductivity (along with pulse duration) and is thus governed by the predominance of either photochemical ablation or photo-thermal ablation, which then relates the laser-induced ablation to optical and mechanical parameters of the target material (Desbiens and Masson, 2007).

Both types of ablation play an important role in UV laser irradiation but differ depending on the polymer material. The irradiation induces a pressure jump in polymer materials and the creation of small molecules. UV lasers are used for the ablation of polymer materials, but are also useful for the transfer or deposition of polymer or other sensitive materials via pulsed laser deposition (Lippert, 2009).

6.3.2 Laser-assisted deposition methods

In addition to simply removing the material from a substrate via laser ablation, there are processes (laser chemical vapour deposition, laser-induced transfer methods, pulsed laser deposition, etc.) where the laser is used for the patterned deposition of materials on a substrate. Laser chemical vapour deposition (LCVD) was used for the *in situ* fabrication of micro-lenses with precise control of film properties (Wang *et al.*, 1996). This process takes place in a reaction chamber, which contains precursor gases and the substrate. The laser is used to locally heat the substrate, which subsequently dissociates the gas precursor and a thin film deposits on the substrate. Using multiple beams or a layered approach to build the structure, it is possible to create 3D structures. The deposition rate of LCVD depends linearly on the precursor gas pressure and the laser power density and decreases with increasing scanning speed. The deposition rate can be adjusted by these parameters and is much higher than in conventional CVD (Hon *et al.*, 2008). In a similar method (liquid-phase chemical laser-assisted deposition) an XeCl excimer laser irradiation was used for the local deposition of Pd layer from a precursor solution to form mirrors on fibre core end faces. The Pd deposition is followed by electroless copper plating in order to yield the final mirror. A subtractive method for the mirror fabrication was compared to the additive method: the laser was used for removal of the metal film around the fibre core after chemical plating of the fibre end faces, but it was found that the adhesion of the mirrors made by the subtractive method was poor compared to the additive method (Kordás *et al.*, 2002).

Sensitive materials that are easily destroyed by the laser are often embedded in a matrix material that absorbs the laser energy. This method (matrix-assisted pulsed laser deposition, MAPLE) uses a frozen solvent, which is evaporated upon laser irradiation. The material for deposition is evaporated together with the matrix and deposits on a receiving substrate. Without masks, the material deposition is unpatterned, hence another method, laser-induced forward transfer method (LIFT) is often used for laser-assisted patterning (Fig. 6.9). Pulsed laser deposition methods, ablation mechanisms and applications are discussed in Schneider and Lippert (2010). In the LIFT process, the laser energy is absorbed in a thin film on a transparent substrate, which leads to evaporation of the transfer material. Subsequently, the evaporated material precipitates on a second receiving substrate, which faces the first substrate in close (micrometer) vicinity. For soft-matter materials such as polymers or biological compound materials, a direct contact between the substrates was found to yield the best transfer results regarding resolution and defined edges of transferred pixels (Palla-Papavlu *et al.*, 2010a, 2010b).

LIFT can be achieved with various types of lasers (UV excimer lasers, Nd:YAG, Ar-ion lasers, fs lasers). The transfer materials are often sensitive to oxygen or humidity, thus requiring a vacuum or inert gas setup. Originally used for the patterned transfer of metal films, it can be applied for a variety of materials including oxides and biomaterials or even more complex

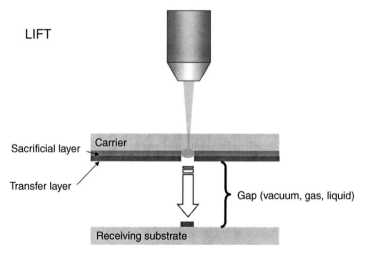

6.9 Scheme of laser-induced forward transfer.

multilayer systems such as a polymer light emitting diode pixel (Shaw Stewart *et al.*, 2011) or organic thin film transistors (Rapp *et al.*, 2011). Such sensitive materials or materials which are transparent to the incident laser or easily destroyed by the incident laser, can be transferred by using an energy absorbing sacrificial layer (dynamic release layer (Shaw Stewart *et al.*, 2010) between the transfer material and the carrier, which promotes the release of the material. Additionally, the temporal shape of ultrafast laser pulses influences the LIFT process and the achievable resolution on the receiving substrate, which is attributed to fast electron and lattice interactions. It was shown that fs pulses with a short separation (less than 500 fs) show large impact of the deposited pixel size, while the covered area stays constant for longer pulse separations up to 10 ps (Papadopoulou *et al.*, 2010).

Using microsphere arrays as micro-lenses, parallel material transfer (parallel LIFT) with an unfocused laser beam can be achieved (Othon *et al.*, 2008). The polystyrene beads are on top of a transparent substrate (quartz glass) and focus the incident light onto the single or multilayered transfer material, which is on the other side of the substrate. Thus, micron to submicron holes can be written into the films and corresponding dot patterns on the receiving substrate.

6.3.3 Laser interference lithography

Instead of scanning a focused laser beam through a photosensitive material, laser interference lithography (LIL) is very well suited for the fabrication of various types of nanoscale structures (Xie *et al.*, 2008). It has been used for the fabrication of nanostructures in various fields and with many different technical approaches and is comprehensively reviewed in the literature (Jang *et al.*, 2007; Xia *et al.*, 2010). Here, only some basic issues of this technology are summarized.

LIL is capable of structuring rather large areas in a single shot exposure or limited exposure time without defects and without scanning, but is limited to periodic patterns. The fabrication of a 4×4 cm^2 photonic crystal using a holographic element in combination with a mask is reported in Zhang *et al.* (2006). The large holographic element comprises three separate gratings (each rotated by 120°) and generates four transmitted beams (first diffraction order from each grating plus transmitted centre beam). The incident laser beam is expanded to a diameter of 20 cm and the exposure of the positive type resist for the photonic crystal fabrication is 3 min.

The laser provides light of defined wavelength, polarization and coherence, thus enabling coherent superposition of multiple laser beams, whereas the experimental conditions such as laser fluence, film thickness, angle of incidence and polarization of the beams directly correlate to the fabricated patterns. The periodicity (or the lattice constant of the unit cell) of the pattern is determined by the difference between the wave vectors of the interfering beams and is thus proportional to the wavelength of the laser and the angle between the interfering beams, which are crucial and limiting parameters for the achievable spatial resolution. The shape of the image formed within a unit cell of the pattern is influenced by the polarization of the beams and its position within the unit cell depends on the initial phase difference of the beams. The laser intensity, exposure time and development procedure also have an impact on the final pattern shape. Depending on the number of beams (N), angle between the beams and polarization, 1D, 2D and 3D (maximum $N - 1$) periodic patterns can be fabricated over a fairly large area in a single exposure step. Multiple exposure steps with rotation and translation of a 1D phase mask facilitate complex 3D patterns such as woodpile structures with three beams (Xu *et al.*, 2010). The coherence length limits path differences in the optical setup and determines also the maximal area that can be processed in a single exposure step. The coherence length of the used laser determines whether wave front splitting or amplitude splitting of the laser beam (Marconi and Wachulak, 2010) is used in the optical setup. The beam superposition leads to the generation of stable interference patterns, which can be used for patterning films of (usually) positive type photoresist (Ellman *et al.*, 2009), negative type resist (Stankevicius *et al.*, 2011), TiO_2 gel films (Wang *et al.*, 2011), hybrid organic–inorganic sol–gel materials (Della Giustina *et al.*, 2011), biomimetic tissue (Daniel, 2006), as well as PEDOT-PSS (Lasagni *et al.*, 2009), a conducting polymer, which is important for organic (opto)electronics (cf. Fig. 6.10).

Unwanted reflections from the substrate surface may degrade the interference patterns and lead to undercut phenomena, which can be reduced by anti-reflection coating of the substrate or thermal post-exposure treatment of the sample to induce thermal diffusion and redistribution of the photo active compound, which is applied in order to improve sidewall profiles and line edge quality and eliminate undercut features of fabricated structures. This is crucial if the fabricated structures are used as molds for subsequent nano-imprint lithography (Jang *et al.*, 2010).

Various types of laser sources such as Nd:YAG lasers at 266 nm (fourth harmonic) or 355 nm (third harmonic) (Lasagni *et al.*, 2011) are used for interference and holographic lithography, for which some use a laboratory setup with a more exotic wavelength in the EUV (46.9 nm) based on a capillary discharge-pumped excitation scheme that produces an intense amplification by the excitation of Ar^{+8} ions (Marconi and Wachulak, 2010). Regarding costs, large area LIL with semiconductor lasers seems more attractive. The AlInGaN laser has a rather low price and a long coherence length, which is a prerequisite for processing large sample areas (Byun and Kim, 2010). Recently, LIL was combined with multiphoton polymerization

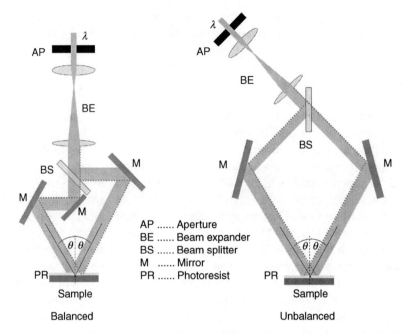

6.10 Dual beam laser interference lithography – optical setup for balanced and unbalanced transverse beam coherence. A balanced setup requires the same number of reflections in the two interferometric paths in order to overlay same transverse beam parts.

in a four-beam setup for the fabrication of micro-lenses. The four beams were generated using a diffractive optical element and a diaphragm to remove undesired laser light from the optical path. The negative type resist was exposed to multiple laser pulses, which facilitates much faster processing as compared to multiphoton-based direct laser writing, which is a sequential voxel-by-voxel buildup process. Appropriate hatching or stepping of the exposed area enables the processing over large areas and reduction of the structure degradation at the edge of the exposed area due to the spatial intensity profile of the laser beams (Stankevicius *et al.*, 2011).

Since LIL requires multibeam arrangements and stable control of laser parameters for intricate intensity patterns of interfering laser beams, sophisticated and complex optical setups are required. The use of the phase mask (which actually can be fabricated by LIL) technology correlated to the desired structure is a more stable experimental approach rather than large optical setups that suffer from alignment inaccuracies or vibrations. The diffracted beams from such a mask are inherently phase-locked and provide a stable beam superposition (Xu *et al.*, 2010). Nevertheless, multiple-beam LIL is compatible with automatic and cost-effective industrial processes beyond manual laboratory configurations and usage. An interference module facilitates the automatic setting of beam number, angles of incidence and polarization in order to obtain different interference patterns. The exposure area can be stepped over larger sample areas such as 3 inch Si wafers (Rodriguez et al., 2009). Originally used for regular 2D patterns, interference lithography is increasingly applied to 3D structures such as photonic bandgap structures with increasing structural complexity. Photonic applications taking advantage of the LIL capabilities and the combination of LIL with replication methods involve the patterned arrangement of quantum dots (Lu et al., 2009), where a three-beam configuration LIL is used for the fabrication of pore structures in SU-8, which are subsequently immersed in a solution containing CdSe/ZnSe quantum dots. The quantum dots are then dragged by capillary forces into the pores yielding a patterned luminescent architecture that could be useful for LED-based lighting applications. Further optical applications involve the fabrication of structures for enhanced light out-coupling from light emitting diodes (Kim K.-R. et al., 2011) as well as broadband anti-reflective coatings in the visible spectral range with sub-wavelength conical structures to reduce the reflection well below 1% (Ting et al., 2009) or with an extended wavelength range (250–1200 nm) and wide incidence angles up to 50° (Chen et al., 2010). Regarding such photonic fabrication processes, LIL is often used for the generation of regular master structures that are subsequently used as a mold for the fabrication of a photonic crystal-like structure diode by nano-imprint lithography on light emitting or on flexible substrates such as foils (e.g. PET films), which facilitates roller lamination, exposure through the substrate and the use of Ni shims (roller imprinting).

A simple optical setup such as Lloyd's mirror setup for dual beam interference requires a sufficient temporal and spatial coherence of the laser beams, since the wave front of the incident beam is divided and coherently added with itself after travelling different optical paths. Folding of the wave front at the mirror requires good transverse and longitudinal spatial coherence. An interferometric setup using a beam splitter or a phase mask for the generation of the interfering beams is more relaxed in terms of spatial coherence, since two phase-locked replicas of the wave front are divided in amplitude at the splitting element and each section of the interference pattern is achieved by the superposition of the same beam section and is therefore spatially coherent (Marconi and Wachulak, 2010).

The use of micro-prism arrays in combination with interference lithography enables the fabrication of nanoscale sub-structures in micro-structures due to the complex interference patterns of the beams, which are refracted from each micro-prism. With this method, micro-particles with nanoscale sub-structures can be fabricated and subsequently released from a substrate. Such particles, showing enhanced fluorescence sensitivity, can be suspended in solutions and used for sensing (Lee et al., 2009).

6.4   Laser lithography based on one-, two- or multiple-photon absorption

The accurate fabrication of 3D structures at high spatial resolution requires alternative approaches to photolithography without the need for vertically stacking multiple planar layers with a finite thickness. A stacked 3D structure that is fabricated by a sequence of exposure steps suffers generally from limited alignment accuracies of the masks between the subsequent exposure steps. Furthermore, the creation of smooth 3D topologies requires a local variation of the exposure dose. While direct beam writing methods are more flexible than mask-based fabrication due to their inherent capability of varying the exposure dose as a function of the beam position, such methods are often used for the generation of a continuous relief in the target material (Fu and Ngoi, 2001). Although a local variation of the exposure dose in photolithography might be achieved by the use of expensive gray-tone masks with a sophisticated transmission profile, it is not an easy task, even for binary masks, to fabricate a mask with the desired transmittance and requires sophisticated pixel coding in order to achieve the desired number of gray levels. Nevertheless, gray-scale lithography can be applied to the fabrication of smooth micro-optical elements such as lenses (Cui *et al.*, 2003).

A special technique of direct write laser lithography that overcomes these problems is 3D laser lithography (3D-LL) based on multiphoton absorption, which is reviewed in this chapter. Recently, Misawa and Juodkazis (2006) edited a very comprehensive book about 3D laser micro-fabrication. 3D-LL is a true 3D method that has definitely reached a level of technical perfection over the last few years and may potentially replace other direct writing methods, such as electron beam lithography in a wide field of applications. 3D-LL based on two- or multiphoton polymerization has been applied as a versatile tool in micro- and nano-fabrication (Anscombe, 2010; Farsari *et al.*, 2010; Fourkas and Baldacchini, 2004; Maruo and Fourkas, 2008; Ostendorf and Chichkov, 2006; Schmidt *et al.*, 2007a, 2007b; Serbin *et al.*, 2004; Sun and Kawata, 2004; Yi *et al.*, 2004a, 2004b). The 3D-LL technique has been used in the following ways: for 3D structures with arbitrary shapes (Fig. 6.11), with freely moving components (Sun *et al.*, 2000) such as photo-driven micro-pumps and micro-sensors (Lin *et al.*, 2004), micro-needles, periodic and scaffold structures for photonic and biomedical applications (Ovsianikov, 2007a, 2007b), designable refractive micro-lenses (Guo *et al.*, 2006) and pyramids (Satzinger *et al.*, 2008), embedded photonic structures such as line gratings (Guo *et al.*, 2003), 3D optical memory (Nakahama *et al.*, 2005; Walker and Rentzepis, 2008), waveguides (Langer and Riester, 2007; Stampfl *et al.*, 2009), waveguides in a flexible PDMS matrix (Infuehr *et al.*, 2007), waveguides coupled to photonic crystal structures (Serbin and Gu, 2006), diffractive optical

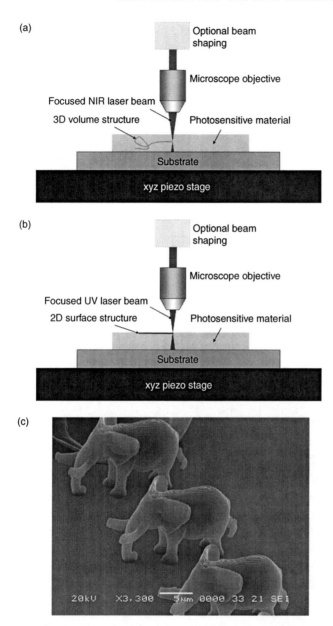

6.11 Comparison between lithography based on multiphoton (a) and single-photon (b) absorption. The material is transparent to the laser wavelength and focusing beneath the surface is possible. Hence, 3D scanning of the laser focus through the volume generates arbitrary intricate 3D structures (c).

elements (Chen *et al.*, 2007), photonic bandgap structures, photonic hetero-structures and meta-materials (von Freymann *et al.*, 2010), optical cloaking (Ergin *et al.*, 2010), polymer templates for metallic structures (Gansel *et al.*, 2009; Rill *et al.*, 2008), flexible structures for biological cell culture studies (Klein *et al.*, 2010), micro-replication of biological architectures for cellular scaffolds or custom tissue replacements (Nielson *et al.*, 2009), and *in vivo* processing of scaffolds with embedded living organisms (Torgersen *et al.*, 2010).

3D-LL is a laser-based photolithographic method, which is used for true 3D microscale patterning (Fig. 6.11). A 3D patterning method is important for the development of micro- and nanoscale systems because it enables new approaches to the fabrication and integration of complex shaped features into the micro- and nano-cosmos. The fabrication of miniaturized photonic systems requires a true 3D method, which avoids geometrical and alignment restrictions usually imposed by planar and sequential processing methods.

There are several ways of building a 3D micro-structure by means of laser-based lithographic methods. A common approach is micro-stereo-lithography (μ-SL) (Neumeister *et al.*, 2008), where a 3D structure is built layer-by-layer in a photosensitive resin. The structure in each layer is built either by scanning the focus of a UV laser or by projecting patterned UV light via masks or a digital mirror device onto the material. This method is compatible with specially tailored materials with tunable properties (Stampfl *et al.*, 2008). Usually lasers with a wavelength within the absorption range of the photosensitive materials are used, but μ-SL with enhanced spatial resolution takes advantage of multiphoton-based exposure with suitable materials (Houbertz *et al.*, 2010).

Since only one layer is exposed at a time, the process of alignment and exposure has to be repeated for each layer in order to obtain a 3D structure, which is represented by a layer stack. After a thin layer of photosensitive material is generated, the layer is exposed to light, whereas the light is absorbed within a thin layer of the material (several microns) starting from the surface. Generally, these methods use light sources that are matched to the absorption spectrum of the photosensitive material, and are hence based on single-photon processes. The 3D structure is built as a layer stack in a repetitive process by vertically translating the material photoreactor. This imposes several restrictions on the structure regarding the 3D design (geometrical restrictions due to layer-by-layer processing) and the structure resolution. The vertical resolution is limited by the achievable thickness of the individual layers. The lateral resolution is determined from the spot size of the light on the layer.

A very powerful method is the 3D-LL based on multiphoton absorption. Especially the use of femtosecond lasers in combination with polymer materials offers several benefits for 3D micro- and nano-fabrication. The generation of 3D polymer structures is based on two-photon polymerization, which happens via two-photon absorption with subsequent polymerization of the material in the laser focus. The most important technical benefits of this method are true 3D structuring capabilities and high spatial resolution beyond the diffraction limit (e.g. sub 100 nm structures fabricated with a laser wavelength of 800 nm), which is controlled by the number of applied laser pulses and the laser pulse energy. Such benefits enable the micro-fabrication of freely movable structures (Sun *et al.*, 2000) and photonic structures such as prisms and diffractive photonic elements (Ostendorf and Chichkov, 2006). The implementation of this technique is a rather simple lithographic apparatus, which in general does not require special ambient conditions such as vacuum and inert atmosphere or special coating tools for a thin layer application. More details about the technical implementation are provided in the next section.

The technological maturity of ultrafast lasers has advanced two-photon absorption as a realistic exposure mechanism with several technical benefits over single-photon exposure. Femtosecond pulses provide a high peak power, whereas the laser fluence remains below the damage threshold of common photosensitive materials originally not designed for two-photon processes (Wu *et al.*, 1999).

In contrast to conventional μ-SL with UV or VIS lasers, absorption via inter-band transitions involving more than one photon is a key issue regarding 3D-LL. The two- or multiphoton absorption-based lithography takes advantage of the strong confinement of the energy transfer of a tightly focused laser beam to a photosensitive material. The energy transfer is responsible for a modification of the material around the laser focus (Fig. 6.12). The smallest exposed volume element is typically called a voxel (in analogy to the 2D pixel), which represents the smallest building unit of a 3D structure. The in-volume photo-induced modification of the material is described in Section 6.5.

Due to the unique properties of two-photon-absorption (TPA)-based 3D-LL, an increasing number of groups deal with 3D fabrication of polymer structures for various applications. Therefore, numerous comprehensive review papers can be found in the literature about the recent progress in multiphoton micro-fabrication with femtosecond lasers (Fourkas and Baldacchini, 2004; LaFratta *et al.*, 2007; Maruo and Kawata, 1998), and direct laser-written 3D polymer templates for advanced nanostructures for photonics (von Freymann *et al.*, 2010).

A related topic, which exploits the 3D features of the two-photon technology is 3D data storage. Taking advantage of 3D multilayer storage would enable capacities of 10 Tbyte on a DVD-size disk (Walker and Rentzepis, 2008). The recording of optical data realized by photodegradation of

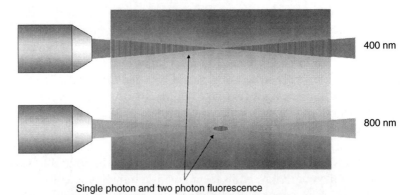

400 nm

800 nm

Single photon and two photon fluorescence

6.12 Two-photon fluorescence indicating the energy transfer from the laser to the material. The string confinement around the laser focus enables the high spatial resolution and the three-dimensional structuring.

fluorescent chromophores at a spatial resolution of less than 1 μm in three dimensions has been successfully demonstrated by Nakahama *et al.* (2005).

6.4.1 Two-photon absorption – team play of photons

Seventy years ago, Maria Göppert-Mayer discussed the fundamentals of two-photon absorption in her PhD thesis (Göppert-Mayer, 1931). However, the experimental proof was demonstrated three decades later in 1961 (Kaiser and Garrett, 1961). Phenomena including more than one photon (multiphoton absorption and multiphoton ionization) occur in materials at high intensities of the incident radiation. With the simultaneous presence of a large number of photons in the same place, it may happen that a photosensitive entity absorbs multiple photons with energies below the energetic gap between the ground and the excited state (Fig. 6.13). At high laser intensities, the absorption coefficient of a material becomes intensity dependent (Boyd, 2003), which leads to a modified Beer-Lambert absorption law and a non-linear dependence of absorbed energy on the incident laser intensity. Thus, a localization of the energy transfer to the material around the laser focus takes place, even for radiation that is spectrally located in a transparent region of the photosensitive material. The nonlinear intensity dependence facilitates the high achievable spatial resolution and true three-dimensional structuring capability (Fig. 6.14). 3D-LL based on the simultaneous absorption of multiple ($n > 1$) photons was first proposed by Maruo *et al.* (1997) in photo-polymerizable resin. The resin is hardened by a polymerization process, which is triggered by two-photon absorption. Small 3D features with a lateral and axial (with respect to the laser beam) resolution of 0.62 μm and

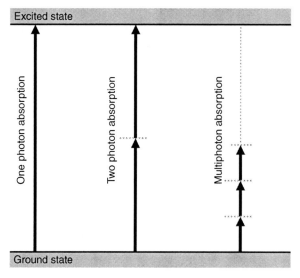

6.13 One- (SPA), two- (TPA) and multiphoton (MPA) absorption between ground and the excited state.

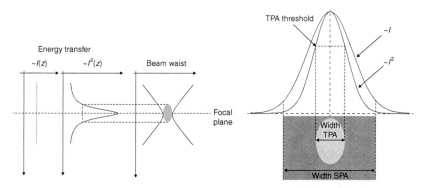

6.14 Nonlinear absorption enhances spatial resolution of TPA process as compared to SPA. The nonlinear energy transfer decreases faster in radial direction than a linear process and peaks at the focal plane of a Gaussian beam. Hence, the TPA process strongly confines energy transfer axially and radially to the voxel size. See also Maruo and Kawata (1998).

2.2 μm are immediately fabricated by scanning the laser focus through the resin volume. The main limiting factor of the achievable resolution is attributed to spherical aberration at the resin–air interface (Maruo and Kawata, 1998). Hence, this method circumvents drawbacks of other rapid prototyping methods based on a sequential layer-by-layer approach, because the laser focus can be arbitrarily scanned in three dimensions through the volume of the photosensitive resin and more structure-related scanning strategies (including non-horizontal scanning approaches, cf. Section 6.4.3 and Section 6.6.1) can be developed. A short description of the method can be found in the literature (Gebeshuber *et al.*, 2010).

The underlying mechanisms of photo-polymerization including the rate of free electron generation via avalanche mechanisms and nonlinear absorption mechanisms at tight focusing conditions are considered important regarding chemical bond breaking, photo-initiation and structural material changes. The difference between high and low irradiance as well as the difference between high and low repetition regime is related to the material cooling mechanisms (Malinauskas *et al.*, 2010b).

Multiphoton processes are technically exploited for spectroscopy, microscopy, and induced damage of various optical materials. Furthermore, laser-induced material modification by multiphoton absorption is of growing interest regarding 3D micro- and nano-fabrication, which is discussed later in this chapter.

The future development of micro- and nano-fabrication of micro- or nano-sized devices or devices with micro- and nano-features, which is an important task of nano-photonics, is expected to be essentially influenced by two-photon polymerization (TPP). TPP is an important method of multiphoton-based laser micro- and nano-fabrication and should play an important role in producing polymer-based optoelectronic and MEMS devices, since it offers fabrication of 3D structures below the diffraction limit (Witzgall *et al.*, 1998). It is also expected that this method will penetrate many scientific research fields due to ongoing research in materials, optics and fabrication of functional devices. From the materials point of view, there is a need for highly efficient two-photon reactive materials, which will reduce costs for equipment, and a need for functional chromophores as a bonded component for advanced material with specially tailored properties. The optical point of view deals with adapting the voxel size and shape and a mechanism for a parallel production system, such as micro-lens arrays, which increase the number of foci (Sun and Kawata, 2004).

6.4.2 Technical implementation of 3D-LL based on TPA

There are some comprehensive reviews of two-photon-based micro-fabrication (e.g. Maruo and Fourkas, 2008; Sun and Kawata, 2004), which include also an overview of the development of laser lithography setups (Sun and Kawata, 2004). Basically, it is possible to provide the photons from two different laser beams in a 'crossbeam' arrangement. The advantages are the use of two different wavelengths (colors) and a better longitudinal resolution, but the alignment of such an optical system is difficult because the two beams have to overlap in space and time in the photosensitive sample (Sun and Kawata, 2004). Therefore, a setup similar to a laser scanning microscope is preferable. A single laser beam is tightly focused into the photosensitive resin and both photons are provided from the single laser beam. One advantage is the rather simple experimental setup although a two-color setup similar to the crossbeam setup with improved longitudinal resolution is not possible. There are other means (e.g. shaded ring filter [SRF], phase masks, beam shaping) to improve longitudinal/axial resolution for a single beam setup. Diffractive elements have also been introduced in the optical path, which decrease laterally the size of the point spread function of the focused laser beam in order to increase the resolution of the two-photon polymerization process (Wei *et al.*, 2011). Another way of setting up multiphoton lithography is by raster scanning a focused fs laser beam over a digital mirror device (DMD) that represents a computer-controlled intrinsically aligned reflective photo mask. The reflection pattern from the DMD is imaged onto a photosensitive material by a high NA objective. A 2D pattern can be fabricated easily and a layer stacking of multiple patterns is fast and simple due to the inherent alignment of the DMD reflection patterns (automated mask sequence) and synchronized plane-to-plane movement of the laser voxel (Nielson *et al.*, 2009).

A typical experimental or lab-scale setup of a 3D-LL fabrication system comprises generally a pulsed (fs, ps) laser source, part or sample handling and positioning, a beam delivery system, which consists of transmission optics, focusing optics, scanning systems or fiber delivery. Femtosecond laser pulses are popular material processing tools regarding two-photon-based lithography, although they are expensive. Femtosecond lasers provide high peak power at moderate pulse energy and average power, which is beneficial for reducing thermal damage to materials.

The high achievable resolution of the two-photon absorption-based lithographic process requires high resolution positioning of the sample or the laser focus. The implementation is either via high-precision linear stages, galvano-scanners, piezo stages or a combination of multiple positioning units. A common combination uses the long travel ranges of high-precision linear stages in order to handle large samples and a fast scanning of the laser focus with a galvano-scanner unit. A large structure can be realized by stitching of multiple smaller structures. Another combination takes advantage of the large scanning ranges of motorized stages with lower positioning accuracy that carry a high-precision piezo stage with a smaller scanning range but a much better positioning accuracy. For large-scale samples and applications that require a spatial resolution in the micrometer range, a sample motion with the motorized stages is done and stitching is not necessary. Photonic applications usually require high spatial resolution, which needs piezo stages for sample positioning.

Galvano-mirror scanning has a fast scanning speed, but suffers from a small fabrication area and distortions by the focusing lens. In contrast, piezo stage scanning is slower, but benefits from a larger fabrication area and better fabrication uniformity due to the absence of lens distortions (Yi and Kong, 2007).

The fabrication over large areas with piezo stages can be optimized by applying a continuous scanning method. This method takes advantage of the slow response of the piezo system at higher scanning speeds and is preferable to a point-to-point motion and a corresponding dwell time for exposure. A stable fabrication window with adequate stage settings enables uniform motion of the stage without errors considering the stage characteristics and hence provides optimized resolution and uniformly exposed material at minimized processing times (Lim *et al.*, 2008).

With respect to a large format fabrication based on two-photon polymerization in suitable materials, the laser writing process must be optimized regarding the fabrication time. Laser power, repetition rate, sample scan speed and voxel size are not independent from each other and from the material. The optimal parameters must be determined from calibration experiments that screen different exposure (scan speed versus laser power) and determine the size of the inscribed voxels by scanning electron microscopy in order to account for a sufficient polymerization of the material at a given scan speed (Liu *et al.*, 2010).

3D-LL for TPA is usually implemented with a titanium-sapphire (Ti:S) laser, which emits radiation in the near-infrared (NIR) range, typically 750–850 nm. These systems consist typically of a Ti:S oscillator or a frequency doubled fiber laser, that provide a pulse train at a repetition rate at several tens of MHz (80–100 MHz) and a pulse width of approximately 80–150 fs. Ti:S oscillators are often used to induce TPA in polymer materials since they provide enough power. In order to increase the variety of materials, an amplification stage can be implemented. This results typically in higher peak power of the provided laser pulses, but at a lower repetition rate in the kHz regime. Long cavity lasers provide pulses at higher power than oscillators at a reasonable fast repetition rate of typically 10–20 MHz. The Ti:S crystal as gain medium is well suited for the generation of fs pulses due to its wide spectral emission, but it is also reported that lasers with other active materials such as Yb:KGW (ytterbium doped potassium-gadolinium tungstate crystals) are used for two-photon polymerization in polymer materials (Malinauskas *et al.*, 2010a). The second harmonic of such a laser is in the visible range, which is beneficial regarding the resolution limit of the method. A good spatial beam profile of the emitted TEM_{00} mode is essential for tight focusing of the laser. Fiber lasers offer a nearly perfect profile, which is generally better than from amplified Ti:S systems and better suited for the highest achievable resolution.

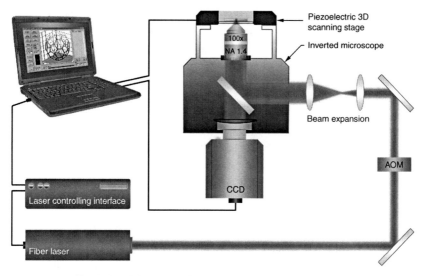

6.15 Typical writing setup fort two-photon-based 3D-LL. Reproduced with permission from Nanoscribe GmbH.

Although a 3D-LL apparatus can be built up in a modular way by combining the above-mentioned components, a compact setup is advantageous regarding the spatial resolution, stability, repeatability and quality of the 3D micro- and nanostructures. Today, there are systems available that are built on a microscope or inverted microscope (Fig. 6.15) taking advantage of a stable beam delivery, sample positioning and visual inspection or live monitoring of the fabrication process. Another advantage is the combined positioning of the sample with a motorized stage and a high-precision piezo stage. Additionally, different dimensions and scales of structures can be addressed by using different high NA (up to NA = 1.4 with oil immersion objectives) or low NA objectives. A system that is equipped with a motorized objective revolver enables automatic change of the fabrication resolution by simply changing the objective.

The increasing interest of industry and research pushes the number of commercially available lithographic systems for a routine fabrication of 3D micro- and nanostructures from several suppliers. The lithographic systems have outgrown the lab-only use and provide reliable tools for micro- and nano-rapid prototyping. Advanced systems offer a user-friendly control setup with automation of the writing process (remote control and programming, automatic sample exchange/handling) and CAD interface in combination with optimized optics for compensation of laser beam distortions (temporal as well as spatial pulse profile) and focusing. Optional optical features such as shaping of the voxel volume are available by means of filters (e.g. shaded ring filter), which change the point spread function of the beam

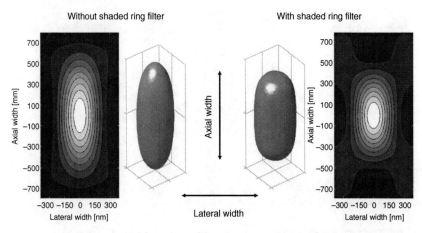

6.16 Point spread function of focal volume. Lines of constant intensity show the shape and size of the focal volume without (lhs) and with (rhs) shaded ring filter (SRF). A reduction of >28% of the axial elongation is achieved. Reproduced by courtesy of Erik Waller and Georg von Freymann, TU Kaiserslautern.

and reduce the axial elongation of the voxel (Fig. 6.16), important for photonic applications or applications, where structuring close at the resolution limit is required (e.g. 3D meta-materials). Beside beam shaping of the laser focus by cylindrical telescopes (Cerullo *et al.*, 2002), axicons (Winfield *et al.*, 2007) and slits (Ams *et al.*, 2005) a dual beam approach for voxel shaping and enhanced resolution is reported (Fischer *et al.*, 2010).

The structuring process is optimized via various computer-controlled writing parameters such as the laser power, scan speed, distance between individual voxels, acceleration, deceleration of the positioning system and synchronized ramping of the laser power.

The offered flexibility facilitates the generation of 2D and 3D structures in photosensitive materials with high and low refractive index for photonic applications, whereas positive-tone and negative-tone materials are available for two-photon-based direct laser writing. Via polymer templates, 3D structures can be molded in materials that are not suitable for direct laser writing.

Recently, direct laser writing of three-dimensional submicron structures using a continuous wave (cw) laser at 532 nm was demonstrated (Thiel *et al.*, 2010). Using such a laser at moderate powers (some 10 mW for commercially available photoresists such as SU-8 or IP-L and IP-G) significantly reduces the cost and increases the reliability and stability of advanced optical lithographic systems.

Recently, large-scale patterning based on two-photon absorption in inorganic–organic hybrid materials was realized (Houbertz *et al.*, 2010). A setup

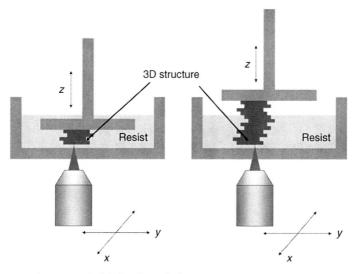

6.17 Large-scale fabrication of 3D structures according to Houbertz *et al.* (2010).

similar to micro-stereolithography was applied. The setup is inverted, that is, the laser is focused from the bottom side into a material bath with an immersed sample holder. The structure is generated head over heels and structures with dimensions of several mm up to 2 cm at high spatial resolution were processed (Fig. 6.17).

6.4.3 Technical benefits of 3D-LL based on TPA

Applications that require a true 3D fabrication method for intricate structures at high resolution (sub-diffraction limit resolution) with great flexibility in design benefit from 3D-LL based on TPA. As a rapid prototyping method it facilitates the direct conversion from a CAD model to a physical structure with reliable and reproducible results in a single exposure step. True 3D offers the adaptation of the laser focus scanning strategy according to the geometry, that is, slicing, hatching and contouring of a 3D model is not restricted to planes parallel to the substrate (horizontal slices). Vertical and horizontal scanning are possible in the same way. Similarly, vertical and horizontal stitching can be made for the direct writing of large and tall structures over larger areas or volumes. The vertical limit is mainly given by the focal length of the objective. Therefore, 3D-LL is situated between high resolution processes such as electron beam lithography or focused ion beam lithography and processes with worse resolution such as micro-stereolithography. The simple apparatus without vacuum requirements facilitates a faster and higher throughput than, for example, e-beam lithography and hence has the potential for mass production. Upon availability of efficient material the costs of the equipment are further reduced due to the possibility to use cheaper laser systems for triggering the TPA process. The material development supports the trend to faster writing speeds and larger sample areas. Technical implementation of machine vision enables the processing of pre-configured substrates in order to align laser-inscribed structures into micro-systems. Depending on the optical configuration of the apparatus, both transparent and opaque substrates can be handled. For this purpose, 3D-LL is of growing interest in industry and research.

The above-mentioned benefits can be technically exploited to make superior micro- and nanostructures. There are several benefits of using two-photon absorption-based photo-polymerization rather than a one-photon-based polymerization: due to the localization of the laser–matter interaction around the laser focus, the polymerization is confined to a small volume inside the material volume (resin) and can be triggered within the bulk of the resin. Consequently, a 3D scanning of the laser focus across the resin yields a true 3D structure (no sequential layering) instead of a layer-by-layer processing initiated by a one-photon-based process. Additionally, the quenching of radicalized molecules and hence inhibited polymerization is avoided, since there is no direct contact with ambient oxygen inside the resin volume. Furthermore, two-photon absorption-based polymerization results in higher spatial resolution and smaller structure features as compared to one-photon absorption-based polymerization. The threshold of the two-photon absorption enables sub-diffraction limit feature sizes by controlling the laser power energy and the number of applied laser pulses per voxel. This yields very smooth surfaces as required by micro-optical components.

The achievable resolution of a structure depends on various parameters such as material chemistry, laser parameters (beam profile, pulse width) and writing parameters. An estimation of the achievable resolution is given by the investigation of the voxel size as a function of laser and writing parameters. The size and shape of the voxel can be studied by SEM characterization of the voxels. The shape of the voxel reveals additionally the internal structure of the focal volume of the laser, because it is sensitive to aberrations introduced by misalignment of the optical setup. The width of a voxel can be estimated from the intensity distribution from a focused Gaussian beam and the dependence of the squared intensity of the polymerization and easily measured by SEM characterization (Yi *et al.*, 2005), but for the measurement of the vertical size of the voxel, an isolated voxel lying on the substrate is required. Since the voxel needs to be attached to the substrate in order to avoid being washed away during the development it is partly truncated and does not show its full shape. The ascending scan method circumvents this problem and avoids truncation of voxels and measurements on isolated voxels (Sun *et al.*, 2002). The ascending scan method moves the laser focal position with respect to the substrate surface starting from inside the substrate. As the focus moves away from the interface, the voxel is less truncated and more weakly attached to the substrate. If the laser focus moves further, the voxel becomes detached and floats away during development. Some of the voxels that are barely attached will tilt over and exhibit the full longitudinal or axial elongation (Fourkas and Baldacchini, 2004; LaFratta *et al.*, 2007). This method enables the voxel characterization (size and shape) as a function of exposure time and laser power and shows the influence of laser beam properties such as the polarization on the voxel formation or the numerical aperture of the focusing optics (Sun *et al.*, 2003). The smaller a voxel gets, the more technically challenging is the application of the ascending scan method and the suspending bridge method can be used instead (Sun and Kawata, 2004). Instead of isolated voxels by single point/single pulse exposure, a suspended line is written between two posts, whereas the line width and height corresponds to the voxel dimensions. This method is also commonly used for determination of achievable resolution for various materials and systems.

In a polymerization process initiated by radicals, the voxel size can be estimated from a rate equation and the intensity distribution of a focused Gaussian beam (lateral and along the optical axis), assuming that polymerization starts when a certain threshold of generated radicals is exceeded upon incident laser pulses (Wu *et al.*, 2006b).

Typical negative-tone photoresists have voxel with sub-diffraction diameters in the range 0.2–0.5 µm for a numerical aperture NA = 1.25 up to 2.4–4 µm for a NA = 0.25. The control over the laser power and the power of the focusing optics (numerical aperture) facilitates the generation of complex 3D structures at different length scales in the same structure (Malinauskas *et al.*, 2010a). In addition to the numerical aperture of the focusing optics and the laser wavelength the laser pulse width shows an effect on the achievable fabrication resolution. A study of the voxel size as a function of the pulse width showed a decreasing voxel size with a stretched pulse length (up to picoseconds) (Kong *et al.*, 2007).

Although usually laser wavelength in the NIR enables the realization of sub-diffraction-limit features in photo-polymers with high reproducibility, a laser wavelength in the VIS is preferable to an NIR wavelength regarding the achievable resolution (Malinauskas *et al.*, 2010a; Thiel *et al.*, 2010).

Nevertheless, it is possible to achieve a voxel size of approximately $\lambda/50$ with a 780 nm laser for a two-photon process by tuning the laser power and the scan speed. Additionally, repolymerization of the material leads to further reduction of the feature size of written lines down to just 15 nm (Tan *et al.*, 2007).

The size of the voxel can be attributed to different interaction volumes, which are governed by different mechanisms and which are not known to the same extent. The smallest volume is determined by the threshold behavior of the two-photon process and depends on the minimum required photo-initiator concentration. All mechanisms contribute simultaneously to the achievable voxel shape and size. The technical interaction volume is determined by the hardware of the apparatus and can be optimized by using high NA optics, stabilization of the laser, and a high-precision positioning system, while the chemical interaction is influenced by material properties such as the reaction kinetics and the diffusion of initiator molecules and is more difficult to minimize (Houbertz *et al.*, 2010).

6.5   Material modification aspects

Normally, ultraviolet (UV), visible (VIS) or infrared (IR) light is used to initiate photo-polymer reactions, but sources at a shorter wavelength or electron beam lithography have also been utilized. Since the development of the basic chemical and physical reactions of organic-based materials that are initiated by electromagnetic radiation, the related photo-polymer technology has been used in a wide variety of practical applications. The relation between photo-polymer chemistry and applications can be described by the exposure mode, the mechanism of the photo-polymer reaction and the visualization method, which is applied to detect changes that were induced by the exposure. The exposure mode differentiates between an imaging (patterned or 'pattern-wise') and non-imaging (e.g. flood) way of exposing the photo-polymer. An imaging exposure is a patterned exposure, which is either by optically projecting a mask onto the material or by scanning a focused light beam over the material for forming a well-defined pattern in the photo-polymer. A non-imaging exposure is, for example, a flood exposure of the photo-polymer or an exposure through a shadow mask without any optics. The mechanisms involved in photo-polymer reactions comprise photo-polymerization (crosslinking of monomers and oligomers to form higher molecular weight material), photo-crosslinking (reaction of unsaturated moieties, which are attached to an organic polymer), photomolecular reaction (reaction of small molecules, which modify the embedding polymer matrix), photodegradation (fragmentation of light sensitive linkages and degradation of the polymer into smaller units) and a photo-thermal reaction (physical changes such as laser ablation). All these reactions form detectable changes in organic materials, which can be visualized by an adequate visualization method. The light-induced changes are, for example, solubility, adhesion, color change, phase change, refractive index or electrical conductivity (Pfeiffer, 1997).

Upon exposure by light or laser, a latent image is formed in the photosensitive material, which must be developed. In most applications, the following different classes of visualization are used, which depend on the property change of the photo-polymer: if the solubility of the material is changed either by photo-polymerization or photo-crosslinking, which both transform an initially soluble material into an insoluble, or by photodegradation or photomolecular reaction, which both can make an insoluble material more soluble in an adequate solvent, the exposed material can be separated from the unexposed material and the latent image is made visible. In a similar way, other properties such as adhesion modulation, color changes, phase changes, changes of the refractive index or the electrical conductivity can be used to visualize the pattern in the photo-polymer (Pfeiffer, 1997).

In most of the described photonic applications in this chapter, the change in solubility and/or the refractive index is used for either the direct inscription of embedded photonic structures or the forming of a freestanding 2D or 3D shape after separation of exposed and unexposed material. Depending on the material system the polymerization happens during laser exposure via radically reacting photo-initiators (liquid resins such as acrylate-based systems or organically modified ceramics). Other materials such as SU-8 generate an acid and require a post-exposure baking in order to polymerize. Since the polymerization usually leads to a change of the refractive index, cationic systems such as SU-8, which have only a negligible change of refractive index upon laser exposure, allow a more flexible exposure strategy due to negligible perturbation of the laser focus due to index modulation (Ostendorf and Chichkov, 2006). The maintained transmittance and refractive index of the material is an important prerequisite for a true 3D scanning of the laser focus through the material and hence optimized exposure strategies for a time optimized fabrication of complex structures.

It was shown that 3D structures can be fabricated by using a thermo-sensitive resin in a micro-fabrication system. The nonlinear thermally triggered photo-polymerization shows similar capabilities for 3D fabrication such as the two-photon absorption-based lithography (Yamakawa et al., 2004).

6.5.1 Towards high efficient two-photon photo-initiators

An efficient photochemistry is a prerequisite for the application of a wide range of applications regarding two-photon-based laser lithography. An efficient photo-initiator in a suitable polymer matrix has a great impact on the fabrication result (Fig. 6.18). The development of efficient materials aims at improving the fabrication efficiency of 3D nano/micro-fabrications based on two-photon polymerization (TPP) and the development of organic molecules with large TPA cross-sections (Blanche et al., 2002).

(a)

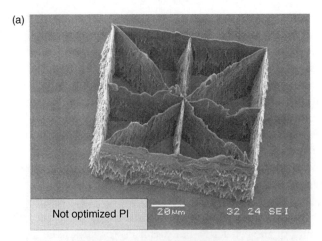

(b)

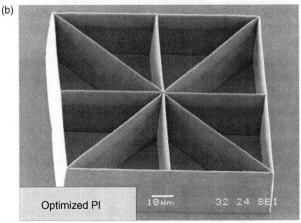

6.18 (a) Result with a non-optimized photo-initiator; (b) result with an optimized photo-initiator. Exposure conditions are the same for both structures.

High TPA cross-sections are directly related to the molecule structure (Lee *et al.*, 2008). Different donor–acceptor functionalities in cross-conjugated polymers allow investigation of the structure–activity relationship (Pucher *et al.*, 2009). An efficient two-photon photo-initiator is technically very important, because it enables photochemistry far below the damage threshold of the material and therefore a wide power range for tailoring the micro-structure dimensions. In addition, the more efficiently a material can be structured, the less optical power is required for initiating the two-photon absorption process and hence the less expensive is the technical setup, since cheaper laser systems (cw or ns) instead of femtosecond laser systems can be used.

It was found that the spatial resolution in pinpoint photo-polymeriza-tion of radical-type resins is influenced by the temperature of the resin. A temperature decrease leads to a less pronounced radical diffusion, while a temperature increase leads to a voxel size reduction due to enhanced chain termination. In addition to the influence of the temperature on the polymerization effect, heated photonic structures showed a blue shift of the reflection spectra due to shrinking of the structure. Heating is considered to increase the robustness and facilitates a tuning of the working wavelength of photonic structures (Takada *et al.*, 2008).

The development of photo-polymers for rapid prototyping is not restricted to optical systems, but is also of growing interest for biological applications. Structures can be made of newly developed acrylate-based biocompatible and biodegradable formulations for cellular implants with various rapid prototyping techniques such as digital light processing (DLP), micro-stereolithography (μ-SL) and rapid prototyping based on two-photon absorption photo-polymerization techniques. A new cross-conjugated pho-to-initiator for TPA was found recently, which enables micro-fabrication in biocompatible monomer formulation based on trimethylolpropanetriacry-late (TTA) at a very low photo-initiator concentration of 0.005 wt% (Liska *et al.*, 2007). Epoxy/acrylate resins can be processed by two-photon-induced crosslinking of the acrylate groups and subsequent thermally linking the epoxy groups, without the need of directly writing into the epoxy system (Winfield and O'Brien, 2010).

The selective chemical functionalization of 3D polymer micro-structures was demonstrated and used for the fabrication of a functional micro-induc-tor (Farrer *et al.*, 2006). Both acrylate and methacrylate monomers were used for 3D micro-fabrication in a sequential laser writing process. At increased laser power it was possible to create also methacrylic structures of simi-lar quality as compared to acrylic structures. Due to the cross-reaction of methacrylates and acrylates it was possible to fabricate structures that are partially composed of acrylic and methacrylic polymer. Subsequent treat-ment added amines only to the acrylic portion of the structure, onto which other materials can be deposited. For example, a selective metallization can be achieved. This approach presents new possibilities for multiphoton-based micro-fabrication in areas such as photonic crystals, meta-materials or biosensing.

The good performance of 1,5-diphenylpenta-1,4-diyn-3-one (DPD) as photo-initiator for radical polymerization motivated the investigation of several donor substituted derivatives. Because of to the D-π-A-π-D sys-tem of these compounds two-photon-induced 3D photo-polymerization experiments were performed. It was found that the dimethylamino deriva-tive N-DPD showed outstanding performance with respect to two-photon-induced photo-polymerization even at very low concentrations as compared to single-photon initiators such as Irgacure 369. Test structures were fabricated in a polymer matrix (1:1 blend of commercial trimethylolpropane triacrylate and ethoxylated trimethylolpropane triacrylate) that contains the N-DPD photo-initiator and compared to similar structures in the same polymer matrix but with the Irgacure 369 photo-initiator. Subsequent to fabrication, the structures were investigated by means of SEM. Taking into account the worse efficiency of Irgacure 369, a higher initiator concentration (approximately 1–1.5 wt%) and higher laser power was required for structuring. Nevertheless, no structures at sufficient quality could be processed in the material with Irgacure 369 in contrast to structures with very thin (300 nm) and smooth surfaces and high quality in the same matrix material with a lower concentration (0.025 wt%) of the very efficient photo-initiator N-DPD (Heller *et al.*, 2007).

SU-8 is a commercially available, negative-tone photoresist for high aspect ratio micro-patterns in MEMS technology. It contains an epoxy resin and a photo-initiator, which are both dissolved in gamma-butyrolactone (GBL). A more detailed description of the composition and properties can be found in the literature (Teh *et al.*, 2005). Upon irradiation by UV light, the photo-initiator generates a photoacid with a concentration that is proportional to the irradiation dose (latent image). A post-exposure bake is required for a chain reaction that converts the latent image into a crosslinking density. The degree of crosslinking determines the solubility in the developer. SU-8 was used for laser-based 3D micro-fabrication (Lim *et al.*, 2008), fabrication of 3D photonic crystal templates (Hermatschweiler *et al.*, 2007; Tétreault *et al.*, 2006; von Freymann *et al.*, 2010), biomimetics patterns (Lasagni *et al.*, 2010), micro-fluidics (Liu *et al.*, 2010), etc. The latent image has a negligible refractive index change and hence subsequent exposure steps do not influence each other (Deubel *et al.*, 2004). SU-8 remains solid during the writing process. Since SU-8 is capable of forming thick films, up to mm height, high aspect ratio structures can be fabricated even with TPP. For this purpose, low numerical aperture focusing facilitates the exposure across the thick film and hence rapid processing at high lateral resolution. Structures with an aspect ratio up to 50 have been fabricated in SU-8 (Teh *et al.*, 2005).

Organically modified ceramics (ORMOCER®s) are a class of inorganic–organic hybrid materials that have attracted attention in micro-optical (e.g. waveguides) and micro-electrical applications (e.g. dielectrics). They exhibit important properties such as low attenuation at standard telecom wavelengths (especially at 1310 and 1550 nm), a precisely tunable refractive index within a specific range (1.44–1.56 at 633 nm), surface planarization, no parasitic polymerization outside the exposed area and long-term stability (Houbertz *et al.*, 2003). Such materials can be processed by mask aligner UV light exposure or by combining lithography, reflow and UV molding. Two-photon polymerization induced by 3D-LL based on TPA was used for the direct micro-fabrication from nm to mm scales with high precision and high optical quality (Houbertz, 2005). Organically modified ceramics are also used for the fabrication of movable complex 3D assemblies in one lithographic step using a UV laser. In order to reduce the penetration depth of the laser and to increase feature resolution, the material was mixed with azo phenyl UV absorbers (Overmeyer *et al.*, 2011).

Acrylate-based transparent photo-polymers are suitable for micro-fabrication, whereas the photo-polymerization of the acrylate system transforms the reactive liquid into a solid polymer network. The polymerization is initiated by free radical species, which are generated upon two-photon absorption from a visible laser beam and well confined to the laser focus. Good adhesion, low shrinkage and a non-toxic removal of unexposed material are additional benefits (Nguyen *et al.*, 2005).

The fabrication of 3D structures in a high-index material is accomplished either by the fabrication of a polymer templates with subsequent double-inversion or by high-index materials that can be directly patterned, such as chalcogenide glasses (Busch *et al.*, 2007; Wong *et al.*, 2006).

The full photonic bandgap of a 3D photonic structure requires high-index materials (refractive index $n \sim 2$–3) with a sufficient contrast to the ambient medium (generally air). TiO_2 has a high refractive index ($n \sim 2.4$) and high transparency in the visible spectra and is therefore very promising for photonic applications. A direct 3D patterning with fs lasers is possible with a specially developed photosensitive sol–gel-based spin-coatable TiO_2 resist, which has a refractive index of 1.68 and can be increased to > 2 by heating the film to temperatures between 400°C and 500°C due to a phase change of TiO_2. The heating of fabricated TiO_2 structures is challenging due to the strong shrinkage of the film, which results in deformation and delaminating of the structure from the substrate, although no fragmentation of the structure was observed (Passinger *et al.*, 2007).

6.6   Device design, fabrication and applications

Usually 3D-LL machining systems use femtosecond lasers for the fabrication process. Such lasers offer sufficiently high pulse peak power for initiating a two-photon transition in the photosensitive material and moderate average energies, which avoids destruction of the surrounding material. A common system is a Ti:S laser system. The wavelength of such a laser is in the NIR (typically 750–850 nm) and is located in a transparent region of commonly used photosensitive materials.

In the past decades, many applications in the fields of electronic materials, printing materials, optical and electro-optical materials, fabrication of devices and materials, adhesives and sealants, coatings and surface modifications were generated (Pfeiffer, 1997).

Many manufacturing techniques have been developed and implemented to fabricate a wide range of micro-optical products. The challenges of the micro-optics business are diverse and tend to resist a widely accepted manufacturing process such as has been implemented for CMOS fabrication. Many of the challenges that have been addressed with various solutions include optical waveband of operation from DUV through LWIR, material systems, cost of manufacturing for the intended application space, feature sizes based on device functionality, and fabrication technology based on the manufacturing volume. Some of the technologies to be discussed include device patterning by e-beam lithography, optical lithography, gray-scale photolithography, direct CNC machining and micro-polishing, and plastic replication for applications such as micro-optics, micro-lenses, diffractive optics and refractive optics (Borek *et al.*, 2007).

Infrared femtosecond laser pulses are ideal for the fabrication of 3D structures in transparent media. Due to the low absorption cross-section, two or more photons are necessary for absorption. This multiphoton effect limits the affected volume to the focal area allowing for sharp features on the order of the wavelength of light. One possible multiphoton reaction is the photo-destruction (ablation, decomposition, etc.) or photo-polymerization of materials. Using these techniques, 3D photonic components can be realized. A photonic bandgap template has been created with monodisperse polystyrene (PS) spheres (diameter ~ 624 nm). To optimally focus inside the bulk, an index matching material must be infiltrated. By using a photosensitive material, two-photon polymerization can be used to harden the material surrounding the spheres and insert defects inside the bulk. With proper placement of defects, 3D photonic components, that is, waveguides, splitters and filters, can be created (Boyle *et al.*, 2007).

6.6.1 Optical design aspects

Nowadays, there are many versatile tools commercially available for design and simulation of optical components, as well as imaging or non-imaging systems. Pushed by the rapidly increasing computing power of desktop computers, optical modeling software became very powerful and can be used by everyone. The simulation tools are not restricted to simulate single optical components; the tools are capable of calculating the performance of a complete optical system with realistic material properties and a sufficient number of rays in a reasonable time for realistic predictions of system performance. In combination with optimization algorithms and tolerancing features, it is possible to design and create a virtual optical system that is already very close to the prototype, thus reducing costs drastically. Therefore, a close connection between simulations and rapid prototyping is considered crucial for photonic development. Once an optical design is made (including manufacturing tolerances and physical properties of the optical materials), normally a prototype of the designed optical system is required. Regarding microscale or nanoscale optics, 3D-LL is a very suitable method for rapid prototyping and designable nano-fabrication, which gives high resolution and the smoothest surfaces not only due to the small achievable voxel size but also due to the freedom in adjusting the voxel overlap and filling with liquid or partly polymerized resin with a surface tension-based self-smoothing effect (Zhang *et al.*, 2010). Although there are many other laser-based rapid prototyping tools, none of them exhibits the required resolution. Usually, optical design programs support CAD data exchange, which can be used for transferring the CAD data into the lithographic apparatus.

There are manifold ray tracing tools for geometrical optics, tools for wave optics or tools for the design and optimization of photonic components such as diffractive optical elements, gratings, laser cavities, photonic crystals, waveguides, thin films, etc. These tools follow various mathematical approaches depending on their application focus. A main difference is the calculation of propagating rays (geometrical optics) or the calculation of propagating fields (wave optics). Some tools have hybrid algorithms and combine ray propagation with wave optics in order to account for coherent effects in macroscopic optical systems.

Optical simulations

The dimensions of optically relevant features in a photonic system determine the type of optical simulation that has to be performed for a system analysis or design. While ray tracing based on geometrical optics is well suited for the simulation of systems with dimensions larger than the wavelength such as illumination systems, imaging and non-imaging systems, it cannot be used for micro- and nano-photonics. Small-scaled features of photonic structures require a full vectorized solution to the Maxwell equations in order to cover all observed optical phenomena. Geometrical ray tracing requires fully featured non-sequential ray tracing, optimization algorithms and stray light analysis and is developed for the rapid development of refractive optics. Wave-optical effects including evanescent coupling, plasmonics and coherent superposition of polarized electromagnetic fields demand tools based on, for example. the beam propagation method (BPM) or the finite difference time domain (FDTD) algorithm. BPM cannot handle backwards propagating fields and therefore is not suited to account for reflections at system boundaries or refractive index mismatches. The FDTD algorithm is well established and is ideal for calculations of structured interfaces and refractive/diffractive systems with feature sizes in the order of the wavelength or even smaller. It calculates the decay of an injected electromagnetic pulse into the simulation region. It can be used for the design of gratings, coupling structures, color generating structures, dipole emission close to dielectric interfaces and nanoparticles, photonic bandgap structures, etc.

Modern simulation tools facilitate the CAD design of structures, include material databases and support the parameterization of systems for tolerance analysis and optimization. Subsequent to the system design, the simulation tool exports the structural data to a CAD compatible format that can be further processed by lithographic machining. Depending on the tool, post-processing of data is required.

While the design of microscopic refractive optical structures such as lenses is straightforward and can mostly be done by geometric ray tracing, the design of structures with sub-wavelength features is more demanding and usually incorporates different algorithms and approximations. Limitations imposed by fabrication should also be considered in the design phase, since the structure eventually needs to be realized (Mait *et al.*, 1998).

Diffractive optical elements with a high efficiency require a multilevel representation of the designed phase structure (Fig. 6.19). A phase structure with 2^N binary levels requires N separate masks, exposure and machining steps for its fabrication. For $N = 4$, the first machining (ion or laser etching) is done to a depth $d/2^N$ and $d/2^{N-1}$ for the second machining step. Generally, the last etching step has to be done to half of the desired peak-to-peak depth of the structure (Goodman, 1996). Here, 3D-LL can be used to directly write a sawtooth structure without the need of multiple sequential processing (Fig. 6.20).

The optical design based on the representation of the electromagnetic field with all its properties such as spatial and temporal distribution, coherence, spectral distribution and polarization is essential for optical simulations that take into account effects that originate from the electromagnetic nature of light. An optical system is divided into its components and the light field is propagated through the system using appropriate propagation operators and assumptions. The optical design strongly relies on the right choice of the propagation operator and knowledge of its limitations in order to avoid numerical and physically non-existing artifacts. Today there are commercial software packages that facilitate the optical design without troubling too much about numerical issues and support the optical engineer with many tools for optimal design (Wyrowski and Schimmel, 2006, 2007).

Rapid prototyping aspects

Recent advances in multidimensional structures have been demonstrated that serve as the basis for three-dimensional photonic bandgap materials, meta-materials, optical cloaks, highly efficient low-cost solar cells, and chemical and biological sensors. The state-of-the-art design and fabrication of

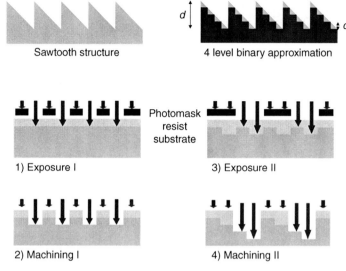

d $d/2^N$

Sawtooth structure

4 level binary approximation

Photomask
resist
substrate

1) Exposure I

3) Exposure II

2) Machining I

4) Machining II

6.19 A diffractive sawtooth structure (a blazed grating element) and its 4-level binary equivalent. A mask-based lithographic fabrication process with a positive tone resist for 4 ($N = 2$, see text) height levels requires two masks with different patterns.

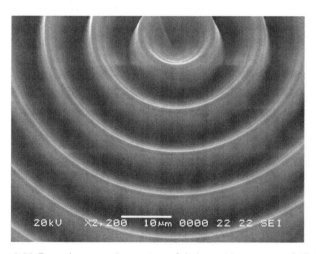

ZØkU   X2,2ØØ   1Øμm ØØØØ 22 22 SEI

6.20 Round sawtooth structure fabricated by means of 3D-LL.

multidimensional architectures for functional optical devices are covered and the next steps for this important field are described in Arpin *et al.* (2010).

3D-LL can be easily integrated into a photonic structure development process. CAD data from an optical design are translated into a physical structure via 3D-LL rapid prototyping. By means of subsequent characterization

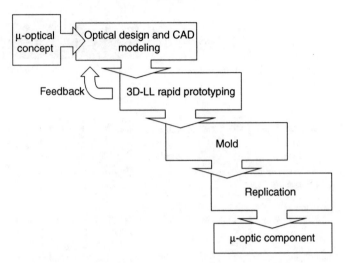

6.21 Design process of photonic components involving 3D laser prototyping.

an optimized prototype can be generated in an iterative way. Afterwards, large-scale replication of the component may start (Fig. 6.21).

Beside the technical improvements of lithographic apparatus and the progress in material development, the writing strategy must be considered. Although the scanning of the laser focus in all three dimensions offers many ways of fabricating a structure directly from a CAD model, there are some ways that yield better results than others. The strategy of horizontally contouring is well suited for most shapes although other geometrical effects must be taken into account. Structures with many vertical lines can also be scanned vertically.

As a general rule, the voxel size correlates directly with applied scanning speed and numerical aperture of the focusing optics and correlates inversely with applied laser power (Zhou *et al.*, 2008). This can be exploited for additional design features, because a very small distance between neighbouring voxels leads to a very even and smooth surface. A large distance may lead to the writing of separated, not connected voxels. With the right choice of voxel separation, a corrugated surface with a period smaller than the voxel diameter can be obtained (Woggon *et al.*, 2009). Hence, the shape of the voxel gives additional freedom in structure design (cf. Fig. 6.22).

Although the voxel diameter correlates to the achievable line width and resolution, post-processing of the fabricated structure such as O_2 plasma ashing is used to decrease the line width of laser-written structures from 280 nm down to 60 nm (Park *et al.*, 2006a).

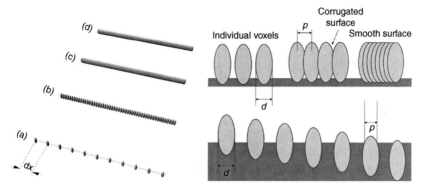

6.22 Influence of voxel size and separation on surface smoothness. At an appropriate separation between individual voxels, it is possible to write corrugated surfaces with a periodicity *p* smaller than the voxel diameter *d*. The same effect is achieved, when the largest part of the voxel is inside the substrate and structures such as lenses are written with only the topmost part of the voxel (diameter *p*).

Structural and shape enhancement methods

A simple representation of a structure by a (watertight) mesh requires additional hatching of the surfaces prior to the fabrication. If a free-standing structure is generated by exposure with subsequent removal of unexposed material (for the sake of simplicity a negative-tone photoresist is considered), the structure must be attached to the substrate in order to prevent floating and losing the structure during the development step.

The photochemistry and the change of optical properties of the photosensitive material must be considered. An immediate change of the refractive index may influence the laser beam by aberrations and focusing through already exposed regions can have a negative impact on resolution and voxel shape and may lead to unwanted results. Usually, the scanning of the laser focus is done in a defined way similar to the stacking known from stereolithography, either from the substrate or towards the substrate in planes parallel to the substrate surface (horizontal slicing). The thickness of each slice depends on the local slope of the structure and hence a uniform layer thickness may lead to a large number of slices. A small slope requires thinner slices, while steep slopes require fewer slices. For an efficient writing it is beneficial to divide a structure into subregions and each region is sliced with a layer thickness depending on the slope characteristics of each region. This subregional slicing reduces substantially the number of required slices and hence the fabrication time. The structural data of a shape can be obtained in different ways, according to the features of the designed model. Although the data represent the same shape, the fabricated structures may differ regarding the quality of the structure (Park *et al.*, 2005).

Especially for larger structures it is advantageous to write only the shell of the structure, since exposing the whole volume would drastically increase the processing time. After removal of the unexposed material, the inner part of the structure may be UV cured by flood exposure, which solidifies the whole structure. A stress in the material may lead to fracture of the thin shell and the liquid resin inside the shell runs out, which leads to a destroyed structure. Writing at increased laser power leads to a larger voxel and hence reinforced shell but also to a reduced resolution. Therefore, multipath scanning is proposed for a shell reinforcement of the shell without losing spatial resolution and physical strength enhancement of the structure in order to avoid collapsing. In multipath scanning an offsetting contour to the original contour is obtained by translating the shell contour a distance towards its inside (Yang *et al.*, 2007).

Among other rapid prototyping (RP) techniques that directly create a physical structure from a CAD model, 3D-LL shows a few unique benefits over other RP methods like micro-stereolithography or 3D printing. Due to the dependence of the energy transfer and hence the polymerization rate on the squared laser intensity, a much higher spatial resolution can be achieved. Since the structure is inscribed into the volume of the polymer material, there is no need for a layer-by-layer buildup of the 3D structure, hence there are nearly no geometrical limitations on the shape and the vertical (along the laser beam axis) resolution is not limited by the layer thickness of applied material as in stereolithography. Additional optics, which modify the transmitted laser beam profile, facilitate shaping of the focal volume of the laser beam in order to reduce the axial elongation of the beam even at weaker focusing with low numerical aperture objectives. It was shown that cylindrical telescopes (Cerullo *et al.*, 2002), anamorphic beam shaping (Osellame *et al.*, 2004) or a slit aperture (Ams *et al.*, 2005) in the beam path are suitable to create a rather spherical volume instead of an elongated focal volume. This is especially important for the fabrication of symmetric multimode waveguides with a diameter of tens of microns (Schmidt *et al.*, 2007a).

6.6.2 3D-LL based fabrication of photonic structures

3D-LL has proven to be suitable for the fabrication of diverse optical and photonic structures, which are presented in the following section. A clear advantage over other fabrication method is the true 3D writing, which facilitates the generation of monolithic optical components in a single exposure step and the three-dimensional integration of such structures in micro-systems with very small dimensions.

Optical waveguides

Nonlinear interaction between the tightly focused laser beam and the material enables the deposition of energy in hot plasma of free electrons leading to a permanent change of the refractive index due to structural material modifications. 3D-LL based on two-photon absorption has been used for direct inscription of waveguides in glasses (Osellame *et al.*, 2004), in organic–inorganic hybrid materials (Schmidt *et al.*, 2007a, 2007b), in a PDMS matrix (Infuehr *et al.*, 2007) and modified porous silica films (Krivec *et al.*, 2010). Additionally, direct laser writing is suited for repairing polymer waveguides in specially functionalized polymer materials that increase locally the refractive index upon laser heating. It enables the control over the power splitting ratio in waveguide branches via laser-induced change of the waveguide width at the splitter or fanout junction (Srisanit *et al.*, 2005).

The possibility to change the refractive index of a material by exposure to intense laser pulses motivates the direct inscription of waveguides into transparent materials, whereas the change of the refractive index might be positive (increase of refractive index) or negative (decrease of refractive index), depending on the target material (Krol *et al.*, 2004).

The illumination of semiconductor nanocrystal doped glasses with fs laser pulses induced a fairly high permanent change of the refractive index ($1.8 \cdot 10^{-2}$) as compared to undoped glasses, which circumvents a usually observed drawback of directly written embedded photonic structures (Martinez-Vazquez *et al.*, 2007).

It is possible to inscribe the waveguide in a longitudinal and in a transverse writing scheme (Fig. 6.23). In the longitudinal setup, the laser beam is focused in the material and the waveguide axis is orientated along the propagation axis of the laser beam. In contrast, in the transverse setup, the writing direction of the waveguide is perpendicular to the laser beam propagation axis. Although the longitudinal writing yields symmetric waveguides, the length of the waveguide is limited by the focal length of the focusing optic. Transverse writing is more flexible regarding the waveguide shape and the focal length of the optics limits only the inscription depth of the waveguide. Hence, transverse writing is the common technique to fabricate waveguides by means of laser fabrication. Unfortunately, without additional optical means, simple focusing of the laser beam in transverse writing suffers from the strong asymmetric shape of the focal volume in lateral and axial direction with respect to the beam propagation direction and hence very asymmetric waveguide cross-sections, which are disadvantageous for light coupling.

Introducing an asymmetry into the laser beam cross-section by astigmatically shaped beams (Osellame *et al.*, 2004) or slit beam shaping (Ams *et al.*, 2005) before focusing removes the asymmetry of the waveguide

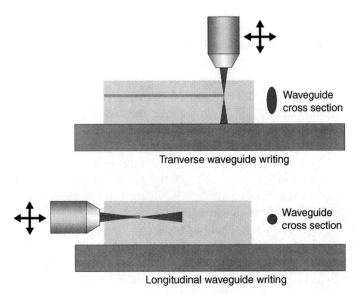

6.23 Longitudinal and transverse waveguide writing scheme.

cross-section in the transverse writing scheme. This kind of beam shaping by a cylindrical telescope allows control of both cross-sectional shape and size of directly inscribed waveguides, because of control of the independent and individual beam waist in two orthogonal directions and the introduction of an astigmatic shift of the corresponding focal planes (Cerullo *et al.*, 2002).

In contrast to the common concept of out-of-plane coupling of light into a planar waveguide based on a 90° deflection of the light via 45° mirrors or the use of immersed secondary emitters (fluorescent molecules) into waveguides (Steindorfer *et al.*, 2010), 3D-LL facilitates the direct integration of photonic structures into printed circuit boards (PCBs) by means of *in situ* fabrication and alignment of optical interconnects (Fig. 6.24) (Langer and Riester, 2007). Such boards are capable of transmitting signals at a data rate of more than 7 Gbits/s at a bit error ratio of 10^{-9} (Schmid *et al.*, 2009). Recently, PCBs with a more complicated optical interconnection system were fabricated that had two optical interconnects crossing each other at a distance of several tens of microns but did not intersect each other. A data rate of up to 8.25 Gbit/s at a bit error ratio of 10^{-9} was achieved. It was found that the data signal between the optical interconnects did not interfere and no optical cross-talk occurred. It was also shown that in contrast to an electrical signal line the optical data line was not disturbed by electrical noise that was added to the data signal (Langer *et al.*, 2011).

The waveguide alignment becomes an intrinsic part of the fabrication process. Provided that there is a suitable material that shows a sufficiently large

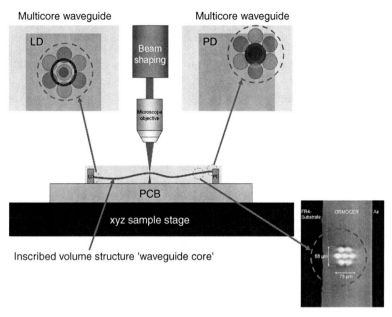

6.24 In situ coupling concept of laser-written embedded multicore optical interconnects.

refractive index change upon laser irradiation such as specially designed acrylate-based ORMOCER® (Houbertz *et al.*, 2010), working optical interconnections were fabricated and integrated into an optical layer of a multilayer PCB (Schmidt *et al.*, 2007a, 2007b). Optical interconnects exhibit a larger data transmission bandwidth and are robust against electrical noise. Their use is advantageous for high data rate applications, where a reliable connection even over long distances is required.

The fabrication of waveguides embedded in flexible matrix such as PDMS is based on selective photo-polymerization of monomers, which are swollen by photoreactive monomer formulations prior to the laser writing process. The reactive monomers contain acrylates or vinyl compounds with higher refractive index than PDMS. The monomer can be polymerized by two-photon absorption and the uncured monomer can be removed by evaporation at elevated temperatures leading to a local change of the refractive index (Satzinger *et al.*, 2008).

The optical loss of a waveguide is a key parameter of the waveguide quality. Non-uniformity of the induced refractive index change may lead to increased optical losses of laser-fabricated waveguides. By splitting a single laser pulse into a pair of pulses at a delay in the femto- to picosecond range, the optical loss may be significantly reduced as compared to a single pulse exposure. This is due to a pre-conditioning of the optical material caused by a partial melting or softening of the material in the focal volume (Nagata *et al.*, 2005).

Direct femtosecond laser-written waveguides were realized in freshly deposited thin chalcogenide glass films (Zoubir *et al.*, 2004). The inscription happened at laser intensities below the ablation threshold and induced an increase of the refractive index at the waveguide location. Raman spectra revealed permanent structural change of the glass matrix due to modifications in the molecular arrangement of the film upon femtosecond laser irradiation.

A fast and cheap approach to waveguide fabrication is the laser-induced densification of Ti-doped SiO_2 sol–gel films using a continuous wave Ytterbium fiber laser at 1070 nm and subsequent etching in order to remove unexposed material. Laser irradiated areas exhibit a lower etch rate. Due to internal stress, the laser power is limited to a certain fabrication window, whereas the laser power has influence on the width of the fabricated waveguides (Li *et al.*, 2011).

Micro-optics

The fabrication of micro-optical features on non-planar surfaces facilitates novel hybrid optical elements. The challenging technical implementation requires the development of special laser lithographic equipment with auto-focusing and tilting options for substrate mount and beam delivery. The more complex scan motion and the mapping of the design data onto the curved substrate enables the fabrication of diffractive optical elements on top of biconvex lenses for imaging corrections (Radtke and Zeitner, 2007), or a micro-lens array on a concave lens, which can be combined with pinholes on a convex surface forming a spherical compound eye objective (Radtke et al., 2007).

In addition to directly inscribed waveguides, 3D-LL has been applied to the fabrication of polymer-based micro-lenses and micro-pyramids (Satzinger *et al.*, 2008). Such structures represent optical features that can be directly integrated into a system or be replicated by imprint lithography via a PDMS mold. 3D-LL is capable of processing micro-optics on top of devices such as organic light emitting diodes or photo detectors for specially designed light coupling features and radiation characteristics. Femtosecond-based laser lithography is also capable of fabricating optical, fluidic, electronic and plasmonic elements in photosensitive glasses (Foturan). Even the fabrication of complex integrated systems with mirrors and a chamber containing a laser dye is reported (Cheng *et al.*, 2008).

Micro-lenses in high-index GeO_2–SiO_2 glass films are fabricated by a combined process of laser writing and pattern transfer via plasma etching. For this purpose, a resist coats the glass film. The 3D shape of the lens is inscribed into the volume of the resist and subsequently transferred to the glass film with a CHF_3 and O_2 plasma treatment, yielding high-index glass lenses (Nishiyama *et al.*, 2009). With the same approach it is possible to fabricate optical structures on non-planar inorganic substrates. It is demonstrated that a diffractive Fresnel lens can be processed on top of a refractive micro-lens. The combined process in yields a diffractive–refractive hybrid lens (Mizoshiri *et al.*, 2009).

Gratings

Photonic structures such as gratings can be rapidly fabricated by laser interference lithography, where multiple laser beams are overlapped in a photosensitive material. The spatial intensity distribution of the interfering beams is translated into a physical structure of the photoresist. Upon removal of the unexposed material, a photonic relief structure is obtained. This method is well suited for the generation of large area defect-free periodic structures, but does not allow individual shaping of single grating lines. However, this is possible by direct laser writing of the lines, which facilitates ultimate freedom in designing the gratings and the fabrication of embedded gratings in semiconductor nanocrystal doped glasses (Martinez-Vazquez *et al.*, 2007). Upon appropriate selection of the voxel overlap, structures with a periodicity smaller than the voxel diameter can be written, which was demonstrated with two-photon polymerization of ORMOCER® for the fabrication of a distributed feedback resonator with a grating period of 400 nm and a modulation depth of 40 nm for an optically pumped organic laser based on Alq3:DCM gain medium (Woggon *et al.*, 2009). Additionally by changing voxel overlap and design, it is possible to combine gratings with different grating constants into a single hybrid grating structure, which could be used for color-separating purposes of transmitted light. Upon fs laser exposure (790 nm, 130 fs), embedded diffraction gratings in BK-7 glass are directly inscribed. The high peak intensity of the focused laser beam generates a plasma, which leads to a graded bulk index modification ranging from 2×10^{-3} up to 1.5×10^{-2}. The dimensions of the plasma formation are between 400 nm and 4 µm (Park *et al.*, 2011).

The interference of fs laser pulses in a two-photon sensitive material is a hybrid structuring method that takes advantage of the periodic intensity modulation of laser interference lithography and the 3D features of two-photon-based 3D-LL. The optical arrangement requires the splitting of the laser beam and the introduction of a delay line for one laser pulse in order to control the spatial and temporal overlapping of the short pulses within the bulk material. The two-photon interaction between laser and material facilitates the inscription of a structure inside the bulk material and induces a local refractive index modulation. This yields embedded phase structures such as linear gratings without a surface relief (Guo *et al.*, 2003). The interference of fs laser pulses shows a pulse width dependence of formed thin film surface relief Au–Cr gratings. Shorter pulse duration (25 fs) is preferable in terms of fabrication quality, since shorter pulses slightly above the ablation threshold exhibit smaller ablation fringes and hence higher spatial resolution (Wang *et al.*, 2009).

Phase lenses and diffractive optical elements (DOEs)

Diffractive phase lenses with up to eight levels were fabricated by two-photon-based laser lithography (Chen *et al.*, 2007). In contrast to Fresnel zone plates, which have alternating transparent and opaque zones, a phase lens is completely transparent. The alternating zones are defined either by different refractive index or step heights, which introduce the required phase shift for focusing transmitted light. Therefore, phase lenses offer usually a much higher efficiency than Fresnel zone plates. Depending on the number of levels the performance of phase lenses changes.

Alternatively to waveguides, DOEs have been designed and used for planar photonic devices and optical interchip interconnections of large-scale integration chips and proved to be suitable for a high degree of integration (Takamori *et al.*, 2003).

The maskless approach of 3D-LL is ideally suited for the fabrication of DOE with binary levels or continuous grey-level encoding (256 phase levels) (Jia *et al.*, 2007). From optical simulation, a phase transmission is generated that produces the desired intensity distribution in the image plane behind the DOE. From this data, a bitmap file, which represents the height profile pattern, is generated. The vectorized bitmap file data are used for controlling the sample motion of the apparatus (Fig. 6.25).

Fabrication of photonic crystals

A photonic crystal is the optical analogy to a crystal lattice, where atoms or molecules are periodically arranged and the periodic potential introduces gaps into the energy band structure of the crystal. Due to Bragg-like diffraction, electrons are forbidden to propagate with certain energies through the crystal and at a sufficiently strong potential the gap extends to all possible directions and a complete bandgap is built. In a photonic crystal, the periodic potential is replaced by the geometrical periodic arrangement of dielectric media. This leads to similar phenomena for photons in such a crystal as for electrons in a lattice. The optical phenomena are due to scattering at the periodic interfaces of the media that constitute the crystal. The optical properties of photonic crystal structures are influenced by the lattice period and structure and the effective refractive index, which is also modified by external agents. Thus, photonic crystals are used in various types of sensor

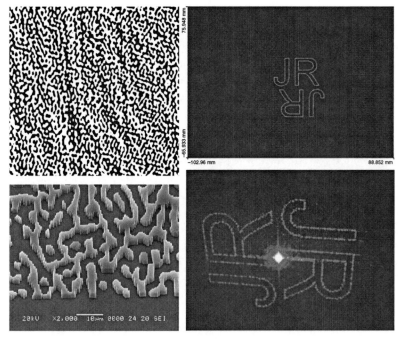

6.25 Top left: binary structural data of DOE; top right: visual representation of simulation result; bottom left: SEM image of fabricated DOE structure; bottom right: transmission of fabricated binary DOE.

applications, such as temperature sensors, humidity sensors, gas sensors, oil sensors, chemical sensors, biosensors, etc. A comprehensive overview of photonic crystal applications in sensorics is given in Nair and Vijaya (2010). Provided that the dielectric constants of the media differ sufficiently from each other and the absorption of the media is negligible, such a photonic crystal exhibits a photonic band structure that defines allowed and forbidden propagation states of photons within the volume of the crystal. A comprehensive treatment of photonic crystals can be found in Joannopoulos *et al.* (2008). Fabrication methods and applications of photonic crystals are presented in Thylén *et al.* (2004). It is evident that the influence of periodic dielectric media on light propagation enables an advanced use of light. Photonic crystal films, particles, fibers and photonic crystals in optofluidics facilitate novel applications not only in photonics, but also in bioassays (Zhao *et al.*, 2010). Hence, suitable and powerful fabrication and replication methods are required. Photonic crystals could be beneficial for applications involving light emitting diodes and waveguides, where localization of light and modification of radiation behavior (Florescu *et al.*, 2010) are required. A 3D photonic structure on a Si wafer, which is compatible with materials and standard micro-electronics fabrication methods, was realized more than 10 years ago (Lin *et al.*, 1998). The 3D woodpile structure is made of poly-crystalline Si and showed a transmission stop band in the infrared (10–14.5 µm) with a strong attenuation of light in this spectral region. Such a structure demonstrates the possibility to fabricate large-scale Si photonic crystals acting, for example, as filters in devices such as Si waveguides or photo detectors. Although the above-mentioned fabrication method is compatible with standard micro-electronics methods and takes advantage of the high refractive index of Si, it is a complex process to achieve the vertical topology of the 3D lattice structure. It is built by the repetitive deposition and etching of multiple dielectric films. For this reason, other methods for building complex 3D shapes were investigated. A rather straightforward approach to the fabrication of photonic crystals is direct inscription by ultrafast laser pulses (direct laser writing – DLW). Again, the benefits of a nonlinear interaction based on two- or multiphoton absorption between the laser and the dielectric material enables the direct 3D structuring of the material inside its volume. The 3D scanning of the laser focus through the bulk of the material and the associated modification of the material (photodegradation or photo-polymerization, etc.), which changes locally the refractive index and/or the solubility, create directly a photonic bandgap structure. Subsequent development removes either the exposed (positive-tone photoresist) or the unexposed (negative-tone photoresist) material. Important features of the lithographic method for the fabrication of photonic crystals are (a) good long-range periodic ordering; (b) intentionally engineered non-periodic defects; (c) high dielectric contrast; and (d) controllable feature size. DLW based on two-photon photo-polymerization has proved suitable for the fabrication of photonic crystal lattices with submicron features and arbitrary geometry (Kaneko *et al.*, 2003). DLW is well suited for the fabrication of long-range periodic structures in photoresists (although the writing time may become an issue) and the introduction of structural defects in the periodic arrangement. Conventional photoresists with a refractive index around $n = 1.5$ suffer from low index contrast upon exposure and hence a complete bandgap is generally not achieved. Therefore, either higher-index organic, inorganic or hybrid materials are required that still can be structured by means of DLW. Other approaches such as atomic layer deposition or chemical vapor deposition use the DLW fabricated resist structures as 3D templates for transforming these structures into materials that are not suited for direct processing by DLW or structures that require a higher refractive index contrast than provided from the polymer structures. Alternative methods include inversion (Hermatschweiler *et al.*, 2007) or double-inversion (Tétreault *et al.*, 2006) of the templates with another high-index material such as Si. DLW written polymer templates are coated by a thin SiO_2 film. After infiltration with silicon, etch of the SiO_2 layer and calcination of the polymer, an inverse silicon structure is obtained. These advanced methods yield structures with a complete near-IR photonic bandgap or chiral polymeric photonic crystals, helical photonic crystals with polarization stop bands, and near-IR 3D photonic quasi-crystals. These types of structures and related fabrication methods are comprehensively reviewed in von Freymann *et al.* (2010).

Recently, metallic structures were fabricated from polymer templates by means of chemical-vapor deposition of Ag (Rill *et al.*, 2008) or electroplated Au filling (Gansel *et al.*, 2009).

Another approach towards the generation of metallic structures is the direct inscription of metallic structures in a polymer matrix, which is reported in Shukla *et al.* (2010). A two-photon-based *in situ* reduction of an Au salt and photo-polymerization of SU-8 produces gold nanoparticle doped polymer lines. The high loading of the film with the gold precursor yielded high quality metallic structures consisting of approximately 10 nm Au nanoparticles.

The direct fabrication of 3D, high-index-contrast woodpile nanostructures by DLW in chalcogenide glasses has shown a complete bandgap (Wong *et al.*, 2006). The high index of the material required a partial compensation of optical aberrations for the writing in non-index-matched materials and a highly selective etchant for the accurate fabrication of the structure.

Another fabrication method of photonic crystals is based on the interference of multiple laser beams, also referred to as holographic fabrication. Multiple laser beams, which are aligned at certain angles of incidence, create a complex interference pattern, which exposes a photosensitive material. A comprehensive survey of nano-fabrication of photonic crystal structures can be found in the literature (Xia *et al.*, 2010). Three-dimensional holographic patterning can be achieved also with a phase mask and a single laser beam. The phase mask creates multiple beams of a single incident beam that are phase locked to each other. Thus, the experimental setup is simpler as compared to a multiple beam setup and the sample quality is generally improved. Photonic woodpile structures with 3D periodicity were fabricated in a single beam 1D phase mask dual-exposure experiment, where the phase mask had been rotated and translated between the two exposure steps (Xu *et al.*, 2010).

Furthermore, it is known that the introduction of controlled defects into photonic crystals is necessary in order to add functionality to the structures, which is comparable to that of dopants in semiconductors (cf. Braun *et al.*, 2006).

Holographic lithography based on multiple beam interference can be combined with direct two-photon laser writing to produce photonic crystals containing precisely localized truly 3D structural modifications. This process has sufficient resolution to create embedded waveguide and micro-cavity structures by selectively modifying individual unit cells in the bulk of a 3D photonic crystal (Scrimgeour *et al.*, 2006).

2D and 3D periodic structures with submicron features can be fabricated over large areas by interfering two laser beams with a wavelength in the absorption band of the photoresist, whereas the angle between the two beams facilitates the control over the periodic arrangement of the structure. 3D structures can be fabricated in a layer-by-layer approach, whereas different exposure conditions (line width of interference pattern, rotation of the sample for hexagonal patterns) for each layer facilitate a periodic but complex architecture (Lasagni *et al.*, 2010).

Multiple beams for interference lithography can be generated by a conformal phase mask, which is fabricated by an imprinted phase-shift structure on top of a photoresist. This so-called proximity-field nanopatterning (PnP) takes advantage of diffraction from the phase mask and generates a 3D intensity distribution in the underlying photoresist. Advantages of PnP over interference lithography with multiple beam is the single beam exposure of a single self-aligned diffractive structure and reduced requirements on the coherence of the laser due to the proximity of the photoresist to the conformal mask. Hence, this method is also compatible with two-photon-based exposure of thick polymer films (Jeon *et al.*, 2006). Two-photon-based exposure provides an increased range of 3D structure geometries and generates better images in the photoresist due to the squared intensity dependence of energy transfer from the laser to the material (George *et al.*, 2009).

A review of PBG fabrication methods can be found in the literature (Lopez, 2003). Self-assembly for photonic structures has proven suitable (Galisteo-López *et al.*, 2010). Pulsed fs laser irradiation resulted in self-organized formation of surface nano-gratings in soda-lime glass with a periodicity well below the laser wavelength, whereas the periodicity and orientation of the grating depended on the number of laser pulses and the laser polarization (Ahsan *et al.*, 2011). However, self-assembly of colloidal nanoparticles may lead to unwanted defects such as missing particles or dislocations. Multiphoton polymerization can be used for direct pattern generation inside a 3D colloidal crystal (Lee *et al.*, 2002; Pruzinsky and Braun, 2005). The pattern represents defect sites within the colloidal crystal. This method is considered less time consuming since the whole crystal does not need to be fabricated by multiphoton polymerization and the colloidal neighborhood stabilizes the inscribed defects.

Photonic crystals (PC) or photonic bandgap (PBG) structures can be processed in a wide range of materials by DLW (Deubel *et al.*, 2004, 2006). They can be fabricated via optical damage in inorganic glasses, organic glasses, two-photon solidification in photo-curing resins with woodpile (Deubel *et al.*, 2004) or spiral (Thiel *et al.*, 2007) architectures. Another method is holographic lithography by multiple-beam interference for the generation of periodic light intensity patterns in 2D and 3D. Here, a photosensitive material is exposed to the interference pattern of multiple beams and subsequent silicon replication. In this context, shrinkage of the polymeric templates is found to be the most prominent deviation and requires pre-compensation (Meisel *et al.*, 2006). Beside the low refractive index contrast, the deformation of photonic structures due to polymer shrinkage is considered mainly as contributing to the absence of a photonic bandgap in polymer structures. Hence a shape pre-compensation during the structure design is required and showed impact on the photonic bandgap for logpile structures (Sun *et al.*, 2004) and photonic structure with a unit cell resembling the diamond lattice (Kaneko *et al.*, 2003).

DLW facilitates the direct fabrication of 3D–2D–3D photonic crystal heterostructures (Deubel *et al.*, 2006), which consist of a sandwiched 2D waveguiding structure between a top and bottom cladding that is represented by a 3D PBG structure. The top and bottom PBG structure prevents optical out-of-plane losses of the 2D structure. DLW fabricates a polymeric template, which is replicated via silicon double inversion in order to get a corresponding high-index structure (Tétreault *et al.*, 2006) with 3D square spiral architecture and a photonic stop gap centered at 2.5 µm (Seet *et al.*, 2008).

Circular dichroism (different absorption for left and right circularly polarized light) is known from optically active chiral molecules and can be observed in dielectric chiral structures, which are, regarding losses, superior to metallic chiral structures. Similar to woodpile structures, layer-by-layer 3D chiral photonic crystals can be fabricated by DLW and work as thin film optical isolators (Thiel *et al.*, 2007).

The threshold behavior of TPP can be exploited for structuring without the need of a shutter for the laser. At a given laser pulse energy the exposure dose and hence the effect of photo-polymerization can be controlled by the scan velocity of the laser focus with respect to the sample. First, the feature size can be controlled within the fabrication window of the material by adapting the scan velocity and sub-diffraction limited features can be fabricated. Second, above a certain velocity the exposure dose is insufficient to initiate photo-polymerization and discontinuous features can be written by simply changing the scan speed from below to above the threshold velocity (material and velocity dependent shutter mechanism) (Teh *et al.*, 2004). This approach avoids the use of blocking the laser and simplifies a continuous scan strategy.

6.7   Conclusions and future trends

6.7.1   Conclusions

Since its invention in 1960, the laser proved its capability as a superior light source for scientific research and industrial applications. The laser provides a unique coverage of intensity (from highest power for laser fusion processes and material processing to lowest powers for laser therapy or opto-electronic components) and wavelength ranges (from deep UV to IR) due to the existence of manifold gain media, combined with properties such as long coherence length and polarization. Many modes of operation exist, from continuous wave to shortest pulses via Q-switching or mode-locking, and as a consequence of the manifold features of the laser there are manifold macroscopic and microscopic applications, such as applications in material science, photonics, µ-fluidics, medical applications, biotechnology, and as described in this chapter as a versatile lithographic tool.

The improvements in laser technology moved many applications from research labs into industrial and commercial applications. The laser proved compatible with established tools for optical lithography and pushed essentially optical lithography to lower and lower limits in terms of optical resolution. Laser interferometric lithography or a direct writing laser pen, combined with sophisticated scanning and masking methods, achieved 3D structuring, which is most important for the growing complexity of micro- and nano-systems. Owing to the achievable high intensity, nonlinear material aspects came into the focus of research and material development supported the establishment of laser writing methods based on the absorption of multiple photons. This method is also capable of fabricating 3D structures but without the use of masks and hence a rather simple optical setup. The three-dimensional nature is an intrinsic property of the laser writing process, which is accompanied by diffraction-unlimited spatial resolution. With increasing efficiency of photosensitive materials, the writing process is increasingly faster.

Nowadays, there are tools for 3D-LL for automated processing available that are equipped with many technical features such as auto-focus and automated tilt correction and work with transparent and opaque substrates. This technological maturity promises a bright future to direct laser writing methods and does not require preliminary steps for setting the laser focus location as reported in Park *et al.* (2006b). The realization of novel ideas with laser rapid prototyping is rather straightforward and simple with suitable laser tools. This holds true for nanoscale rapid prototyping, which is important for photonics. The promising combination of laser fabrication with other nano-processing widens the fields of applications and may set some future trends.

6.7.2   Future trends

Since DLW is a process that generates one structure after the other, highly efficient two-photon materials are a prerequisite for increasing the processing speed. Faster writing and parallel writing with multiple foci becomes possible and makes two-photon-based 3D-LL attractive to industrial or commercial applications. The combination with soft-lithography (PDMS molding) and the development of new materials that expand today's available photosensitive materials is considered another step towards increasing technological exploitation of this tool.

Sophisticated optical setups applying more than one laser beam are about to be explored. A second annular shaped laser beam, which is used for photo-depletion/deactivation is guided collinearly to the writing laser beam and increases the lateral resolution of the DLW process, where a $\lambda/20$ resolution is achieved (Li *et al.*, 2009). NIR or visible light-based nano-lithography that reaches the same resolution targets (~ 30 nm) as compared to next generation lithography is very promising (Fourkas, 2010).

In order to increase writing speed and decrease the costs of the time consuming sequential writing process, micro-lens arrays can be used for parallel writing of identical structures. Provided that the laser delivers enough power, the laser beam can be expanded and focused on the sample with a micro-lens array (MLA). MLAs as big as 1 cm^2 with micro-lenses at a period of 50 μm and a focal length in the μm range in order to produce numerous focused micro beams for parallel writing were used for the fabrication of phase masks (Huang *et al.*, 2010). Such phase masks may be used in other lithographic methods: intricate polymer surface relief masks were fabricated by means of holographic lithography applying a phase mask. The structures were replicated by soft lithography and subsequently used as holographic elements for single laser beam exposure for the fabrication of 2D and 3D photonic elements (Goldenberg *et al.*, 2010).

The direct laser printing of pixels useful for organic lighting or displays seems very interesting for future fabrication. The use of laser writing in a reel-to-reel tool would be a pathway to flexible and large-scale fabrication.

The increasing computational power of desktop PCs is advantageous for optical simulations and the more and more sophisticated technical implementation of laser lithography is a path towards optical systems of growing complexity, which makes light the most versatile means of material processing and information technology.

6.8   References

Ahsan Md, S., Kim, Y. G. and Lee, M. S. (2011), 'Formation mechanism of nanostructures in soda–lime glass using femtosecond laser', *Journal of Non-Crystalline Solids*, **357**, 851–857.

Ams, M., Marshall, G. D., Spence, D. J. and Withford, M. J. (2005), 'Slit beam shaping method for femtosecond laser direct-write fabrication of symmetric waveguides in bulk glasses', *Optics Express*, **13**(15), 5676–5681.

Anscombe, N. (2010), 'Direct laser writing', *Nature Photonics Technology Focus*, **4**, 22–23.

Arpin, K. A., Mihi, A., Johnson, H. T., Baca, A. J., Rogers, J. A., Lewis, J. A. and Braun, P. V. (2010), 'Multidimensional architectures for functional optical devices', *Advanced Materials*, **22**, 1084–1101.

Basting, D., Pippert, K. and Stamm, U. (2002), 'History and future prospects of excimer laser technology', RIKEN Review No. 43, LPM2001: Focused on *2nd International Symposium on Laser Precision Microfabrication*, May 16–18, 2001, Data Storage Institute, National University of Singapore, Singapore.

Bäuerle, D. (2008), *Laser: Grundlagen und Anwendungen in Photonik, Technik, Medizin und Kunst*. Weinheim: Wiley-VCH Verlag.

Bettiol, A. A., Ansari, K., Sum, T. C., van Kan, J. A. and Watt, F. (2004), 'Fabrication of micro-optical components in polymer using proton beam writing', *Proceedings of SPIE*, **5347**, 255–263.

Blanche, P. A., Kippelen, B., Schülzgen, A., Fuentes-Hernandez, C., Ramos-Ortiz, G., Wang, J. F., Hendrickx, E. and Peyghambarian, N. (2002), 'Photorefractive polymers sensitized by two-photon absorption', *Optics Letters*, **27**(1), 19–21.

Borek, G., Weissbrodt, P., Schrenk, M. and Cumme, M. (2007), 'Challenging micro-optical applications demand diverse manufacturing solutions', *Proceedings of SPIE*, **6462**, 64620X 1–11.

Boyd, R. W. (2003), *Nonlinear optics*, 2nd edn. Burlington, MA: Academic Press.

Boyle, M., Neumeister, A., Kiyan, R., Reinhardt, C., Stute, U., Chichkov, B., Wohlleben, W. and Leyrer, R. J. (2007), 'Production of 3D photonic components with ultrafast micromachining', *Proceedings of SPIE*, **6462**, 646212 1–9.

Braun, A., Zimmer, K., Hösselbarth, B., Meinhardt, J., Bigl, F. and Mehnert, R. (1998), 'Excimer laser micromachining and replication of 3D optical surfaces', *Applied Surface Science*, **127–129**, 911–914.

Braun, P. V., Rinne, S. A. and García-Santamaría, F. (2006), 'Introducing defects in 3D photonic crystals: State of the art', *Advanced Materials*, **18**, 2665–2678.

Busch, K., von Freymann, G., Linden, S., Mingaleev, S. F., Tkeshelashvili, L. and Wegener, M. (2007), 'Periodic nanostructures for photonics', *Physics Reports*, **444**, 101–202.

Byun, I. and Kim, J. (2010), 'Cost-effective laser interference lithography using a 405 nm AlInGaN semiconductor laser', *Journal of Micromechanics and Microengineering*, **20**, 1–6.

Callegari, V. (2009), 'Fabrication of photonic elements by focused ion beam (FIB)', dissertation, ETH No. 18558, Zürich.

Cerullo, G., Osellame, R., Taccheo, S., Marangoni, M., Polli, D., Ramponi, R., Laporta, P. and De Silvestri, S. (2002), 'Femtosecond micromachining of symmetric waveguides at 1.5 μm by astigmatic beam focusing', *Optics Letters*, **27**(21), 1938–1940.

Chen, Q.–D., Wu, D., Niu, L.–G., Wang, J., Lin, X.–F., Xia, H. and Sun, H.–B. (2007), 'Phase lenses and mirrors created by laser micronanofabrication via two-photon photo-topolymerization', *Applied Physics Letters*, **91**, 171105.

Chen, Y.–P., Chiu, H.–C., Chen, G.–Y., Chiang, C.–H., Tseng, C.–T., Lee, C.–H. and Wang, L. A. (2010), 'Fabrication and measurement of large-area sub-wavelength structures with broadband and wide-angle antireflection effect', *Microelectronic Engineering*, **87**, 1323–1327.

Cheng, Y., Xu, Z., Xu, J., Sugjioka, K. and Midorikawa, K. (2008), 'Three-dimensional femtosecond laser integration in glasses', *The Review of Laser Engineering*, **36**, 1206–1209.

Chiu, C.-C. and Lee, Y.-C. (2011), 'Fabrication of aspheric micro-lens array by excimer laser micromachining', *Optics and Lasers in Engineering*, **49**, 1232–1237.

Chong, T. C., Hong, M. H. and Shi, L. P. (2010), 'Laser precision engineering: From micro-fabrication to nanoprocessing', *Laser & Photonics Reviews*, **4**(1), 123–143.

Cui, Z., Du, J. and Guo, Y. (2003), 'Overview of greyscale photolithography for micro-optical elements fabrication', *Proceedings of SPIE*, **4984**, 111.

Daniel, C. (2006), 'Biomimetic structures for mechanical applications by interfering laser beams: More than solely holographic gratings', *Journal of Materials Research*, **21**(8), 2098–2105.

Debaes, C., Van Erps, J., Vervaeke, M., Volckaerts, B., Ottevaere, H., Gomez, V., Vynck, P., Desmet, L., Krajewski, R., Ishii, Y., Hermanne, A. and Thienpont, H. (2006), 'Deep proton writing: A rapid prototyping polymer micro-fabrication tool for micro-optical modules', *New Journal of Physics*, **8**, 270–288.

Della Giustina, G., Zacco, G., Zanchetta, E., Gugliemi, M., Romanato, F. and Brusatin, G. (2011), 'Interferential lithography of Bragg gratings on hybrid organic–inorganic sol–gel materials', *Microelectronic Engineering*, **88**, 1923–1926.

Desbiens, J.-P. and Masson, P. (2007), 'ArF excimer laser micromachining of Pyrex, SiC and PZT for rapid prototyping of MEMS components', *Sensors and Actuators A*, **136**, 554–563.

Deubel, M., von Freymann, G., Wegener, M., Pereira, S., Busch, K. and Soukoulis, C. M. (2004), 'Direct laser writing of three-dimensional hotonic-crystal templates for tele-communications', *Nature Materials*, **3**, 444–447.

Deubel, M., Wegener, M., Linden, S., von Freymann, G. and John, S. (2006), '3D–2D–3D photonic crystal heterostructures fabricated by direct laser writing', *Optics Letters*, **31**(6), 805–807.

Dyer, P. E., Farley, R. J., Giedl, R. and Karnakis, D. M. (1996), 'Excimer laser ablation of polymers and glasses for grating fabrication', *Applied Surface Science*, **96–98**, 537–549.

Elliott, D. J. and Ferranti, D. C. (1989), 'Sub-micron lithography at 248nm and 193nm excimer laser wavelengths', *Microelectronic Engineering*, **9**, 59–63.

Ellman, M., Rodríguez, A., Pérez, N., Echeverria, M., Verevkin, Y. K., Peng, C. S., Berthou, T., Wang, Z., Olaizola, S. M. and Ayerdi, I. (2009), 'High-power laser interference lithography process on photoresist: Effect of laser fluence and polarisation', *Applied Surface Science*, **255**, 5537–5541.

Ergin, T., Stenger, N., Brenner, P., Pendry, J. B. and Wegener, M. (2010), 'Three-dimensional invisibility cloak at optical wavelengths', *Science*, **328**, 337–339.

Farrer, R. A., LaFratta, C. N., Li, L., Praino, J., Naughton, M. J., Saleh, B. E. A., Teich, M. C. and Fourkas, J. T. (2006), 'Selective functionalization of 3-D polymer microstructures', *Journal of the American Chemical Society*, **128**, 1796–1797.

Farsari, M., Vamvakaki, M. and Chichkov, B. N. (2010), 'Multiphoton polymerization of hybrid materials', *Journal of Optics*, **12**, 124001.

Fay, B. (2002), 'Advanced optical lithography development, from UV to EUV', *Microelectronic Engineering*, **61–62**, 11–24.

Fischer, J., von Freymann, G. and Wegener, M. (2010), 'The materials challenge in diffraction-unlimited direct-laser-writing optical lithography', *Advanced Materials*, **22**, 3578–3582.

Florescu, M., Busch, K. and Dowling, J. P. (2007), 'Thermal radiation in photonic crystals', *Physical Review B*, **75**, 201101 R.

Flury, M., Benatmane, A., Gérard, P., Montgomery, P. C., Fontaine, J., Engel, T., Schunck, J. P. and Fogarassy, E. (2003), 'Excimer laser ablation lithography applied to the fabrication of reflective diffractive optics', *Applied Surface Science*, **208–209**, 238–244.

Fourkas, J. T. (2010), 'Nanoscale photolithography with visible light', *Physical Chemistry Letters*, **1**, 1221–1227.

Fourkas, J. T. and Baldacchini, T. (2004), 'Three-dimensional nanofabrication using multiphoton absorption', *Dekker Encyclopedia of Nanoscience and Nanotechnology*, DOI: 10.1081/E-ENN 120021631.

Fritze, M., Tyrrell, B. M., Astolfi, D. K., Lambert, R. D., Yost, D.-R. W., Forte, A. R., Cann, S. G. and Wheeler, B. D. (2003), 'Subwavelength optical lithography with phase-shift photomasks', *Lincoln Laboratory Journal*, **14**(2), 237–250.

Fu, Y. and Ngoi, B. K. A. (2001), 'Investigation of diffractive-refractive microlens array fabricated by focused ion beam technology', *Optical Engineering*, **40**, 511–516.

Galisteo-López, J. F., Ibisate, M., Sapienza, R., Froufe-Pérez, L. S., Blanco, Á. and López, C. (2010), 'Self-assembled photonic structures', *Advanced Materials*, **20**, 1–40.

Gansel, J. K., Thiel, M., Rill, M. S., Decker, M., Bade, K., Saile, V., von Freymann, G., Linden, S. and Wegener, M. (2009), 'Gold helix photonic metamaterial as broadband circular polarizer', *Science*, **325**, 1513–1515.

Gattass, R. R. and Mazur, E. (2008), 'Femtosecond laser micromachining in transparent materials', *Nature Photonics*, **2**, 219–225.

Gebeshuber, I. C., Belegratis, M. and Schmidt, V. (2010), 'Emerging nanopatterning methods'. In Guston, D. and Golson, J. G. (eds.), *Encyclopedia of nanoscience and society*. California: Sage Publications, pp. 184–185.

Geisler, S., Bauer, J., Haak, U., Stolarek, D., Schulz, K., Wolf, H., Meier, W., Trojahn, M., Matthus, E., Beyer, H., Old, G., Marschmeyer, S. and Kuck, B. (2008), 'Double exposure technology for KrF lithography', *EMLC 2008 – 24th European Mask and Lithography Conference*, VDE VERLAG, Berlin Offenbach, pp. 62–70.

George, M. C., Nelson, E. C., Rogers, J. A. and Braun, P. V. (2009), 'Direct fabrication of 3D periodic inorganic microstructures using conformal phase masks', *Angewandte Chemie*, **121**, 150–154.

Gläbe, R. and Riemer, O. (2010), 'Diamond machining of micro-optical components and structures', *Proceedings of SPIE*, **7716**, 771602-1.

Goldenberg, L. M., Gritsai, Y., Sakhno, O., Kulikovska, O. and Stumpe, J. (2010), 'All-optical fabrication of 2D and 3D photonic structures using a single polymer phase mask', *Journal of Optics*, **12**, 1–7.

Goodman, J. W. (1996), *Introduction to Fourier optics*, 2nd edn. New York: McGraw-Hill.

Göppert-Mayer, M. (1931), 'Über Elementarakte mit zwei Quantensprüngen', *Annalen der Physik*, **401**(3), 273–294.

Guo, H., Jiang, H., Luo, L., Wu, C., Guo, H., Wang, X., Yang, H., Gong, Q., Wu, F., Wang, T. and Shi, M. (2003), 'Two-photon polymerization of gratings by interference of a femtosecond laser pulse', *Chemical Physics Letters*, **374**, 381–384.

Guo, R., Xiao, S., Zhai, X., Li, J., Xia, A. and Huang, W. (2006), 'Micro lens fabrication by means of femtosecond two photon photopolymerization', *Optics Express*, **14**(2), 810–816.

Guo, X. W. and Dong, Q. (2010), 'Rapid fabrication of micro optical elements using DMD-based maskless lithography technique', *Advanced in Materials Research*, **146–147**, 143–146.

Heller, C., Pucher, N., Seidl, B., Kalinyaprak-Icten, K., Ullrich, G., Kuna, L., Satzinger, V., Schmidt, V., Lichtenegger, H. C., Stampfl, J. and Liska, R. (2007), 'One- and two-photon activity of cross-conjugated photoinitiators with bathochromic shift', *Journal of Polymer Science: Part A: Polymer Chemistry*, **45**, 3280–3291.

Hermatschweiler, M., Ledermann, A., Ozin, G. A., Wegener, M. and von Freymann, G. (2007), 'Fabrication of silicon inverse woodpile photonic crystals', *Advanced Functional Materials*, **17**, 2273–2277.

Hirai, Y., Inamoto, Y., Sugano, K., Tsuchiya, T. and Tabata, O. (2007), 'Moving-mask UV lithography for 3-dimensional positive-and negative-tone thick photoresist microstructuring', Solid-State Sensors, Actuators and Microsystems Conference, 2007. *Transducers*, 545–548. DOI:10.1109/SENSOR.2007.4300188.

Hon, K. K. B., Li, L. and Hutchings, I. M. (2008), 'Direct writing technology – Advances and developments', *CIRP Annals – Manufacturing Technology*, **57**, 601–620.

Houbertz, R. (2005), 'Laser interaction in sol–gel based materials – 3-D lithography for photonic applications', *Applied Surface Science*, **247**, 504–512.

Houbertz, R., Domann, G., Cronauer, C., Schmitt, A., Martin, H., Park, J.-U., Fröhlich, L., Buestrich, R., Popall, M., Streppel, U., Dannberg, P., Wächter, C. and Bräuer, A. (2003), 'Inorganic–organic hybrid materials for application in optical devices', *Thin Solid Films*, **442**, 194–200.

Houbertz, R., Steenhusen, S., Stichel, T. and Sextl, G. (2010), 'Two-photon polymerization of inorganic-organic hybrid polymers as scalable technology using ultra-short laser pulses'. In Duarte, F. J. (ed.), *Coherence and Ultrashort Pulse Laser Emission*. Rijeka, Croatia: Intech, pp. 583–608.

Huang, Z., Lin, Q. Y. and Hong, M. (2010), 'Phase shift mask fabrication by laser microlens array lithography for periodic nanostructures patterning', *JLMN – Journal of Laser Micro/Nanoengineering*, **5**(3), 233–237.

Ihlemann, J. and Rubahn, K. (2000), 'Excimer laser micro machining: Fabrication and applications of dielectric masks', *Applied Surface Science*, **154–155**, 587–592.

Infuehr, R., Pucher, N., Heller, C., Lichtenegger, H., Liska, R., Schmidt, V., Kuna, L., Haase, A. and Stampfl, J. (2007), 'Functional polymers by two-photon 3D lithography', *Applied Surface Science*, **254**, 836–840.

Jahns, J., Cao, Q. and Sinzinger, S. (2008), 'Micro- and nanooptics – An overview', *Laser Photonics Review*, **2**(4), 249–263.

Jang, H. S., Kim, G. H., Lee, J. and Choi, K. B. (2010), 'Eliminating the undercut phenomenon in interference lithography for the fabrication of nano-imprint lithography stamp', *Current Applied Physics*, **10**, 1436–1441.

Jang, J.-H., Ullal, C. K., Maldovan, M., Gorishnyy, T., Kooi, S., Koh, C. Y. and Thomas, E. L. (2007), '3D micro- and nanostructures via interference lithography', *Advanced Functional Materials*, **17**, 3027–3041.

Jeon, S., Malyarchuk, V., Rogers, J. A. and Wiederrecht, G. P. (2006), 'Fabricating three dimensional nanostructures using two photon lithography in a single exposure step', *Optics Express*, **14**(6), 2300–2308.

Jia, B., Serbin, J., Kim, H., Lee, B., Li, J. and Gu, M. (2007), 'Use of two-photon polymerization for continuous gray-level encoding of diffractive optical elements', *Applied Physics Letters*, **90**, 073503.

Joannopoulos, J. D., Johnson, S. G., Meade, R. D. and Winn, J. N. (2008), *Photonic crystals: Molding the flow of light*. Princeton: Princeton University Press.

Kaiser, W. and Garrett, C. G. B. (1961), 'Two-photon excitation in CaF2:Eu2+', *Physical Review Letters*, **7**, 229–231.

Kaneko, K., Sun, H.-B., Duan, X.-M. and Kawata, S. (2003), 'Submicron diamond-lattice photonic crystals produced by two-photon laser nanofabrication', *Applied Physics Letters*, **83**(11), 2091–2093.

Kazansky, P. G., Qiu, J., Shimotsuma, Y., Bricchi, E. and Hirao, K. (2004), 'Femtosecond laser nano-structuring of transparent materials', *Proceedings of SPIE*, **5399**, 88–95.

Kim, K.-R., Jeong, H.-W., Lee, K.-S., Yi, J., Yoo, J.-C., Cho, M.-W., Cho, S.-H. and Choi, B. (2011), 'Rapid laser fabrication of microlens array using colorless liquid photopolymer for AMOLED devices', *Optics Communications*, **284**, 405–410.

Kim, S. H., Lee, K.-D., Kim, J.-Y., Kwon, M.-K. and Park, S.-J. (2007), 'Fabrication of photonic crystal structures on light emitting diodes by nanoimprint lithography', *Nanotechnology*, **18**, 55306.

Kim, S.-K., Oh, H.-K., Jung, Y.-D. and An, I. (2011), 'Advanced lithography simulation for various 3-dimensional nano/microstructuring fabrications in positive- and negative-tone photoresists', *Journal of Nanoscience and Nanotechnology*, **11**, 528–532.

Klein, F., Striebel, T., Fischer, J., Jiang, Z., Franz, C., von Freymann, G., Wegener, M. and Bastmeyer, M. (2010), 'Elastic fully three-dimensional microstructure scaffolds for cell force measurements', *Advanced Materials*, **22**, 868–871.

Kondo, T., Yamasaki, K., Juodkazis, S., Matsuo, S., Mizeikis, V. and Misawa, H. (2004), 'Three-dimensional microfabrication by femtosecond pulses in dielectrics', *Thin Solid Films*, **453–454**, 550–556.

Kong, H. J., Yi, S. W., Yang, D.-Y., Lee, K.-S., Kim, J.-B., Lim, T.-W. and Kim, S. (2007), 'Pulse-width dependency of the fabricating resolution of the two-photon absorption photo-polymerization', *Proceedings of SPIE*, **6462**, 646202-1.

Kordás, K., Pap, A. E., Lyöri, V., Uusimäki, A., Vähäkangas, J. and Leppävuori, S. (2002), 'Mirror fabrication on optical fibres using maskless excimer laser-assisted methods', *Surface and Coatings Technology*, **155**, 285–288.

Krivec, S., Matsko, N., Satzinger, V., Pucher, N. U., Galler, N., Koch, T., Schmidt, V., Grogger, W., Liska, R. and Lichtenegger, H. (2010), 'Silica-based, organically modified host material for waveguide structuring by two-photon-induced photopolymerization', *Advanced Functional Materials*, **20**, 1–9.

Krol, D. M., Chan, J. W., Huser, T. R., Risbud, S. H. and Hayden, J. S. (2004), 'Fs laser fabrication of photonic structures in glass: The role of glass composition', *Proceedings of SPIE*, **5662**, 30–39.

Kunz, T., Stebani, J., Ihlemann, J. and Wokaun, A. (1998), 'Photoablation andmicrostructuring of polyestercarbonates and their blends with a XeCl excimer laser', *Applied Physics A*, **67**, 347–352.

LaFratta, C. N., Baldacchini, T., Farrer, R. A., Fourkas, J. T., Teich, M. C., Saleh, B. E. A. and Naughton, M. J. (2004), 'Replication of two-photon-polymerized structures with extremely high aspect ratios and large overhangs', *Journal of Physical Chemistry B*, **108**, 11256–11258.

LaFratta, C. N., Fourkas, J. T., Baldacchini, T. and Farrer, R. A. (2007), 'Multiphoton fabrication', *Angewandte Chemie International Edition*, **46**, 6238–6258.

Langer, G. and Riester, M. (2007), 'Two-photon absorption for the realization of optical waveguides on printed circuit boards', *Proceedings of SPIE*, **6475**, 64750X-1.

Langer, G., Satzinger, V., Schmidt, V., Schmid, G. and Leeb, W. R. (2011), 'PCB with fully integrated optical interconnects', *Proceedings of SPIE*, **7944**, 794408, DOI:10.1117/12.873744.

Lasagni, A. F. and Menéndez-Ormaza, B. S. (2010), 'Two- and three-dimensional micro- and sub-micrometer periodic structures using two-beam laser interference lithography', *Advanced Engineering Materials*,**12**(1–2), 54–60.

Lasagni, A. F., Hendricks, J. L., Shaw, C. M., Yuan, D., Martin, D. C. and Das, S. (2009), 'Direct laser interference patterning of poly3, 4-ethylene dioxythiophene-polystyrene sulfonate (PEDOT:PSS) thin films', *Applied Surface Science*, **255**, 9186–9192.

Lasagni, A. F., Roch, T., Langheinrich, D., Bieda, M. and Wetzig, A. (2011), 'Large area direct fabrication of periodic arrays using interference patterning', *Physics Procedia*, **12**, 214–220.

Lee, K.-S., Kim, R. H., Yang, D.-Y. and Park, S. H. (2008), 'Advances in 3D nano/microfabrication using two-photon initiated polymerization', *Progress in Polymer Science*, **33**, 631–681.

Lee, S.-K., Park, H. S., Yi, G.-R., Moon, J. H. and Yang, S.-M. (2009), 'Holographic fabrication of microstructures with internal nanopatterns using microprism arrays', *Angewandte Chemie*, **121**, 7134–7139.

Lee, W., Pruzinsky, S. A. and Braun, P. V. (2002), 'Multi-photon polymerization of wave-guide structures within three-dimensional photonic crystals', *Advanced Materials*, **14**, 271–274.

Li, A., Wang, Z., Liu, J. and Zeng, X. (2011), 'Low cost fabrication of SiO_2 optical wave-guides by laser direct writing on Ti-doped sol–gel films', *Optics and Lasers in Engineering*, **49**, 351–355.

Li, L., Gattass, R. R., Gershgoren, E., Hwang, H. and Fourkas, J. T. (2009), 'Achieving /20 resolution by one-color initiation and deactivation of polymerization', *Science*, **324**, 910–913.

Lim, T. W., Son, Y., Yang, D.-Y., Kong, H.-J., Lee, K.-S. and Park, S. H. (2008), 'Highly effective three-dimensional large-scale microfabrication using a continuous scanning method', *Applied Physics A*, **92**, 541–545.

Lin, C., Wang, I., Bouriau, M., Casalegno, R., Andraud, C. and Baldeck, P. L. (2004), 'Two-photon induced polymerization of photo-driven microsensors', *Proceedings of SPIE*, **5516**, 52–62.

Lin, S. Y., Fleming, J. G., Hetherington, D. L., Smith, B. K., Biswas, R., Ho, K. M., Sigalas, M. M., Zubrzycki, W., Kurtz, S. R. and Bur, J. (1998), 'A three-dimensional photonic crystal operating at infrared wavelengths', *Nature*, **394**, 251–253.

Lippert, T. (2009), 'UV laser ablation of polymers: From structuring to thin film deposition'. In Miotello, A. and Ossi, P. M. (eds.), *Laser–surface interactions for new materials production tailoring structure and properties*, Springer Series in Material Chemistry Vol. 130. Berlin: Springer Verlag, pp. 141–175.

Liska, R., Schuster, M., Inführ, R., Turecek, C., Fritscher, C., Seidl, B., Schmidt, V., Kuna, L., Haase, A., Varga, F., Lichtenegger, H. and Stampfl, J. (2007), 'Photopolymers for rapid prototyping', *Journal of Coatings Technology and Research*, **4**(4), 505–510.

Liu, Y., Nolte, D. D. and Pyrak-Nolte, L. J. (2010), 'Large-format fabrication by two-photon polymerization in SU-8', *Applied Physics A*, **100**, 181–191.

Logunov, S., Dickinson, J., Grzybowski, R., Harvey, D. and Streltsov, A. (2011), 'Laser-induced swelling of transparent glasses', *Applied Surface Science*, **257**, 8883–8886.

Lopez, C. (2003), 'Materials aspects of photonic crystals', *Advanced Materials*, **15**(20), 1679–1704.

Lu, C., Zhou, J., Lipson, R. H. and Ding, Z. (2009), 'Simple method to fabricate large scale quantum dot architectures', *Materials Letters*, **63**, 563–565.

Maiden, A., McWilliam, R., Purvis, A., Johnson, S., Williams, G. L., Seed, N. L. and Ivey, P. A (2005), 'Nonplanar photolithography with computer-generated holograms', *Optics Letters*, **30**(11), 1300–1302.

Mait, J. N., Prather, D. W. and Mirotznik, M. S. (1998), 'Binary subwavelength diffractive-lens design', *Optics Letters*, **23**(17), 1343–1345.

Makarovsky, O., Kumar, S., Rastelli, A., Patanè, A., Eaves, L., Balanov, A. G., Schmidt, O. G., Campion, R. and Foxon, C. T. (2010), 'Direct laser writing of nanoscale light-emitting diodes', *Advanced Materials*, **22**, 3176–3180.

Malinauskas, M., Purlys, V., Rutkauskas, M., Gaidukeviciute, A. and Gadonas, R. (2010a), 'Femtosecond visible light induced two-photon photopolymerization for 3D micro/nanostructuring in photoresists and photopolymers', *Lithuanian Journal of Physics*, **50**(2), 201–207.

Malinauskas, M., Zukauskas, A., Bickauskaite, G., Gadonas, R. and Juodkazis, S. (2010b), 'Mechanisms of three-dimensional structuring of photo-polymers by tightly focussed femtosecond laser pulses', *Optics Express*, **18**(10), 10209–10221.

Marconi, M. C. and Wachulak, P. W. (2010), 'Extreme ultraviolet lithography with table top lasers', *Progress in Quantum Electronics*, **34**, 173–190.

Martinez-Vazquez, R., Osellame, R., Cerullo, G., Ramponi, R. and Svelto, O. (2007), 'Fabrication of photonic devices in nanostructured glasses by femtosecond laser pulses', *Optics Express*, **15**(20), 12628–12635.

Maruo, S. and Fourkas, J. T. (2008), 'Recent progress in multiphoton microfabrication', *Laser Photonics Review*, **2**(1–2), 100–111.

Maruo, S. and Kawata, S. (1998), 'Two-photon absorbed near-infrared photopolymerization for three-dimensional microfabrication', *Journal of Microelectromechanical Systems*, **7**(4), 411–415.

Maruo, S., Nakamura, O. and Kawata, S. (1997), 'Three-dimensional microfabrication with two-photon-absorbed photopolymerization', *Optic Letters*, **22**, 132–134.

Meisel, D. C., Diem, M., Deubel, M., Pérez-Willard, F., Linden, S., Gerthsen, D., Busch, K. and Wegener, M. (2006), 'Shrinkage precompensation of holographic three-dimensional photonic-crystal templates', *Advanced Materials*, **18**, 2964–2968.

Misawa, H. and Juodkazis, S. (eds.) (2006), *3D laser microfabrication: Principles and applications*. Weinheim: Wiley-VCH.

Mizoshiri, M., Nishiyama, H., Nishii, J. and Hirata, Y. (2009), 'Silica-based micro-structures on nonplanar substrates by femtosecond laser-induced nonlinear lithography', *Journal of Physics: Conference Series*, **165**, 012048. DOI: 10.1088/1742–6596/165/1/012048.

Mosher, L., Waits, C. M., Morgan, B. and Ghodssi, R. (2009), 'Double-exposure grayscale photolithography', *Journal of Microelectromechanical Systems*, **18**(2), 308–315.

Murray, A. K. and Dickinson, M. R. (2004), 'Tissue ablation-rate measurements with a long-pulsed, fibre-deliverable 308 nm excimer laser', *Lasers in Medical Science*, **19**, 127–138.

Mutapcic, E., Iovenitti, P. and Hayes, J. P. (2005), 'A 3D-CAM system for quick pro-totyping and microfabrication using excimer laser micromachining', *Microsystem Technologies*, **12**, 128–136.

Nagata, T., Kamata, M. and Obara, M. (2005), 'Optical waveguide fabrication with double pulse femtosecond lasers', *Applied Physics Letters*, **86**, 251103.

Nair, R. V. and Vijaya, R. (2010), 'Photonic crystal sensors: An overview', *Progress in Quantum Electronics*, **34**, 89–134.

Nakahama, T., Yokoyama, S., Miki, H. and Mashiko, S. (2006), 'Control of multiphoton process within diffraction limit space in polymer microstructures', *Thin Solid Films*, **499**(1–2), 406–409.

Neumeister, A., Himmelhuber, R., Materlik, C., Temme, T., Pape, F., Gatzen, H. and Ostendorf, A. (2008), 'Properties of three-dimensional precision objects fabricated by using laser based micro stereo lithography', *JLMN – Journal of Laser Micro/Nanoengineering*, **3**(2), 67–72.

Nguyen, L. H., Straub, M. and Gu, M. (2005), 'Acrylate-based photopolymer for two-photon microfabrication and photonic applications', *Advanced Functional Materials*, **15**(2), 209–216.

Nielson, R., Kaehr, B. and Shear, J. B. (2009), 'Microreplication and design of biological architectures using dynamic-mask multiphoton lithography', *Small*, **5**(1), 120–125.

Nieto, D., Flores-Arias, M. T/, O'Connor, G. M. and Gomez-Reino, C. (2010), 'Laser direct-write technique for fabricating microlens arrays on soda-lime glass with a Nd:YVO4 laser', *Applied Optics*, **49**, 4979–4983.

Nishiyama, H., Nishii, J., Mizoshiri, M. and Hirata, Y. (2009), 'Microlens arrays of high-refractive-index glass fabricated by femtosecond laser lithography', *Applied Surface Science*, **255**, 9750–9753.

Noh, J., Sohn, H., Suh, J., Shin, D. and Lee, J. (2007), 'Fabrication of a rainbow color logo (diffraction grating) using the picosecond laser', *Proceedings of the LPM 2007 – the 8th International Symposium on Laser Precision Microfabrication*, April 24–28, University of Vienna, Vienna, Austria.

Okoshi, M., Iyono, M., Inoue, N. and Yamashita, T. (2009), 'Photochemical welding of silica microspheres to silicone rubber by ArF excimer laser', *Applied Surface Science*, **255**, 9796–9799.

Osellame, R., Cerullo, G., Taccheo, S., Marangoni, M., Polli, D., Laporta, P. and Ramponi, R. (2004), 'Femtosecond laser writing of symmetrical optical waveguides by astigmatically shaped beams', *Proceedings of SPIE*, **5451**, 360.

Ostendorf, A. and Chichkov, B. N. (2006), 'Two-photon polymerization: A new approach to micromachining', *Photonics Spectra*, **40**, 72–79.

Othon, C. M., Laracuente, A., Ladouceur, H. D. and Ringeisen, B. R. (2008), 'Sub-micron parallel laser direct-write', *Applied Surface Science*, **255**, 3407–3413.

Overmeyer, L., Neumeister, A. and Kling, R. (2011), 'Direct precision manufacturing of three-dimensional components using organically modified ceramics', *CIRP Annals – Manufacturing Technology*, **60**, 267–270.

Ovsianikov, A., Chichkov, B., Mente, P. and Monteiro-Riviere, N. A. (2007a), 'Two photon polymerization of polymer–ceramic hybrid materials for transdermal drug delivery', *International Journal of Applied Ceramic Technology*, **4**(1), 22–29.

Ovsianikov, A., Ostendorf, A. and Chichkov, B. N. (2007b), 'Three-dimensional photofabrication with femtosecond lasers for applications in photonics and biomedicine', *Applied Surface Science*, **253**, 6599–6602.

Palla-Papavlu, A., Dinca, V., Luculescu, C., Shaw-Stewart, J., Nagel, M., Lippert, T. and Dinescu, M. (2010a), 'Laser induced forward transfer of soft materials', *Journal of Optics*, **12**, 124014.

Palla-Papavlu, A., Dinca, V., Paraico, I., Moldovan, A., Shaw-Stewart, J., Schneider, C. W., Kovacs, E., Lippert, T. and Dinescu, M. (2010b), 'Microfabrication of polystyrene microbead arrays by laser induced forward transfer', *Journal of Applied Physics*, **108**, 033111.

Papadopoulou, E. L., Axente, E., Magoulakis, E., Fotakis, C. and Loukakos, P. A. (2010), 'Laser induced forward transfer of metal oxides using femtosecond double pulses', *Applied Surface Science*, **257**, 508–511.

Park, J.-K., Cho, S.-H., Kim. K.-H. and Kang, M.-C. (2011), 'Optical diffraction gratings embedded in BK-7 glass by low-density plasma formation using femtosecond laser', *Transactions of Nonferrous Metals Society of China*, **21**, 165–169.

Park, S. H., Lee, S. H., Yang, D.-Y., Kong, H. J. and Lee, K.-S. (2005), 'Subregional slicing method to increase three-dimensional nanofabrication efficiency in two-photon polymerization', *Applied Physics Letters*, **87**, 154108.

Park, S.-H., Lim, T.-W., Yang, D.-Y., Jeong, J.-H., Kim, K.-D., Lee, K.-S. and Kong, H.-J. (2006a), 'Effective fabrication of three-dimensional nano/microstructures in a single step using multilayered stamp', *Applied Physics Letters*, **88**, 203105.

Park, S.-H., Lim, T.-W., Yang, D.-Y., Kong, H.-J., Kim, J.-Y. and Lee, K.-S. (2006b), 'Direct laser patterning on opaque substrate in two-photon polymerization', *Macromolecular Research*, **14**(2), 245–250.

Partel, S., Zoppel, S., Hudek, P., Bich, A., Vogler, U., Hornung, M. and Voelkel, R. (2010), 'Contact and proximity lithography using 193 nm Excimer laser in Mask Aligner', *Microelectronic Engineering*, **87**, 936–939.

Passinger, S., Saifullah, M. S. M., Reinhardt, C., Subramanian, K. R. V., Chichkov, B. N. and Welland, M. E. (2007), 'Direct 3D patterning of TiO_2 using femtosecond laser pulses', *Advanced Materials*, **19**, 1218–1221.

Pätzel, R., Bragin, I., Kleinschmidt, J., Rebhan, U. and Basting, D. (1996), 'Excimer laser with high repetition rate for DUV lithography', *Microelectronic Engineering*, **30**, 165–167.

Pfeiffer, R. W. (1997), 'Application of photopolymerization technology'. In Scranton, A. B., Bowman, C. N. and Pheiffer, R. W. (eds.), *Photopolymerization: Fundamentals and Application*. Washington, DC: American Chemical Society, Chapter 1.

Photonics21 (2010), *Lighting the Way Ahead: Photonics21 Strategic Research Agenda*, 2nd edn. Düsseldorf: European Technology Platform Photonics21.

Pruzinsky, S. A. and Braun, P. V. (2005), 'Fabrication and characterization of two-photon polymerized features in colloidal crystals', *Advanced Functional Materials*, **15**, 1995–2004.

Pucher, N. U., Rosspeintner, A., Satzinger, V., Schmidt, V., Gescheidt, G., Stampfl, J. and Liska, R. (2009), 'Structure–activity relationship in D-TT-A-TT-D-based photoinitiators for the two-photon-induced photopolymerization process', *Macromolecules*, **42**, 6519–6528.

Radtke, D. and Zeitner, U. D. (2007), 'Laser-lithography on non-planar surfaces', *Optics Express*, **15**(3), 1167–1174.

Radtke, D., Duparré, J., Zeitner, U. D. and Tünnermann, A. (2007), 'Laser lithographic fabrication and characterization of a spherical artificial compound eye', *Optics Express*, **15**(6), 3067–3077.

Rapp, L., Nénon, S., Alloncle, A. P., Videlot-Ackermann, C., Fages, F. and Delaporte, P. (2011), 'Multilayer laser printing for organic thin film transistors', *Applied Surface Science*, **257**, 5152–5155.

Reimer, K., Quenzer, H. J., Jürss, M. and Wagner, B. (1997), 'Micro-optic fabrication using one-level gray-tone lithography', *Proceedings of SPIE*, **3008**, 279–288.

Rhee, H.-G. (2010), 'Direct laser lithography and its applications'. In Wang, M (ed.), *Lithography*. Rijeka, Croatia: InTech, pp. 1–16.

Richter, J., Meinertz, J. and Ihlemann, J. (2011), 'Patterned laser annealing of silicon oxide films', *Applied Physics A*, **104**, 759–764.

Rill, M. S., Plet, C., Thiel, M., Staude, I., von Freymann, G., Linden, S. and Wegener, M. (2008), 'Photonic metamaterials by direct laser writing and silver chemical vapor deposition', *Nature Materials*, **7**, 543–546.

Rodriguez, A., Echeverría, M., Ellman, M., Perez, N., Verevkin, Y. K., Peng, C. S., Berthou, T., Wang, Z., Ayerdi, I., Savall, J. and Olaizola, S. M. (2009), 'Laser interference lithography for nanoscale structuring of materials: From laboratory to industry', *Microelectronic Engineering*, **86**, 937–940.

Rothschild, M. (2010), 'A roadmap for optical lithography', *OPN Optics & Photonics News*, **21**, 26–31.

Satzinger, V., Schmidt, V., Kuna, L., Palfinger, C., Inführ, R., Liska, R. and Krenn, J. R. (2008), 'Rapid prototyping of micro-optics on organic light emitting diodes and organic photo cells by means of two-photon 3D lithography and nano-imprint lithography', *Proceedings of SPIE*, **6992**, 699217-1.

Schmid, G., Leeb, W. R., Langer, G., Schmidt, V. and Houbertz, .R (2009), 'Gbit/s transmission via two-photon-absorption-inscribed optical waveguides on printed circuit boards', *Electronics Letters*, **45**(4), 219–221.

Schmidt, V., Kuna, L., Satzinger, V., Houbertz, R., Jakopic, G. and Leising, G. (2007a), 'Application of two-photon 3D lithography for the fabrication of embedded ORMOCER® waveguides', *Proceedings of SPIE*, **6476**, 64760P-1.

Schmidt, V., Kuna, L., Satzinger, V., Jakopic, G. and Leising, G. (2007b), 'Two-photon 3D lithography: A versatile fabrication method for complex 3D shapes and optical interconnects within the scope of innovative industrial applications', *JLMN – Journal of Laser Micro/Nanoengineering*, **2**(3), 170–177.

Schneider, C. W. and Lippert, T. (2010), 'Laser ablation and thin film deposition'. In Schaaf P. (ed.), *Laser Processing of Materials*. Springer Series in Material Science Vol. 139. Berlin: Springer Verlag, pp. 89–112.

Scrimgeour, J., Sharp, D. N., Blanford, C. F., Roche, O. M., Denning, R. G. and Turberfield, A. J. (2006), 'Three-dimensional optical lithography for photonic microstructures', *Advanced Materials*, **18**, 1557–1560.

Seet, K. K., Mizeikis, V., Kannari, K., Juodkazis, S., Misawa, H., Tétreault, N. and John, S. (2008), 'Templating and replication of spiral photonic crystals for silicon photonics', *IEEE Journal of Selected Topics in Quantum Electronics*, **14**(4), 1064–1073.

Serbin, J. and Gu, M. (2006), 'Superprism phenomena in waveguide-coupled woodpile structures fabricated by two-photon polymerization', *Optics Express*, **14**, 3563–3568.

Serbin, J., Ovsianikov, A. and Chichkov, B. (2004), 'Fabrication of woodpile structures by two-photon polymerization and investigation of their optical properties', *Optics Express*, **12**(21), 5221–5228.

Shaw Stewart, J., Lippert, T., Nagel, M., Nüesch, F. and Wokaun, A. (2010), 'Laser-induced forward transfer using triazene polymer dynamic releaser layer'. In Phipps, C. R. (ed.), *AIP Conference Proceedings 1278 – International Symposium on High Power Laser Ablation 2010*. New York: American Institute of Physics, pp. 789–799.

Shaw Stewart, J., Lippert, T., Nagel, M., Nüesch, F. and Wokaun, A. (2011), 'Laser-induced forward transfer of polymer light-emitting diode pixels with increased charge injection', *ACS Applied Material Interfaces*, **3**, 309–316.

Shin, D. S., Lee, J. H., Suh, J. and Kim, T. H. (2006), 'Correction of a coherent image during KrF excimer laser ablation using a mask projection', *Optics and Lasers in Engineering*, **44**, 615–622.

Shukla, S., Furlani, E. P., Vidal, X., Swihart, M. T. and Prasad, P. N. (2010), 'Two-photon lithography of sub-wavelength metallic structures in a polymer matrix', *Advanced Materials*, **22**, 3695–3699.

Srisanit, N., Liu, Z., Ke, X. and Wang, M. R. (2005), 'Laser writing correction of polymer waveguide fanouts', *Optics Communications*, **244**, 171–179.

Stampfl, J., Baudis, S., Heller, C., Liska, R., Neumeister, A., Kling, R., Ostendorf, A. and Spitzbart, M. (2008), 'Photopolymers with tunable mechanical properties processed by laser-based high-resolution stereolithography', *Journal of Micromechanics and Microengineering*, **18**, 125014.

Stampfl, J., Inführ, R., Stadlmann, K., Pucher, N., Schmidt, V. and Liska, R. (2009), 'Materials for the fabrication of optical waveguides with two photon photopolymerization', *Proceedings of the Fifth International WLT Conference on Lasers in Manufacturing*, Munich, June 2009.

Stankevicius, E., Malinauskas, M. and Raciukaitis, G. (2011), 'Fabrication of scaffolds and micro-lenses array in a negative photopolymer SZ2080 by multi-photon polymerization and four-femtosecond-beam interference', *Physics Procedia*, **12**, 82–88.

Steindorfer, M. A., Lamprecht, B., Schmidt, V., Abel, T., Mayr, T. and Krenn, J. R. (2010), 'Light coupling for integrated optical waveguide-based sensors', *Proceedings of SPIE*, **7726**, 77261S-1.

Steingrüber, R., Ferstl, M. and Pilz, W. (2001), 'Micro-optical elements fabricated by electron-beam lithography and dry etching technique using top conductive coatings', *Microelectronic Engineering*, **57–58**, 285–289.

Stratakis, E., Ranella, A., Farsari, M. and Fotakis, C. (2009), 'Laser-based micro/nanoengineering for biological applications', *Progress in Quantum Electronics*, **33**, 127–163.

Sum, T. C., Bettiol, A. A., Venugopal Rao, S., van Kan, J. A., Ramam, A. and Watt, F. (2004), 'Proton beam writing of passive polymer optical waveguides', *Proceedings of SPIE*, **5347**, 160.

Sun, H.-B. and Kawata, S. (2004), 'Two-photon photopolymerization and 3D lithographic microfabrication', *APS*, **170**, 169–273.

Sun, H.-B., Kawakami, T., Xu, Y., Ye, J.-Y., Matuso, S., Misawa, H., Miwa, M. and Kaneko, R. (2000), 'Real three-dimensional microstructures fabricated by photopolymerization of resins through two-photon absorption', *Optics Letters*, **25**(15), 1110–1112.

Sun, H.-B., Maeda, M., Takada, K., Chon, J. W. M., Gu, M. and Kawata, S. (2003), 'Experimental investigation of single voxels for laser nanofabrication via two-photon photopolymerization', *Applied Physics Letters*, **83**(5), 819–821.

Sun, H.-B., Suwa, T., Takada, K., Zaccaria, R. P., Kim, M.-S., Lee, K.-S. and Kawata, S. (2004), 'Shape precompensation in two-photon laser nanowriting of photonic lattices', *Applied Physics Letters*, **85**(17), 3708–3710.

Sun, H.-B., Tanaka, T. and Kawata, S. (2002), 'Three-dimensional focal spots related to two-photon excitation', *Applied Physics Letters*, **80**(20), 3673–3675.

Takada, K., Kaneko, K., Li, Y.-D., Kawata, S., Chen, Q.-D. and Sun, H.-B. (2008), 'Temperature effects on pinpoint photopolymerization and polymerized micronanostructures', *Applied Physics Letters*, **92**, 041902.

Takamori, T., Wada, H., Sasaki, H. and Kamijoh, T. (2003), 'Interchip optical interconnection using planar-type photonic circuit and optoelectronic integrated devices', *Electronics and Communications in Japan*, Part 2, **86**(1), 9–17.

Tan, D., Li, Y., Qi, F., Yang, H., Gong, Q., Dong, X. and Duan, X. (2007), 'Reduction in feature size of two-photon polymerization using SCR500', *Applied Physics Letters*, **90**, 071106.

Teh, W. H., Dürig, U., Drechsler, U., Smith, C. G. and Güntherodt, H.-J. (2005), 'Effect of low numerical-aperture femtosecond two-photon absorption on SU-8 resist for ultrahigh-aspect-ratio microstereolithography', *Journal of Applied Physics*, **97**, 054907.

Teh, W. H., Dürig, U., Salis, G., Harbers, R., Drechsler, U., Mahrt, R. F., Smith, C. G. and Güntherodt, H.-J. (2004), 'SU-8 for real three-dimensional subdiffraction-limit two-photon microfabrication', *Applied Physics Letters*, **84**(20), 4095–4097.

Tétreault, N., von Freymann, G., Deubel, M., Hermatschweiler, M., Pérez-Willard, F., John, S., Wegener, M. and Ozin, G. A. (2006), 'New route to three-dimensional photonic bandgap materials:silicon double inversion of polymer templates', *Advanced Materials*, **18**, 457–460.

Thiel, M., Fischer, J., von Freymann, G. and Wegener, M. (2010), 'Direct laser writing of three-dimensional submicron structures using a continuous-wave laser at 532nm', *Applied Physics Letters*, **97**, 221102.

Thiel, M., von Freymann, G. and Wegener, M. (2007), 'Layer-by-layer three-dimensional chiral photonic crystals', *Optics Letters*, 32(17), 2547–2549.

Thylén, L., Qiu, M. and Anand, S. (2004), 'Photonic crystals – A step towards integrated circuits for photonics', *ChemPhysChem*, **5**, 1268–1283.

Ting, C.-J., Chen, C.-F. and Chou, C. P. (2009), 'Subwavelength structures for broadband antireflection application', *Optics Communications*, **282**, 434–438.

Torgersen, J., Baudrimont, A., Pucher, N., Stadlmann, K., Cicha, K., Heller, C., Liska, R. and Stampfl, J. (2010), 'In vivo writing using two-photon-polymerization', *Proceedings of LPM2010 – The 11th International Symposium on Laser Precision Microfabrication*, June 7–10, 2010, Stuttgart, Germany.

Van Erps, J., Vervaeke, M., Debaes, C., Ottevaere, H., Van Overmeire, S., Hermanne, A. and Thienpont, H. (2010), 'Deep proton writing: A powerful rapid prototyping technology for various micro-optical components', *Proceedings of SPIE*, **7716**, 77160W-1.

Veiko, V. P., Kostyuk, G. K., Meshkovskii, I. K., Chuiko, V. A. and Yakovlev, E. B. (1986), 'Microoptic components formed by local modification of the structure of porous glasses', *Soviet Journal of Quantum Electronics*, **16**, 8.

Voelkel, R., Vogler, U., Bich, A., Pernet, P., Weible, K. J., Hornung, M., Zoberbier, R., Cullmann, E., Stuerzebecher, L., Harzendorf, T. and Zeitner, U. D. (2010), 'Advanced mask aligner lithography: New illumination system', *Optics Express*, **18**(20), 20968–20978.

von Freymann, G., Ledermann, A., Thiel, M., Staude, I., Essig, S., Busch, K. and Wegener, M. (2010), 'Three-dimensional nanostructures for photonics', *Advanced Functional Materials*, **20**, 1038–1052.

Vossmerbaeumer, U. (2010), 'Application principles of excimer lasers in ophthalmology', *Medical Laser Application*, **25**, 250–257.

Walker, E. and Rentzepis, P. M. (2008), 'Two-photon technology: A new dimension', *Nature Photonics*, **2**, 406–408.

Wang, J., Auyeung, R. C. Y., Kim, H., Charipar, N. A. and Piqué, A. (2010), 'Three-dimensional printing of interconnects by laser direct-write of silver nanopastes', *Advanced Materials*, **22**, 4462–4466.

Wang, Q., Zhang, Y. and Gao, D. (1996), 'Theoretical study on the fabrication of a microlens using the excimer laser chemical vapor deposition technique', *Thin Solid Films*, **287**, 243–246.

Wang, X., Chen, F., Liu, H., Liang, W., Yang, Q., Si, J. and Hou, X. (2009), 'Fabrication of micro-gratings on Au–Cr thin film by femtosecond laser interference with different pulse durations', *Applied Surface Science*, **255**, 8483–8487.

Wang, Z., Zhao, G., Zhang, X., Heguang, L. and Zhao, N. (2011), 'Fabrication of two-dimensional lattices by using photosensitive sol–gel and four-beam laser interference', *Journal of Non-crystalline Solids*, **357**, 1223–1227.

Wei, P., Li, N. and Feng, L. (2011), 'A type of two-photonmicrofabrication system and experimentations', *ISRN Mechanical Engineering*, 2011, Article ID 278095, DOI: 10.5402/2011/278095.

Winfield, R. J. and O'Brien, S. (2010), 'Two-photon polymerization of an epoxy–acrylate resin material system', *Applied Surface Science*, **257**, 5389–5392.

Winfield, R. J., Bhuian, B., O'Brien, S. and Crean, G. M. (2007), 'Refractive femtosecond laser beam shaping for two-photon polymerization', *Applied Physics Letters*, **90**, 111–115.

Winfield, R. J., Meister, M., Crean, G. M. and Paineau, S. (2000), 'Excimer laser fabrication of diffractive optical elements', *Materials Science in Semiconductor Processing*, **3**, 481–486.

Witzgall, G., Vrijen, R., Yablonovitch, E., Doan, V. and Schwartz, B. J. (1998), 'Single-shot two-photon exposure of commercial photoresist for the production of three-dimensional structures', *Optics Letters*, **23**(22), 1745–1747.

Woggon, T., Kleiner, T., Punke, M. and Lemmer, U. (2009), 'Nanostructuring of organic-inorganic hybrid materials for distributed feedback laser resonators by two-photon polymerization', *Optics Express*, **17**(4), 2500.

Wong, S., Deubel, M., Pérez-Willard, F., John, S., Ozin, G. A., Wegener, M. and von Freymann, G. (2006), 'Direct laser writing of three-dimensional photonic crystals with a complete photonic bandgap in chalcogenide glasses', *Advanced Materials*, **18**, 265–269.

Worgull, M., Schneider, M., Heilig, M., Kolew, A., Dinglreiter, H. and Mohr, J. (2010), 'Replication of optical components by hot embossing', *Proceedings of SPIE*, **7716**, 771604–1.

Wu, C.-Y., Shu, C.-W. and Yeh. Z.-C. (2006a), 'Effects of excimer laser illumination on microdrilling into an oblique polymer surface', *Optics and Lasers in Engineering*, **44**, 842–857.

Wu, P., Dunn, B., Yablonovitch, E., Doan, V. and Schwartz, B. J. (1999), 'Two-photon exposure of photographic film', *Journal of the Optical Society of America B*, **16**(4), 605–608.

Wu, R., Zheng, Z., Li, H. and Liu, X. (2011), 'Freeform lens for off-axis illumination in optical lithography system', *Optics Communications*, **284**, 2662–2667.

Wu, S., Serbin, J. and Gu, M. (2006b), 'Two-photon polymerisation for three-dimensional micro-fabrication', *Journal of Photochemistry and Photobiology A: Chemistry*, **181**, 1–11.

Wyrowski, F. and Schimmel, H. (2006), 'Elektromagnetisches Optikrechen – eine Einführung', *Sonderdruck aus Photonik*, **38**, 54–57.

Wyrowski, F. and Schimmel, H. (2007), 'Elektromagnetisches Optikrechnen – Lichtausbreitung von rogoros bis geometrisch-optisch', *Sonderdruck aus Photonik*, **39**, 54–57.

Xia, D., Ku, Z., Lee, S. C. and Brueck, S. R. J. (2010), 'Nanostructures and functional materials fabricated by interferometric lithography', *Advanced Materials*, **20**, 1–33.

Xie, Q., Hong, M. H., Tan, H. L., Chen, G. X., Shi, L. P. and Chong, T. C. (2008), 'Fabrication of nanostructures with laser interference lithography', *Journal of Alloys and Compounds*, **449**, 261–264.

Xu, D., Chen, K. P., Ohlinger, K. and Lin, Y. (2010), 'Holographic fabrication of three-dimensional woodpile-type photonic crystal templates using phase mask technique'. In Kim, K. Y. (ed.), *Recent optical and photonic technologies*, Rijeka, Croatia: Intech, pp. 71–88.

Yamakawa, S., Amaya, K., Gelbart, D., Urano, T. and Lemire-Elmore, J. (2004), 'Development of three-dimensional microfabrication method using thermo-sensitive resin', *Applied Physics B*, **79**, 507–511.

Yang, D.-Y., Park, S. H., Lim, T. W., Kong, H.-J., Yi, S. W., Yang, H. K. and Lee, K.-S. (2007), 'Ultraprecise microreproduction of a three-dimensional artistic sculpture by multipath scanning method in two-photon photopolymerization', *Applied Physics Letters*, **90**, 013113.

Yang, H.-K., Kim, M.-S., Kang, S.-W., Kim, K.-S., Lee, K.-S., Park, S. H., Yang, D.-Y., Kong, J., Sun, H.-B., Kawata, S. and Fleitz, P. (2004), 'Recent progress of lithographic microfabrication by the TPA induced photopolymerization', *Journal of Photopolymer Science and Technology*, **17**(3), 385–392.

Yi, S. W. and Kong, H. J. (2007), 'Microfabricated 3-D polymeric structure with SU-8', *Proceedings of SPIE*, **6462**, 646206-1.

Yi, S. W., Lee, S. K., Kong, H. J., Yang, D.-Y., Park, S.-H., Lim, T.-W., Kim, R. H. and Lee, K. S. (2004a), 'Three-dimensional micro-fabrication using two-photon absorption by femtosecond laser', *Proceedings of SPIE*, **5342**, 137–45.

Yi, S. W., Lee, S. K., Cho, M. J., Kong, H. J., Kim, R. H. and Lee, K. S. (2004b), 'Fabrication of PDMS poly-dimethyl siloxane molding and 3D structure by two-photon absorption induced by an ultra fast laser', *Proceedings of SPIE*, **5641**, 227–37.

Yi, S. W., Lee, S. K., Cho, M. J., Kong, H. J., Kim, R. H. and Lee, K. S. (2005), 'Fabrication of 3D micro-structure and analysis of voxel generation by ultra fast laser-induced two-photon absorption', *Proceedings of SPIE*, **5715**, 118.

Yoshino, M., Umeda, H., Tsushima, H., Watanabe, H., Tanaka, S., Matsumoto, S., Onose, T., Nogawa, H., Kawasuji, Y., Matsunaga, T., Fujimoto, J. and Mizoguchi, H. (2010), 'Flexible and reliable high power injection locked laser for double exposure and double patterning ArF immersion lithography', *Proceedings of SPIE*, **7640**, 76402A.

Zhang, X., Liu, S. and Liu, Y. (2006), 'Fabrication of large-area 3D photonic crystals using a holographic optical element', *Optics and Lasers in Engineering*, **44**, 903–911.

Zhang, Y.-L., Chen, Q.-D., Xia, H. and Sun, H.-B. (2010), 'Designable 3D nanofabrication by femtosecond laser direct writing', *Nano Today*, **5**, 435–448.

Zhao, Y., Zhao, X. and Gu, Z. (2010), 'Photonic crystals in bioassays', *Advanced Functional Materials*, **20**, 2970–2988.

Zhou, M., Yang, H. F., Kong, J. J., Yan, F. and Cai, L. (2008), 'Study on the microfabrication technique by femtosecond laser two-photon photopolymerization', *Journal of Materials Processing Technology*, **200**, 158–162.

Zimmer, K. and Böhme, R. (2005), 'Precise etching of fused silica for refractive and diffractive micro-optical applications', *Optics and Lasers in Engineering*, **43**, 1349–1360.

Zimmer, K., Braun, A. and Bigl, F. (2000), 'Combination of different processing methods for the fabrication of 3D polymer structures by excimer laser machining', *Applied Surface Science*, **154–155**, 601–604.

Zimmer, K., Hirsch, D. and Bigl, F. (1996), ,Excimer laser machining for the fabrication of analogous microstructures', *Applied Surface Science*, **96–98**, 425–429.

Zoubir, A., Richardson, M., Rivero, C., Schulte, A., Lopez, C., Richardson, K., Hô, N. and Vallée, R. (2004), 'Direct femtosecond laser writing of waveguides in As2S3 thin films', *Optics Letters*, **29**(7), 748–50.

Laser-induced soft matter organization and microstructuring of photonic materials

L. ATHANASEKOS, University of Patras, Greece and
National Hellenic Research Foundation, Greece, S. PISPAS,
National Hellenic Research Foundation, Greece and
N. A. VAINOS, University of Patras, Greece and
National Hellenic Research Foundation, Greece

Abstract: Laser radiation forces applied in fully transparent, highly entangled semi-dilute polymer solutions generate freestanding, three-dimensional, micro- and, potentially, nano-solids. The underlying phenomena are attributed to a synergy of effects involving the radiation forces exerted by milliwatt laser beams on polymer chains and the entanglement of macromolecules. Most importantly, since the primary stages of formation, the incident optical field is structured and guided by the induced microstructures. This self-confinement enhances the effect and results in great compression of the material, osmotic solvent extraction and, eventually, materials solidification in free space. Structural reversibility verifies the absence of any chemical modification of the material. These innovative concepts are demonstrated through the fabrication of microstructures, including among others plasmonic and fluorescent semiconductor quantum–dot hybrid structures, as well as polymer fibers also drawn by laser radiation forces. The phenomenology of the involved effects is plausibly explained here and further research will resolve the fundamental aspects and lead the way forward to new and emerging concepts for future microfabrication technologies.

Key words: photonic structures, laser radiation forces, materials organization, microstructuring, microfabrication.

7.1 Introduction

Optical radiation exerts forces on matter and leads to the remarkable effects of optical trapping and organization pointing to applications in materials, information technology and biomedical sciences. Originating from the concept of optical trapping in the microscale (Neuman and Block 2004; Stevenson *et al.*, 2010), the present chapter highlights current advances in assembly by laser-induced soft-matter organization. These emerging concepts provide alternative tools for micro- and potentially nano-fabrication and offer a new platform for fundamental investigations and new applications in photonics (Dholakia and Zemanek, 2010). Earlier investigations

238

in this field (Loppinet *et al.*, 2005; Sigel *et al.*, 2002) had shown outstanding results concerning the organization of entangled soft matter upon illumination by a laser beam. These effects in liquid phase could not be fully explained at that stage. Later attempts investigated the formation of optical spatial solitons (Anyfantakis, 2008) and the effect of long time irradiation in the structure formation in polydiene solutions (Anyfantakis, 2010). The chapter presents an overview of laser-induced manipulation in entangled polymer solutions and introduces novel microfabrication applications.

In Section 7.2, the origins and physics of laser radiation forces are reviewed, highlighting the effects of optical forces applied in soft matter. In Section 7.3, polymer material dynamics aspects are discussed introducing polymer entanglement and reptation phenomena. In Section 7.4 of the chapter we overview the formation of microstructures due to radiation forces in several media, such as in bulk solutions, thin films, microfabrication in free space and fiber-drawing. The use of holographic outputs as an aiding tool for surface manipulation and concurrent multiple structure formation and the formation of plasmonic and quantum dot hybrid structures are also discussed. The chapter is summarized in Section 7.5 with the salient ideas of the chapter and comments on the emerging trends of laser-induced soft material organization and microstructuring. In the appendix, a thorough presentation on the material synthesis and characterization is given emphasizing the hybrid material synthesis.

7.2   The origin of radiation forces

Albeit that radiation pressure forces stem directly from Maxwell's equation solutions for electric fields (Maxwell, 1873), it was many years later when a targeted attempt to handle and apply those forces on matter was reported by Ashkin (1970). In particular, the manipulation and trapping of particles by light radiation forces was first proposed by Ashkin (1970). Since then, significant work on the field has been done, evolving the concept of matter organization with the aid of optical tools (Ashkin 1978, 1992; Ashkin *et al.*, 1986; Dienerowitz *et al.*, 2008; Jonas and Zemanek, 2008).

Let us consider a beam propagating in a transparent medium. Once the beam faces an 'obstacle', for example a polymer micelle or a metal nanoparticle in a solution, it is scattered and changes direction (Fig. 7.1). Such a change gives rise to a momentum transfer from the beam to the obstacle. Considering the photon approximation for the light beam, it is obvious that each photon momentum is transferred to the object. According to Newton's Second Law, we get:

$$F = \frac{\Delta p}{\Delta t} \qquad\qquad [7.1]$$

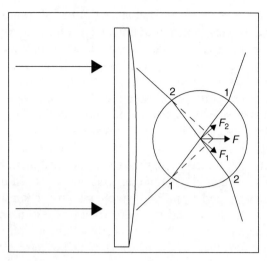

7.1 Schematic representation of optical forces applied in a transparent medium due to refraction.

that is, the rate of momentum change, associated with a change in momentum flux of the beam. It is understood, however, that if the particle is not on the beam propagation axis, that is, it is convergent or divergent, the momentum flux will be decreased.

Depending on the relative size of the particle as compared to the interacting wavelength this light–matter interaction can be approached in two regimes. First, in the Rayleigh regime particle size is much smaller than the wavelength of incident light. In that case, the particle acts as a simple point dipole and the radiation force can be divided in two components: (a) the *scattering force*, associated with the momentum change of the electromagnetic wave due to the scattering by the dipole, and (b) the *gradient force* associated with the Lorenz force acting on the dipole. The second is the Lorenz-Mie regime, where the size of the particle far exceeds the laser wavelength (Cizmar *et al.*, 2010).

Let us consider a particle in a solution that lies in the Rayleigh regime. In this approximation, the scattering and gradient force components are readily separated. As the electric field oscillates harmonically, so does the electric dipole, radiating secondary waves in all directions. Thus, energy flows and the induced scattering force are applied on the dipole, along the light propagation axis. The resulting scattering force holds as (Harada and Asakura, 1996; Nieminen *et al.*, 2007):

$$F_{sc} \approx a|E|^2 \qquad\qquad [7.2]$$

with α being the polarizability of the particle.

In order to evaluate both the scattering and the gradient forces, the polarizability of the dielectric particle needs to be determined. The polarizability can be defined as the ratio of the dipole moment to the applied field. In the case of dielectric particles, as applies in our case, the local polarizability originates mainly from bound electrons. Through the Clausius–Mossotti procedure, it can be shown (Harada and Asakura, 1996) that:

$$a = \frac{8\pi n_{sol} k^4 r_{sph}^6}{c} \left(\frac{m^2 - 1}{m^2 + 2} \right) \qquad [7.3]$$

where n_{sol} is solution/medium refractive index, n_{sph} is the microsphere refractive index, r_{sph} is the microsphere radius and m is the relative refractive index $m = \dfrac{n_{sph}}{n_{sol}}$.

The gradient force stems from the electromagnetic field Lorenz force acting on the dipole (Draine, 1988). It causes the particle to be attracted by high intensity parts of the field, that is, in the focal region, which depends on the gradient of intensity of the incident beam and acts in the direction of the field spatial gradient. The gradient force, being proportional to particle polarizability and the optical intensity gradient, can be expressed (Harada and Asakura, 1996; Nieminen et al., 2007) as:

$$F_{grad} \sim a \cdot \nabla |E|^2 \qquad [7.4]$$

In an attempt to investigate the radiation pressure effects on a particle surrounded by a medium, for example solution, thermal forces have to be taken into account. The irradiating beam causes a temperature gradient in the medium surrounding the particle, due to absorption. This, in turn, results in a thermal force and motion of the particle, called photophoresis. Details on this aspect can be found elsewhere (Greene et al., 1985). In order to minimize or even eliminate such effects, both medium and particles should be transparent to the incident light. Such is the present case using transparent polymer solutions.

In practice, however, the particles move freely in a random manner inside the solution, due to thermal fluctuations and Brownian motion. The thermal energy is given by $k_B T$. The gradient force is the gradient of the so-called trapping potential, which is given by (Nieminen et al., 2007):

$$U \approx -a \cdot |E|^2 \qquad [7.5]$$

If the trapping potential exceeds the thermal kinetic energy, the particles are governed solely by the gradient force. Under these circumstances, the radiation force of the laser beam affects the motion of the particles pulling them into the higher intensity region of the beam, while scattering forces are pushing them along the beam propagation direction.

In the case of irradiating a relatively large sphere having radius of the order of microns, its optical behavior becomes more complex and it can no longer be described by a simple dipole model, but it requires generalization which includes multipole effects. This is achieved within the Lorenz–Mie scattering theory (Metzger *et al.*, 2006) applied appropriately in order to obtain more accurate and generically valid results.

7.3   Organization of entangled polymers and hybrids by laser radiation

7.3.1 Polymer solution dynamics

In a real polymer network of long linear chains, there are a number of topological constraints imposed, due to their inability to cross through one another, commonly known as entanglements. Assume a polymer solution in a good solvent. At low concentrations, the polymer can be considered as isolated coils dynamically positioned very far from each other. As the concentration increases, it reaches a special concentration value, called overlap concentration c^* that equals the concentration inside the coil. Above the overlap concentration, the chains interpenetrate and the solution enters the so-called semidilute regime. In this regime and at small distances, each monomer is surrounded mostly by solvent and the distance between two monomers is quite large (Fig. 7.2a). Here we can introduce the correlation length ξ being the average distance between segments on neighboring chains and which is independent of the degree of polymerization.

In general, the Edwards tube concept of macromolecular motion is considered to be a rather difficult many-body problem. Pierre-Gilles de Gennes succeeded in reducing this many-body problem to the motion of a single chain as it is confined in a tube formed by the surrounding chains (Fig. 7.2b). The simplest tube model was proposed by de Gennes (1971) for the motion of linear entangled polymers, called the reptation model. That model was introduced to explain the dependence of the mobility of a macromolecule on its length. According to the model, the polymer chains are reptating through tubes whose formation and shape is dictated by the neighboring chains and their entanglement points.

7.3.2 Estimations of radiation forces

In order to ensure that radiation forces are capable of organizing the polymer chains to create structures, a feasibility study is necessary. Let us consider a

(a)

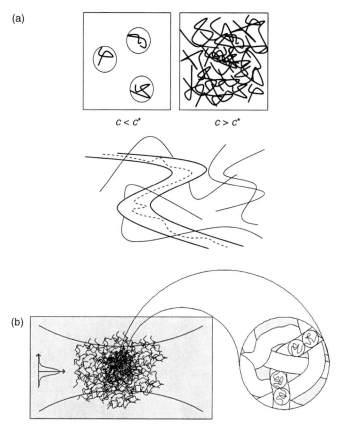

$c < c^*$ $c > c^*$

(b)

7.2 (a) Schematic of dilute (upper left) and semidilute (upper right) solution regimes. (Below) Each polymer is constrained to move within a topological tube due to the presence of the confining surrounding polymers. Within this tube the polymer performs a snake-like reptation motion in the polymer melt. (b) Conceptual view of polymer mesh in the focal point of laser (left) and 'blobs' forming entangled 'tube' in solution (right).

semidilute polyisoprene solution of molecular weight M_w = 1.500 kg/mol in n-heptane at 40%wt which exhibits concentration of c = 0.3 g/cm³ which is much greater than the overlap concentration $c^* \sim 0.008$ g/cm³, indicating a semidilute solution at a high degree of entanglement. In that case, the refractive index of the pure polymer (n_{pol} = 1.52) is higher than the refractive index of the solvent (n_{solv} = 1.388). As a laser beam of relatively large numerical aperture and intensity is incident on a polymer solution, two types of forces act concurrently, namely the scattering and the gradient force, defined previously, giving the resultant optical force that interacts with the semidilute polymer solution. Given M_e the length of macromolecule between adjacent entanglements, a number of $z = \dfrac{M_w}{M_e}$ of about 259 tangles per polymer chain

are expected with $N_e = \dfrac{M_e}{m_{mono}} \approx 85$ monomers per entanglement. We may

thus define here a parameter describing the chain segment between adjacent nodes as a 'suprablob'. The suprablob can be heuristically considered as a spherical nanoparticle composed of a chain section between tangles. It is surrounded by solvent and has a density and refractive index which ranges in between the values of melt and solution average. In analogy to the blob parameter, a diameter of $D \sim (2N_e/c)^{1/3} \sim 7$ nm may be estimated for the given concentration (Fig. 7.2b). The resultant polarizability is thus estimated in the range of $\alpha = 6 \times 10^{-37}$ Fm² and thus, the gradient force F_{gr} can be estimated at about $F_{gr} \sim 2\text{--}3 \times 10^{-18}$ N on each suprablob particle. We note, though, that due to the strong connectivity and high degree of entanglement, the forces are summed as depicted in Fig. 7.3, giving a total force acting on the polymer mesh. This total gradient force far exceeds the Brownian motion force ($F_B \sim 10^{-29}$ N), and thus leads to the evolution of the phenomenon.

As previously mentioned, the gradient forces compel the particles to reconfigure their spatial distribution leading to an organized spatial variation of a refractive index in the solution. This effect is greatly assisted here by the connectivity. By these means, the first condensate is created at the focal point of the objective. Due to its transparency and slightly different refractive index from the surrounding medium it produces a Mie scattering. Several Mie scattering configurations are simulated in Fig. 7.4a to strengthen this perspective. It becomes apparent that since the initial stages of formation a strong

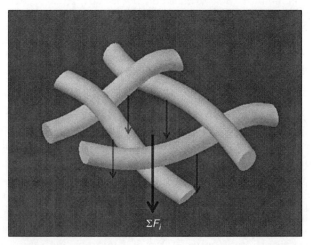

7.3 Radiation forces applied on tube segments leading to local resultant forces ΣF_i.

forward and backward field is produced which enhances further the applied forces and leads to attraction of soft matter. In weak focusing conditions an array of microspherical regions, condensates, is formed in liquid phase since the induced microlens refocuses the incoming light in a neighboring region as depicted in Fig. 7.4b and observed experimentally (Sigel *et al.*, 2002). Each of the induced microlenses acts to refocus the incident field at consecutive positions along the incident beam propagation, resulting in the simultaneous buildup of the array. Considering the gradient forces, the members of this array should be located at local minima of the trapping potential, or otherwise at the foci of the formed beam. Upon prolonged exposure this may result in the gradual buildup of the fiber-like microstructures as demonstrated

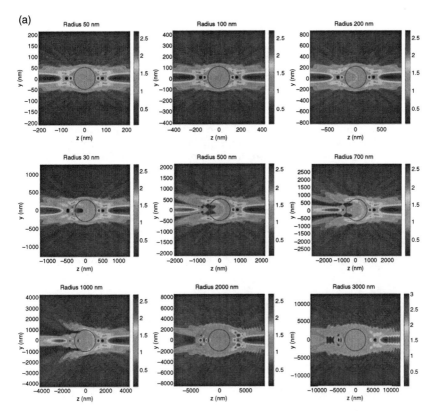

7.4 (a) Simulation of Mie scattering (by uniform spherical polymeric condensates). Light intensity {$I(r)$} distribution of field scattered by spheres of various radii. The incident plane wave has $I = 1$, is right circularly polarized at wavelength $\lambda = 671$ nm and propagates from the left to right toward the +z-direction. The pseudochrome gray scale is absolute with respect to unity. Simulation by electromagnetic Mie scattering methods.

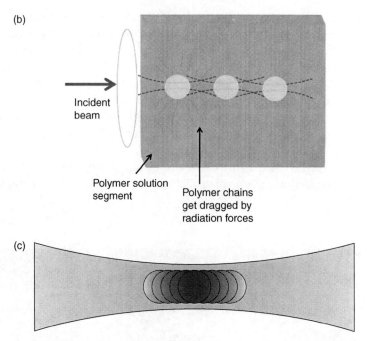

(b)

Incident
beam

Polymer solution
segment

Polymer chains
get dragged by
radiation forces

(c)

7.4 (b) Schematic representation of the structure creation process in
the polymer material. (c) Strongly focusing regime in which a continuous solid structure is produced commencing from forward and backward scattering fields in the nanoscale.

in the next section. Furthermore, under strong focusing conditions the Mie
scattered field and the consequent high field produces strong condensation
at the focal point and a fast attraction of matter to create microstructures as
schematically shown in Fig. 7.4c and presented in the next section.

7.4   Organization and microfabrication by radiation forces: an emerging technology

The creation of polymer condensates and micro-/nanostructures implemented in various forms is discussed here. It includes formations in bulk
solution, thin films, free space and finally the drawing of fibers, each requiring experimental procedures resulting in structures of different shapes and
characteristics.

7.4.1   Material densification in bulk solutions

A focused Gaussian beam emitting in the red is propagating in a semidilute
polymer solution as depicted in the experimental configuration of Fig. 7.5.

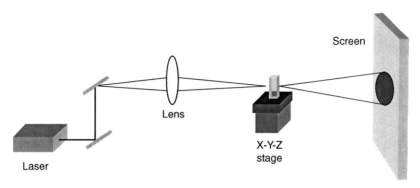

7.5 Experimental setup for structure creation in cuvette. The beam profile is projected onto a screen.

In the experimental arrangement, the cuvette containing the semidilute polymer solution is placed on a translation stage to facilitate positioning. The process is monitored in real time by a CCD camera and the profile of the laser beam is monitored on a screen. The laser beam is allowed to pass through the polymer solution for a time period of a few seconds (Fig. 7.6a) and recorded by the CCD camera. The actual beam profile observed on the screen is Gaussian with a smooth intensity distribution shown in Fig. 7.7a. After a short period, the creation of the condensate in the solution scatters the transmitted light as observed by a gradual change in the beam profile depicted in Fig. 7.7b, 7.7c and 7.7d. The CCD camera records an increase of intensity of the transversely scattered beam propagating in the condensate, as recorded in Fig. 7.6b, indicating lossy waveguiding through the formatted fiber string. This waveguiding phenomenon is induced due to the radial refractive index gradient in the newly formed fiber, owing to the refractive index difference between the pure polymer solute (n = 1.51 for PI) and the solvent (n = 1.37 for n-hexane). In effect, due to radiation forces, the solute polymer chains are dragged along the beam axis, forming a string-like structure along this axis. Concurrently the solvent is forced osmotically to move away and the solvent is replaced by the polymer chains along the beam axis.

To elucidate the process, let us consider the phenomenon through a more extensive perspective for condensate microstructure creation by referring back to the schematic of Figs. 7.4b and 7.4c. The laser beam of some tens of mW power is transmitted through a positive lens and is focused weakly (e.g. by a f = 150–200 cm lens) in the polymer solution producing some kW/cm^2 intensities at focus. The laser radiation and gradient forces localized in the focus point of the lens tend to drag the polymer chains formerly dispersed in the solution in a random manner to form a polymer microsphere

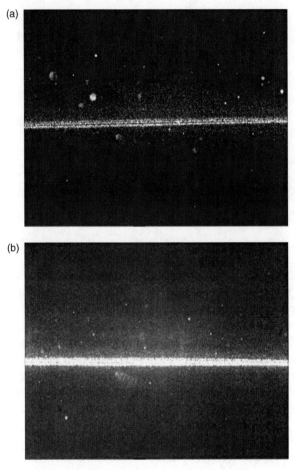

7.6 (a) The laser beam passes through the polymer solution; (b) the
laser beam is waveguided by the formatted structure.

condensate. The latter acts in turn as a second focusing element that forms a
third one, thus consecutively creating a series of microspherical condensates
(Fig. 7.4b). As discussed above this array formation is a simultaneous process. In effect, as a consequence of the accumulation of the microspheres,
a string-like almost macroscopic structure may be formed upon prolonged
exposure, or a direct formation of a fiber string is realized under suitable
conditions. The phenomenon is self-terminated, when no further polymer
chains can join the micro-string, due to inability of radiation forces to drag
efficiently more matter. Macroscopically, the termination of the process is
signaled on the screen where the beam profile becomes stationary, but on
the contrary, it appears frozen after a period of time when steady state of
formation is reached.

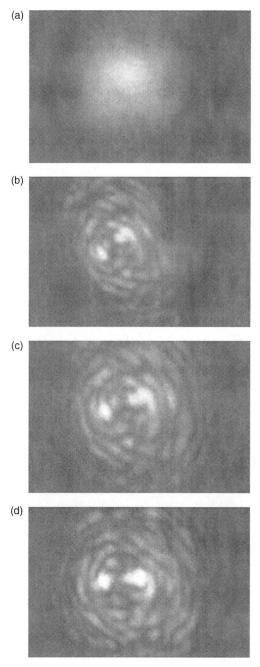

7.7 (a) The laser beam profile upon laser illumination initiation, (b), (c), (d) gradual change of beam profile due to scattering from the created structure.

The quality of the structures can be further improved by controlling several parameters, such as the laser intensity, by the objective focal length, the exposure time, the solution characteristics (concentration, viscosity, etc.) and others.

7.4.2 Microstructure creation in thin films and freestanding structures

Radiation forces can be employed as a manipulation tool for polymer thin films. In the context of the aforementioned scheme, a polymer solution is deposited to form a thin film. In this form the material exhibits a smaller degree of freedom as compared to the bulk solution, owing to the smaller total volume of material available, the strong surface interactions and contact with the environment. The polymer chains are thus not fully free to move under the gradient force, but forces act to manipulate the surface by dragging the polymer and counteracting surface tension thus leading to the formation of elongated microstructures on the thin film surface. The created structures remain localized in a fraction of volume and remain a function of the force field imposed.

Experimentally the deposition of the polymer solution on the glass substrates was facilitated using the doctor blade deposition method. The thickness of the polymer film was controlled by micropositioning the blade and parallel movement in order to produce a uniform spread. The distance between the razor and the substrate is controlling the thickness of the films. The thickness of the samples was in the order of several microns. In a complementary way, thin films may be deposited via spin coating where, depending on the viscosity, the solution needs to be in a rather fluid form.

A typical setup used to controllably organize the polymer material and create solid structures by manipulating the surface of thin films deposited on glass substrates is shown in Fig. 7.8. It consists of a cw laser diode source emitting in the visible, in our case at 671 nm, a microscope objective lens, a series of beam aligning mirrors and a translation stage for micromotion control. A Peltier heater device is placed under the glass substrate for the fast vaporization of the solvent and the concurrent solidification of the solute. A center hole-Peltier module is preferable, due to the fact that the laser beam is allowed to pass through the hole and interact with the thin film, while concurrently the glass substrate is heated in a sought temperature, controlled by a power supply. The writing process is real time monitored via a CCD camera system attached next to the film.

Surface patterning by radiation forces

By applying moderate laser power in the order of some tens mW, corresponding to a few ~ MW/cm^2 at focus, and by a short illumination time

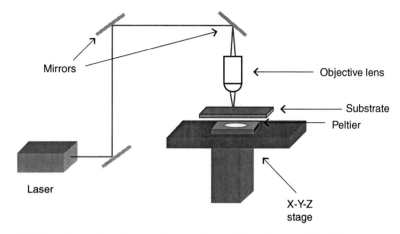

7.8 Experimental setup for the creation of structures in thin films.

duration ranging from a few seconds up to some minutes, a semi-transparent microstructure is created at the focusing point of the objective lens, due to optical radiation forces. For a post-illumination treatment the sample is left on the Peltier device for about 1 h at 60°C and then carefully stored. Exposure duration is of the order of some tens of seconds depending on the concentration and viscosity of the polymer solution, for given laser parameters.

In an attempt to investigate the effect of the beam on the surface morphology two identical films comprising PI-PVP-Au nanoparticles were examined by AFM, one of them being irradiated with a focused laser beam. The results are shown in Fig. 7.9. The effect of the laser beam in the irradiated film becomes apparent, creating a visible recess in the surface of the film. Similar structures were further observed with both profilometer (Tencor Alpha-Step 500IQ) and Atomic Force Microscope as depicted in Fig. 7.10.

In order to examine the effect of focusing on the resultant structures, a number of objective lenses exhibiting different magnification powers (10× and 20×) were used. The resultant structures were found to be highly dependent on the magnification of the lenses, with the structure size being inversely proportional to the magnification power of the lens owing to the focusing strength, thus confirming the theoretical expectations. In particular, in the case of 20× objective lens, it was noticed that only one sphere of large diameter was created, albeit the relevantly high writing duration and full power of laser irradiation. Given the depth of focus, Δf:

$$\Delta f = 2\lambda f_{num}^2,$$

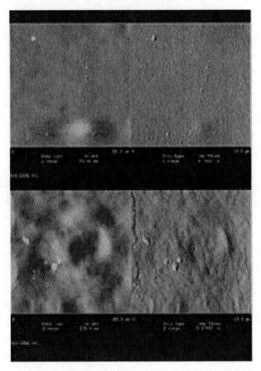

7.9 AFM image of a smooth (up) film surface and rough (down) film surface, indicative of a structure formation on it.

where: $f_{num} = \dfrac{1}{2NA}$, for both 10× (0.25 NA) and 20× (0.4 NA) objectives we have $\Delta f \sim 5$ and 2, respectively. These numbers corroborate the creation of rod-like structures by longer depth of focus imaging, whereas shorter depth of focus produces spherical micropatterns.

Experimentally it has been observed that integrated exposure is a main parameter in structure fabrication. In our attempt to investigate the effect of laser beam exposure in the creation of patterns, the power of the laser beam was changed, by using an optical attenuator, from 150 to 70 mW, as measured with a power meter (Newport model 2832-C Dual Channel). It was found that in order to achieve a structure of the same dimensions, almost double writing time was needed, that is by using a 10× objective in both cases, the exposure time was about 10–15 s at full power of ~ 150 mW, and 30–40 s for the lowest attenuated beam of ~ 70 mW. Thus, the resulting structure was created at longer writing times, indicating the inverse proportionality of the two parameters. The dependence of the structure length on exposure time for given power is shown in Fig. 7.11 and appears almost

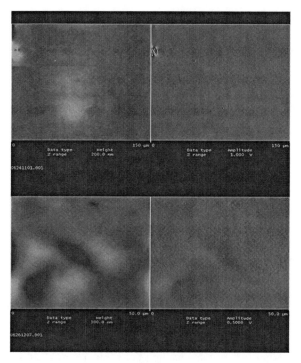

7.10 (Up) and (down): AFM images of film surface showing bulges on top, indicative of structure formation on it.

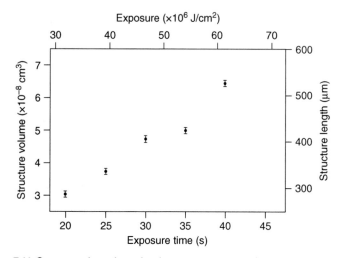

7.11 Structure length and volume v. exposure time graph.

linear. A typical created structure is about 0.4 mm in length and about 10 μm in diameter.

Application of holographic optics for surface microstructuring

Given the phenomenon of structure creation in liquid-phase solution as well as in thin films, someone could try to create multiple structures concurrently by using optical techniques such as holographic masks. The holographic masks were designed by using special software for CGHs holographic laser beam splitters in particular. The fan-out of the masks was formerly designed in CAD software. Then, it was inserted in software for calculation of the sought holographic mask. The calculated mask is again inserted in CAD software for replication in order to increase the overall size without reducing the resolution. The final mask design is transferred onto glass substrate with metal oxides, by using the reactive ion etching (RIE) technique. Several masks were manufactured exhibiting different fan-out configurations as shown in Fig. 7.12.

The holographic masks are designed on the computer using special software for computer generated holograms (CGHs) and a typical micrograph of such a holographic mask produced on oxide coated silica substrate is presented in Fig 7.13.

Surface microstructures by fan-out irradiation were created in thin films of PB (polybutadiene) under similar writing procedures. The experimental

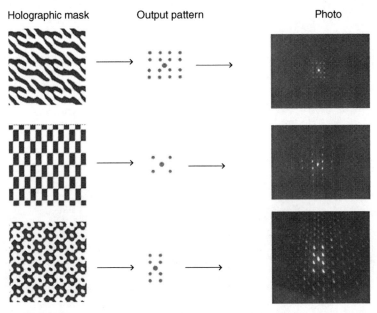

7.12 Several holographic mask patterns and the resulting fan-outs.

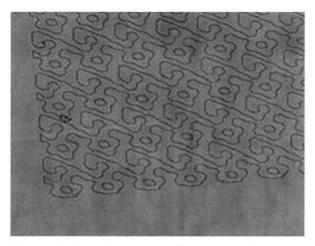

7.13 Detail of a holographic mask used for multiple structure creation.

arrangement of Fig. 7.14a was used and the laser beam was properly manipulated before entering the film by inserting the holographic mask. Fan-out
patterns were transferred onto the film and imaged by a microscope as
shown in Fig. 7.14b, where the features correspond to surface modifications
induced by transferring simultaneously all the fan-out topology. The resulting structures have also been characterized by SEM and the diameter of
spots was found to be in the range of 50–60 μm.

Creation of freestanding structures

Interesting results can be obtained by modifying the orientation of the
thin film with reference to the beam normal. Parallel, perpendicular and
slightly tilted configurations have been used to investigate the dependence
of angle of incidence to the geometrical characteristics and uniformity of
the structures. In a vertical direction, normal to the film, with the deposited
material being under the glass surface, tiny columns of material of the film
appear on the substrate surface, aided minimally by gravity forces. Typical
three-dimensional cylinder rod-like microstructure formed on planar parent substrates are shown in Fig. 7.15a and 7.15b, with structure diameter of
the order of 5–10 μm. Upon illumination, the structures exhibit remarkable
waveguiding properties, though lossy due to scattering effects. As outlined
above, such properties are attributed to the refractive index gradient owing
to the difference between the polymer solute ($n_{PI} = 1.51$) and the solvent
n-heptane ($n_S = 1.39$). In the case of 1,4-polybutadiene the refractive index
variation applies as $n_{PB} = 1.51$ and $n_{decane} = 1.41$.

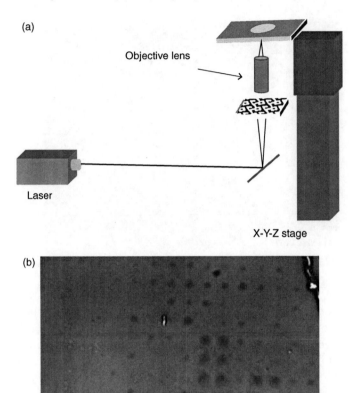

7.14 (a) Setup used for holography-assisted multiple spot patterning in polymer thin film. (b) Scanning electron micrographs of the formation of concurrent multiple spots with the aid of holographic masks.

The length of the structures was found to be in the range of 20–25 μm, the diameter about 3 μm, while the height above the substrate surface was about 300 nm. It should be mentioned at this point that the created structures cannot be extracted from the film due to the strong adhesion forces evolved in the material.

As previously mentioned, the tiny columns can be likened to rods. The rods could not be noticed by observation through an optical microscope due to the orientation of the structures and only the edges were visible. They were short and rather thick in order to be self-standing, static and self-supported. Once the structures get quite long (in the order of mm), they appear to be curved due to gravity and the viscoelastic nature of the material.

(a)

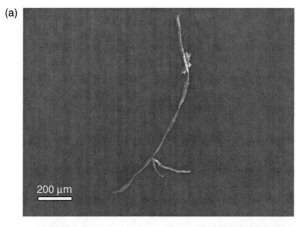

200 μm

(b)

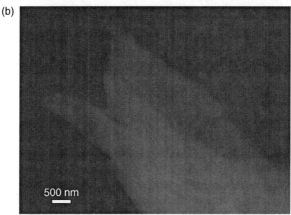

500 nm

7.15 Scanning electron micrographs of (a) a string-like microstructure, (b) the edge of the structure.

The writing/creation of microstructures can be accomplished in almost any form of entangled polymer solution. The aforementioned setup was used for the writing of a structure in a polymer droplet that had just been ejected from a syringe tip, in free space. The resulting structure was then observed through a microscope, as shown in Fig. 7.16.

7.4.3 Plasmonic and fluorescent quantum dots hybrid structures

In a complementary effort, the polymer solution acts as a matrix element where several dopants can be added. Metal Au nanoparticles can be successfully added in the polymer matrix by following the procedures outlined

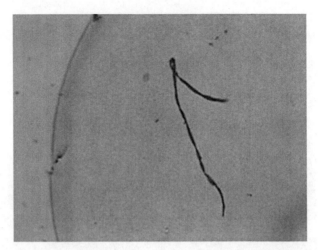

7.16 String-like formation in free space polymer blob (optical microscope).

in the annex. The structures to be created using this solution exhibit remarkable plasmonic characteristics. Furthermore, CdS quantum dots were added in the polymer solution in micellar form. The final solution has a yellowish appearance due to quantum dots. The solution was then deposited onto a glass substrate following the same procedure for microfiber creation in the thin film, and the created microstructures were observed under the microscope. In Fig. 7.17a and 7.17b, two created structures are shown via optical microscope in Fig. 7.17a and the same structures are examined using a fluorescence microscope in Fig. 7.17b. The intense contrast between the polymer thin films and the created structures is apparent due to the fact that the quantum dots were dragged and accumulated at high concentrations during the formation process, being trapped in the net of entangled polymer chains under the exertion of optical forces.

7.4.4 Fiber drawing by laser radiation forces

The experimental setup was further changed in order to examine the feasibility of structure creation in open space. The setup consisted of a laser source, an objective lens and a syringe carrying the polymer solution. The lens used was 10×, the laser operated at full power (150 mW) and the writing process was monitored in real time using a CCD camera. The laser beam was focused in a microdroplet formed at the syringe tip as shown in Fig. 7.18a. Radiation forces are applied and drag a polymer fiber by pulling and solidifying the entangled polymer in free space. This process is self-sustained as feeding comes naturally through the syringe. The phenomenon is

(a)

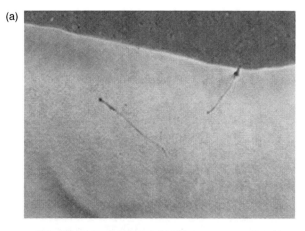

(b)

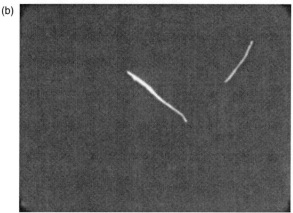

7.17 (a) Optical microscope image of created structures in thin film. (b) Fluorescence microscope image of (a) structures. The structures are glowing.

thus self-conserved resulting in the creation of a flocculent and fully elastic PI fiber rubbery structure. The ring-like appearance shown in Fig. 7.18b is due to self-winding. The total duration of the process is about 10–15 min. It should also be noted here that the whole process is driven by radiation forces and no further operation is needed to assist the fiber drawing process.

The structures created by using that method develop strong electrostatic forces. By approaching a metal tip, the polymer string is strongly affected, moving toward the tip. In addition, the string is highly elastic acting as a spring; the typical structure shown in Fig. 7.19 exhibits a rubbery mechanical response, as expected for the PI material. A close-up of helical structures formed is shown in the optical microscope image of Fig. 7.20.

(a)

(b)

7.18 (a) Real-time image of the fiber-drawing process. (b) Ring-like structure formation at the edge of a syringe needle.

7.4.5 Hybrid structure formation

Hybrid materials have been synthesized as described in the appendix and used in the context of this work. The above microstructure formation process has also been achieved using PI enriched with CdS quantum dots. A typical TEM image of a CdS encapsulated micelle is shown in Fig. 7.21. Both freestanding structure formation and fiber drawing operations have been performed. The structures produced are found to be highly fluorescent as compared to the parent material owing to the strong densification and trapping of quantum dots. In the open space fiber drawing mode, very long strings (~ 10 cm) have been created. The fluorescent objects formed were observed under the fluorescent microscope and spectrally analyzed. A typical fluorescent emission spectrum recorded is shown in Fig. 7.22, with

7.19 String-like structure created in free space. The structure can reach several centimeters in length.

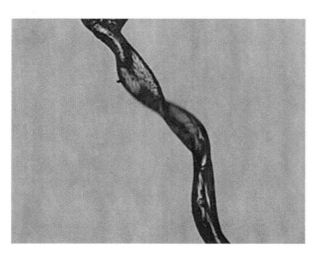

7.20 Close-up of the helix-shaped structure (optical microscope).

the peak at $\lambda_{fluor} \sim 470$ nm indicating a maximum CdS quantum dot size of ~ 4 nm. In a similar operation, the plasmonic behavior of Au enriched material has been verified with typical absorption curves depicted in Fig. 7.23. A typical TEM image of the Au nanoparticles encapsulated micelles is shown in Fig. 7.24. In both cases identical responses with respect to the parent solutions depicted in dotted lines in Figs. 7.22 and 7.23 have been observed with minimal shifts indicating the created denser ambient matrix environment for the nanoparticles and quantum dots. No detrimental effects such as metallization and fluorescence quenching have been observed, verifying that

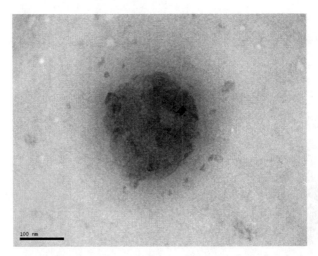

7.21 Transmission electron microscope image of a polymer micelle encapsulating CdS nanoparticles.

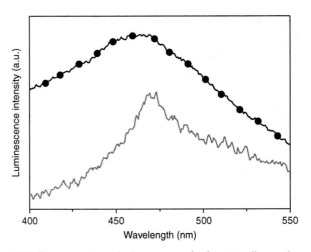

7.22 Fluorescent emission curve of a freestanding polymer structure containing CdS quantum dots. The dotted curve is the spectrum of the polymer solution.

the formation process is dielectrically shielded ensuring absence of aggregation in the final solidified nanocomposites.

7.4.6 Process reversibility and intact materials nature

Significant efforts were devoted to establishing a picture regarding non-reversible chemical processes which may take place during formation of the

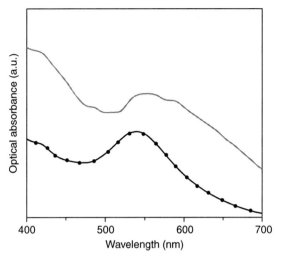

7.23 Plasmonic curve of a freestanding polymer structure containing Au nanoparticles. The dotted curve is the spectrum of the polymer solution.

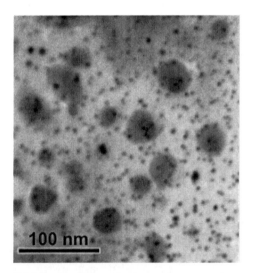

7.24 Transmission electron microscope image of polymer micelles encapsulating Au nanoparticles.

nanostructures, and may compromise reversibility of the formation process. One concern is the chemical stability of polydienes under prolonged and intense irradiation with a laser beam, due to the presence of C=C bonds. Careful NMR and SEC experiments on the polymeric materials isolated after sample irradiation indicated no detectable changes in their chemical structure (i.e. absence of oxidation), as well as the absence of cross-linking or chain scission. The absence of appreciable cross-linking was also verified by dynamic light scattering measurements in solutions of the irradiated material in THF. Reversibility testing of the created structures was performed. Solvent was added to a 3D structure and after stirring for some minutes, the structure disappeared being transformed to the solution previously used. Such findings verifying reversibility in liquid condensates have also been reported by Sigel *et al.* (2002).

7.5   Conclusions and future prospects

The composition of scattering and gradient radiation forces is now offering important tools for the manipulation of inorganic and biological matter by means of particle trapping using optical tweezers and structure actuation. Beyond these applications, the concept of light-induced forces exerted in soft matter at the mesoscale opens up new horizons by allowing efficient density manipulation and the formation of solid objects. Microstructure creation is implemented solely by the above optical means in semidilute, entangled polymer solutions with remarkable results. Pointing to an emerging microfabrication technology, especially suitable for photonics applications, we demonstrated and discussed in this chapter several forms of structures in planar and three-dimensional freestanding micro-objects, in addition to a unique polymer fiber drawing operation, by the sole use of radiation forces.

The observed intriguing phenomena are currently under further study and may be illustrated in terms of a synergy of radiation forces, chain entanglement and optical field structuring and waveguiding, as discussed in this work. The topic offers new scope for fundamental investigations in polymer science and photonics. Emerging concepts on materials microstructuring in three dimensions would take advantage of the chemically inert nature of the process to yield a rich palette of microfabrication tools. These will offer compatibility with both inorganic devices and biological material, leading the way to hybridization and novel lab-on-chip schemes. The ability to tailor surfaces in the nanoscale with reversibility may also offer new tools in optical memory and related applications, and prompt hybrid and unified approaches in photonics, information and biomedical technologies.

7.6   Acknowledgments

This research has been co-financed by the European Union (European Social Fund, ESF) and Greek national funds through the Operational Program 'Education and Lifelong Learning' of the National Strategic Reference Framework (NSRF) — Research Funding Program: Heracleitus II. Investing in knowledge society through the European Social Fund. The authors would like to acknowledge the contributions of Prof. Ioannis Koutselas and Dr Vaggelis Karoutsos in the analysis of results. Useful discussions with Prof Yanopappas, Dr D. Alexandropoulos and M. Vasileiadis are also gratefully acknowledged.

7.7   References

Anyfantakis, M., Loppinet, B., Fytas, G. and Pispas, S. (2008), 'Optical spatial solitons and modulation instabilities in transparent entangled polymer solutions', *Optics Letters*, **23**, 2839–2841.

Anyfantakis, M., Loppinet, B., Fytas, G., Mantzaridis, C., Pispas, S. and Butt, H. J. (2010), 'Experimental investigation of long time irradiation in polydienes solutions: Reversibility and instabilities', *Journal of Optics*, **12**, 124013.

Ashkin, A. (1970), 'Acceleration and trapping of particles by radiation forces', *Physics Review Letters*, **24**, 156–159.

Ashkin, A. (1978), 'Trapping of atoms by resonance radiation pressure', *Physics Review Letters*, **40**, 729–732.

Ashkin, A. (1992), 'Forces of a single-beam gradient laser trap on a dielectric sphere in the ray optics regime', *Biophysical Journal*, **61**, 569–582.

Ashkin, A., Dziedzic, M., Bjorkholm, J. E. and Chu, S. (1986), 'Observation of a single-beam gradient force optical trap for dielectric particles', *Optic Letters*, **11**, 288–290.

Cizmar, T., Davila Romero, L. C., Dholakia, K. and Andrews, D. L. (2010), 'Multiple optical trapping and binding: New routes to self-assembly', *Journal of Physics B: Atomic, Molecular and Optical Physics*, **43**, 102001.

de Gennes, P. G. (1971) 'Reptation of a polymer chain in the presence of fixed obstacles', *Journal of Chemical Physics*, **55**, 572–579.

Dholakia, K. and Zemanek, P. (2010), 'Gripped by light: Optical binding', *Reviews of Modern Physics*, **82**, 1767–1791.

Dienerowitz, M., Mazilu, M. and Dholakia, K. (2008), 'Optical manipulation of nanoparticles: A review', *Journal of Nanophotonics*, **2**, 021875.

Draine, B. T. (1988), 'The discrete-dipole approximation and its application to interstellar graphite grains', *The Astrophysical Journal*, **333**, 848–872.

Gatsouli, K., Pispas, S. and Kamitsos, E. I. (2007), 'Development and optical properties of cadmium sulfide and cadmium selenide nanoparticles in amphiphilic block copolymer micellar-like aggregates', *Journal of Physical Chemistry C*, **111**, 15201–15209.

Greene, W. M., Spjut, R. E., Bar-Ziv, E., Sarofim, A. F. and Longwell, J. P. (1985), 'Photophoresis of irradiated spheres: Absorption centers', *Journal of the Optical Society of America B*, **2**, 998–1004.

Hadjichrisitidis, N., Iatrou, H., Pispas, S. and Pitsikalis, M. (2000), 'Anionic polymerization: High vacuum techniques', *Journal of Polymer Science Part A: Polymer Chemistry*, **38**, 3211–3234.

Harada, Y. and Asakura, T. (1996), 'Radiation forces on a dielectric sphere in the Rayleigh scattering regime', *Optics Communications*, **124**, 529–541.

Jonas, A. and Zemanek, P. (2008), 'Light at work: The use of optical forces for particle manipulation, sorting, and analysis', *Electrophoresis*, **29**, 4813–4851.

Loppinet, B., Somma, E., Vainos, N. and Fytas, G. (2005), 'Reversible holographic grating formation in polymer solutions', *Journal of the American Chemical Society*, **127**, 9678–9679.

Maxwell, J. C. (1873), *A treatise on electricity and magnetism*. Oxford: Clarendon Press.

Meristoudi, A., Pispas, S. and Vainos, N. (2008), 'Self-assembly in solutions of block and random copolymers during metal nanoparticle formation', *Journal of Polymer Science Part B: Polymer Physics*, **46**, 1515–1524.

Metzger, N. K., Wright, E. M. and Dholakia, K. (2006), 'Theory and simulation of the bistable behavior of optically bound particles in the Mie size regime', *New Journal of Physics*, **8**, 139.

Neuman, K. C. and Block, S. M. (2004), 'Optical trapping', *Review of Scientific Instruments*, **75**, 2787–2809.

Nieminen, T. A., Knoner, G., Heckenberg, N. R. and Rubinsztein-Dunlop, H. (2007), 'Physics of optical tweezers', *Methods in Cell Biology*, **82**, 207–236.

Sigel, R., Fytas, G., Vainos, N., Pispas, S. and Hadjichrisitidis, N. (2002), 'Pattern formation in homogeneous polymer solutions induced by a continuous-wave visible laser', *Science*, **297**, 67–70.

Stevenson, D. J., Gunn-Moore, F. and Dholakia, K. (2010), 'Light forces the pace: Optical manipulation for biophotonics', *Journal of Biomedical Optics*, **15**, 041503.

Uhrig, D. and Mays, J. W. (2005), 'Experimental techniques in high-vacuum anionic polymerization', *Journal of Polymer Science Part A: Polymer Chemistry*, **43**, 6179–6222.

7.8   Appendix

7.8.1   Synthesis and characterization of polymer materials and hybrids

High cis 1,4 polybutadiene (PB) sample has been utilized, having a branched molecular architecture. High molecular weight polyisoprene (PI) homopolymer was prepared in-house by the use of anionic polymerization high vacuum techniques (Hadjichrisitidis *et al.*, 2000; Uhrig and Mays, 2005) utilizing home-made glass reactors and break seal techniques. s-BuLi was the initiator used in the non-polar solvent benzene. Polymerization reactions were carried out at room temperature. Under these conditions PI chains of low molecular weight distribution with controlled molecular weights and having high 1,4 microstructure are obtained, as evidenced by size exclusion chromatography determinations and ^{1}H-NMR spectroscopy measurements. The final PI homopolymer was isolated by precipitation in stabilized methanol and dried under vacuum for several days at room temperature. The polyisoprene-b-poly(2-vinyl pyridine) (PI-P2VP) block copolymer also employed as the polymeric functional material in these studies was obtained by an anionic polymerization scheme involving two steps. Isoprene monomer was polymerized first in benzene at room temperature, using s-BuLi as initiator, according to previously described procedures. The living poly(isoprenyl lithium) solution was isolated in a glass ampoule, equipped with a break seal, and was subsequently utilized as the macroinitiator for the polymerization of 2-vinylpyridine. Polymerization of the second monomer was carried out in THF at −78°C. The small amount of benzene present from the preparation of the PI macroinitiator does not perturb the formation of the poly(2-vinyl pyridine block) (P2VP). After formation of the P2VP block, deactivation of the living chains was achieved by degassed methanol. The copolymer solution in THF was concentrated in a rotor evaporator (by distilling about 2/3 of the solvent) since direct precipitation of the copolymer from the THF-rich solution is not quantitative. The pure PI-P2VP diblock copolymer was then isolated in solid form by precipitation in stabilized cold methanol and dried in a vacuum oven at room temperature for several days.

Polymer samples were characterized in terms of molecular weight and molecular weight distribution by size exclusion chromatography (SEC), as well as in terms of PI microstructure and copolymer composition by ^{1}H-NMR spectroscopy. The detailed molecular characteristics of the samples utilized are shown in Table 7.1. No alteration of their chemistry has been observed by the formation operation.

7.8.2   Preparation of PI-P2VP micelles containing Au or CdS nanoparticles

PI-P2VP micelles were prepared in n-heptane which is a selective solvent for the PI block. The solid polymer was directly dissolved in the solvent in a stopper vial in order to give a copolymer concentration $C_{PI\text{-}P2VP} = 1\%$ w/v. The solution was subsequently heated at 60°C for 2 h in order to facilitate complete dissolution of the copolymer and equilibration of the micelles. In this solvent the copolymer forms spherical micelles with P2VP cores and PI coronas. This is evidenced even by the naked eye due to the bluish tint developed in the solution. Quantitative dynamic light scattering measurements after 24 h of solution preparation gave the mean hydrodynamic radius of the micelles ($R_h = 33.6$ nm).

The PI-P2VP micelles were subsequently utilized as nanoreactors for the synthesis of Au and CdS nanoparticles. In the case of Au metal nanoparticle

Table 7.1 Molecular characteristics of the polymers utilized in this study

Sample	M_w (× 10^{-3})	M_w/M_n	Microstructure (PB or PI)	%wt P2VP
PB-390	390	2.5	High cis 1,4	
PI-4	1500	1.07	High cis 1,4	
PI-P2VP-2	55	1.05	High cis 1,4	9

formation, auric acid (as ethanolic solution) was introduced in the PI-P2VP micellar solution in the predetermined stoichiometric amount to the pyridine units (N:Au = 4:1). This high N:Au ratio facilitates quantitative complexation of the gold ions with the nitrogen atoms of the pyridine rings of the copolymer. Complexation of the pyridine units of the P2VP block with the gold ions takes place in the micellar cores which act as nanoreactors for metal nanoparticle nucleation and growth. After allowing for equilibration for 24 h, gold ions were reduced to gold by the use of hydrazine solution in ethanol (hydrazine:Au = 4:1). The nanoconfinement of the micellar cores led to the production of gold metal nanoparticles, as evidenced by UV-vis spectroscopy measurements (Meristoudi *et al.*, 2008). Adsorption maximum for the surface plasmon resonance peak of the produced gold nanoparticles was observed at ca. 540 nm.

A similar preparation protocol was also followed in the second case. Cd^{2+} ions were loaded in the PI-P2VP micellar cores in the form of cadmium acetate (N:Cd^{2+} = 4:1). The mixed solution was allowed under stirring for 24 h. During this period cadmium acetate was completely dissolved, presumably within the micellar cores, due to the chemical affinity of the polar pyridine groups for the Cd^{2+} ions. In the next step, thioacetamide was introduced into the solution of loaded block copolymer micelles. Thioacetamide was utilized as the source for sulfur, through its decomposition at 80°C (S/Cd ratio was kept at 2). The formation of CdS nanoparticles in PI-P2VP micelles was confirmed by UV-vis and fluorescence spectroscopy (Gatsouli *et al.*, 2007).

Inorganic nanoparticle loaded block copolymer micelles were mixed with solutions of PB and PI homopolymers in n-hexane in order to produce hybrid non-covalent interacting homopolymer/micelles mixtures. The similar chemical composition of the micellar corona with the matrix homopolymer enhances such interactions and facilitates decoration of fibrilar polydiene structures, formed upon subsequent exposure of solutions to laser radiation. In this way hybrid nanostructures are produced solely by the laser action. Fluorescent hybrid CdS/PI-P2VP micelles may act also as markers/probes for direct visualization of homopolymer chain ordering.

8

Laser-assisted polymer joining methods for photonic devices

C. H. WANG, Heriot-Watt University, UK

Abstract: This chapter presents the development of laser-based methods for assembly of photonic devices. This includes the development of laser-assisted polymer bonding methods for assembly and packaging of photonic and microelectromechanical systems (MEMS) devices and the associated temperature monitoring method using embedded microsensors. The chapter also covers an overview of the development of a laser microwelding method for assembly of polymer spheres to create photonic bandgap materials for terahertz applications.

Key words: laser assembly, packaging, diode lasers, beam forming, polymer bonding, polymer welding, benzocyclobutene, microelectromechanical systems.

8.1   Introduction

Microscale joining and bonding methods are essential for assembly and packaging of photonic devices and structures for a variety of applications such as coupling of optical fibres to lasers and integration of photonic devices with other microelectronic devices and microelectromechanical systems (MEMS). One well-established application of lasers in assembly of photonic devices is laser welding of metal clips to a metal ferrule on an optical fibre to produce a fibre-coupled semiconductor laser inside a butterfly package. Following the development of the laser welding technology in the 1990s (Jang, 1996), it has become a mature manufacturing method for high speed and high reliability assembly of fibre-coupled laser packages (Jang, 2000; Song *et al.*, 2009; Tan *et al.*, 2005). Lasers are efficient remote heat sources that can deliver a well-controlled beam to an area between parts or components to produce a reliable joint by welding or other thermally induced joining mechanisms. With the rapid development of high-power, high-efficiency diode laser technology and the diode laser pumped solid state lasers, lasers will find many other applications in assembly of photonic devices. This chapter is focused on development of laser-assisted polymer bonding of substrates for photonic and MEMS applications. One potential application of the technology is in encapsulation of photonic components such as lasers, optical detectors, imaging sensors and MEMS devices

269

including optical MEMS. The method offers a potential solution to low cost, low temperature assembly and packaging requirements. The application of the bonding polymer, benzocyclobutene (BCB), in fabrication of semiconductor lasers as a dielectric material for planarization is also covered. The chapter also describes other emerging applications of lasers in assembly of photonic bandgap structures for terahertz applications.

8.2   Properties of benzocyclobutene (BCB) polymers for photonic applications

8.2.1 BCB polymer

BCB represents a family of thermosetting polymers developed by Dow Chemical for microelectronic and photonic applications. Table 8.1 shows a summary of the properties of the BCB polymers. The details of these and further information on the properties of the BCB polymers are available from the manufacturer (Dow Chemical, 2011). The low dielectric constant of the BCB polymers makes them a good dielectric material for fabrication of microelectronic circuits and photonic devices. Polymer materials with a low dielectric constant are essential for high frequency microelectronic and photonic systems. The BCB polymers have good thermal stability and low moisture uptake characteristics. High quality thin films can be produced easily by spin-coating. The BCB polymer films are prepared using B-staged monomer solutions. This means that the curing reaction is already initiated and ready for completion after a film is produced. This can enhance its viscosity stability (Wong, 2000). The photosensitive BCB materials can be patterned by UV lithography and therefore are compatible with the standard microfabrication methods. The photosensitive polymers contain diazo cross-linker moieties. The BCB polymers that are not photosensitive can be patterned using dry etching methods. Therefore the BCB polymers are versatile materials with processing flexibility.

Table 8.1 Summary of properties of BCB polymers

Glass transition temperature	> 350°C
Coefficient of thermal expansion	42 ppm/°C at 25°C
Thermal conductivity	0.293 W/m/K at 24°C
Tensile modulus	2.9 GPa
Volume resistivity	1×10^{19} Ω cm
Dielectric constant	2.5 at 10 GHz
Breakdown electric field	~ 5.3×10^6 V/cm
Refractive index	1.56 at 633 nm
Moisture uptake	< 0.2%

Source: Dow Chemical (2011).

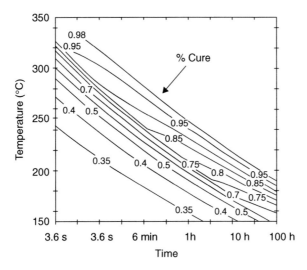

8.1 Curing behaviour of BCB for an oven-based method. (From Dow Chemical, 2009.)

8.2.2 Curing characteristics of BCB polymers

The BCB polymer is a thermal setting resin. It can be cured between 200°C and 350°C. However the curing behaviour is highly dependent on the curing temperature. Figure 8.1 shows the relationship between curing time and temperature for an oven-based curing method (Dow Chemical, 2009). It takes 60 min to achieve over 95% of curing (hard cure) at 250°C, but the curing time can be reduced to a few seconds at curing temperatures above 320°C. This unique property makes it possible to achieve a short processing time provided that a rapid heating method is used such as the laser-based assembly method described in this chapter.

8.3   BCB as a planarization material for fabrication of semiconductor photonic devices

In fabrication of photonic devices such as semiconductor lasers, it is necessary to passivate the ridge waveguide structures to complete the fabrication process. The passivation layer surrounding the ridge waveguide structure is used to provide better confinement of the injected current for laser operation and to suppress the diffraction effects by improving index matching between the waveguide and the air interface. BCB has been used as a planarization material in fabrication of semiconductor lasers. In the work by Wiedmann *et al.* (2001) a BCB polymer was used as the planarization material to passivate InP-based III-V quantum well semiconductors. The active waveguide medium is deposited using an epitaxy method and carried out at wafer scale including additional layers for cladding and ohmic contact. A metal layer is deposited on top of the epitaxy layer for ohmic contact for producing electrodes for the resultant lasers. The multilayer stack is patterned using electron beam lithography and a combination of dry and wet etching methods to obtain freestanding structures including the distributed Bragg reflectors (DBR) to complete the laser fabrication process. A BCB layer is then coated on the etched structures by spin-coating to fill in the vertical gaps in the laser cavities and structures for the DBR reflector. After baking at 250°C for 2 h, the BCB material above the mesa structure is removed by reactive ion etching. Single mode operation of the lasers is achieved at 1.5 μm of wavelength using BCB as the dielectric material for planarization. BCB has been used in a similar manner for fabrication of a ridge waveguide laser operating at 1.3 μm (Yagi *et al.*, 2009). An electrical bandwidth of more than 20 GHz is obtained as a result of reduction of the parasitic capacitance using the BCB buried ridge waveguide structure in the laser. Reliable operation of the laser is demonstrated in accelerated testing for 1200 h at 85°C of temperature and 200 mA of operating current.

In another application of BCB as a planarization polymer for fabrication of semiconductor photonic devices (Demir *et al.*, 2005), a photonic switch is fabricated based on the integration of a mesa photodiode and a quantum well modulator on the same substrate. After selective deposition of the semiconductor layers, a silicon dioxide or silicon nitride based hard mask is produced on the wafer and used to define the areas for device fabrication. The exposed material is etched away vertically to the substrate surface. Selective removal of additional material under the hard mark is useful to create a small degree of undercut in order to produce an extended region of flat BCB surface beyond the ridge structures after the BCB based planarization process. The BCB layer is deposited using spin-coating after fabrication of the device structures under the hard mask. The BCB material seals the structures for photonic devices. After BCB curing the material is etched back to the surface of the hard mask. The advantage of the hard mask approach is that it allows easy control of the process since the hard mask acts as an etch stopper to prevent excessive etching of the BCB material for passivation of the detector and modulator. Using BCB as a planarization material, high yield (> 90%) fabrication, high breakdown voltage and low contact resistance can be achieved.

8.4   Laser-assisted polymer bonding for assembly of photonic and microelectromechanical systems (MEMS) devices

This section describes the development of laser-assisted polymer bonding methods for assembly and packaging of microelectronic, photonic and

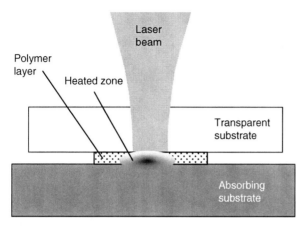

8.2 Schematic illustration of laser bonding of substrates using an intermediate polymer layer.

MEMS devices. An energy efficient diode laser is used to cure a patterned square BCB ring to obtain fast, reliable bonding of glass and silicon substrates for capping or substrate attachment in packaging and integration applications. An accurate temperature monitoring method for the laser bonding process using embedded microsensors has been developed.

8.4.1 Principle of laser-assisted polymer bonding

In the laser-based intermediate polymer layer substrate bonding method, a laser is used to cure a polymer adhesive layer to join two substrates together to form a substrate assembly as illustrated in Fig. 8.2 (Bardin *et al.*, 2007). The laser radiation can be absorbed by one of the substrates or the polymer layer itself to produce the necessary temperature rise to cure the polymer layer to form a strong bond between the substrates. The intermediate layer polymer bonding method allows low temperature and low stress joining of substrates for assembly and packaging of photonic, MEMS and microelectronic devices. It is a flexible approach for bonding of a range of substrate materials for which direct interface bonding is not possible due to incompatibility of the surfaces or requiring extreme high temperature such as silicon direct bonding. Figure 8.3a shows a schematic arrangement for bonding a transparent substrate to a non-transparent substrate while Fig. 8.3b illustrates bonding of two non-transparent substrates. The substrate configuration in Fig. 8.3a can be reversed to have the non-transparent substrate facing the laser beam in the same approach as shown in Fig. 8.3b. A glass cap is commonly used for photonic and optical devices as well as optical MEMS such as micromirrors to provide a window for the input and output light beams. For encapsulation

8.3 Schematic illustration of laser bonding of (a) a transparent cover to an opaque substrate and (b) an opaque cover to a transparent substrate.

of other MEMS devices non-transparent caps are usually used to provide a perfect CTE (coefficient of thermal expansion) match between the substrates thus improving the thermomechanical reliability of the resultant package. For example, a silicon cap is preferred for silicon-based MEMS devices such as accelerometers, gyroscopes and resonators. Silicon is a non-transparent material for laser wavelengths below 1.1 μm.

8.4.2 Beam-forming method

In the laser-assisted polymer bonding method two beam delivery approaches can be used, one based on scanning a focused beam along the polymer bonding track and the second approach based on projection of a laser beam to cover the total area to be processed. When it is not possible to cover the entire bonding area, for example at wafer level processing, each area of the wafer can be processed using the beam project method in sequence to complete wafer level bonding. The beam projection approach is more efficient since an area is processed simultaneously rather than a spot as is the case in the scanning method. In order to make efficient use of laser output and to produce a uniform beam profile, beam shaping of the laser output is essential. Beam shaping is a process of redistributing the irradiance or phase of an optical radiation using a purpose designed optical element with a suitable lens or a number of lenses (Brown *et al.*, 2000). For the laser bonding work described in this section, a diffraction multi-aperture beam integration method is used to convert the laser output from a high-power diode laser into a top-hat beam or a frame shape profile for energy efficient laser bonding (Brown *et al.*, 2000). This method is suitable for beam shaping of the partially coherent radiation from an array of laser bars in the high-power diode laser system.

The principle of the diffraction multi-aperture beam integration method is based on segmentation of the entrance cross-section of a beam using an array of micro-optic elements (e.g. lenslets) produced on a transparent

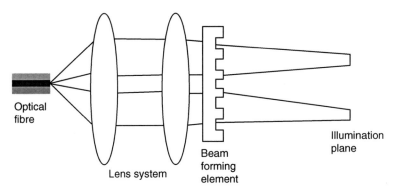

Optical
fibre

Lens system

Beam
forming
element

Illumination
plane

8.4 Schematic drawing of the optical setup for beam shaping.

substrate to direct the beamlets separately onto the target plane. Beam averaging is achieved by integration of the sub-beams at the target plane. Each sub-element applies independent phase correction or change to the respective beamlet. Therefore the beam-shaping element is also called phase-plate. Figure 8.4 shows a schematic illustration of the beam-shaping setup used in the laser-assisted polymer bonding work for packaging of photonic and MEMS devices. The optical output from a fibre-coupled diode laser system is collimated using a telescope module with two lenses. The beam-shaping phase-plate is placed behind the lens for beam integration. It should be noted that the arrangement of the phase-plate and the integrating lens is different from the conventional configuration used for diffracting multi-aperture beam shaping in which the phase-plate is positioned before the integrating lens. The beam-shaping system shown in Fig. 8.4 provides a degree of freedom to change the size of the target beam without significant degradation of beam quality by moving the position of the phase-plate relative to the lens.

A high-power fibre-coupled diode laser at the wavelength of 970 nm is used in the development of a laser-assisted polymer bonding method for assembly of photonic, MEMS and microelectronic devices. The diameter of the optical fibre is 200 μm. A beam delivery module and a beam-forming element are used to produce a suitable beam profile for energy efficient polymer bonding. The beam-shaping method shown in Fig. 8.4 is used to produce top-hat and frame-shape beam profiles for laser bonding of substrate assemblies. Two custom designed beam-shaping phase-plates are used to provide separate beam profiles, one for a top-hat beam profile and the other one for a frame-shape beam profile. The focal length of the integrating lens is 200 mm. The phase-plates are designed and fabricated by Powerphotonic Ltd, Scotland. They are produced on quartz substrates using a CO_2 laser-based micromachining method. Figures 8.5a and 8.5b show the 3D plots

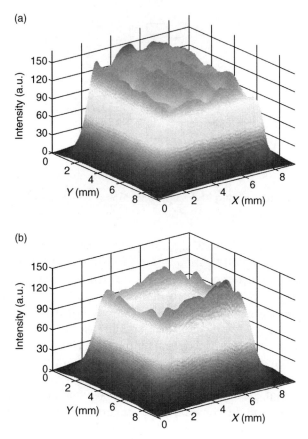

8.5 3D plot of measured beam profiles for (a) top-hat beam and (b) frame shape beam.

of the measured intensity distribution for the top-hat beam and the frame shape beam respectively. The outer dimensions of the beam for laser bonding are about 6 mm × 6 mm for both beam profiles. The ring width of the frame-shaped beam was designed to be 1 mm. The beam size is designed for bonding of substrate assemblies discussed in the following sections.

8.4.3 Laser bonding for assembly of substrates

In the laser bonding method a laser beam is used to produce the necessary temperature increase to cure an intermediate polymer (BCB) layer between two substrates to bond them together to form a microcavity for capping of photonic, sensor and MEMS devices. The BCB polymer bonding rings are produced on caps using photolithography. A BCB film is deposited on a glass or silicon wafer using a spin-coating method. The film is then baked and patterned using UV photolithography. The thickness of the BCB film is about 10 µm. The outer dimensions of the BCB rings are 5.2 mm × 5.2 mm. The track width is 400 µm. The square BCB rings are fabricated on 4 inch diameter glass or silicon wafers. The wafers are diced to produce individual substrates each with a BCB ring for laser bonding to the second substrate to form a microcavity. For substrate attachment, an unpatterned BCB film can be used to provide complete joining of the surface area between two substrates.

A custom designed bonding setup consisting of an XY platform is used to support the substrates to be bonded (Liu *et al.*, 2010). The fibre-coupled laser system and beam-forming optics described in Section 8.4.2 are used to deliver the top-hat or frame shape beam to the substrates to be assembled. The substrates, one with a BCB ring, are aligned and placed on a stainless steel supporting platform. In order to improve thermal efficiency a 0.9 mm thick ceramic plate was inserted between the platform and the substrate. A computer-controlled force applicator (bonding arm) was used to press the substrates together to apply a suitable bonding force/pressure. A window on the bonding arm provides the optical path for the laser beam to reach the substrate assembly to raise the required temperature through the absorption of laser radiation by the silicon substrate.

Figure 8.6 shows an optical picture of laser bonded glass-silicon substrate assemblies using the top-hat beam. It can be seen that defect-free bonding has been achieved. The bonding process for the bonded glass-silicon microcavity shown in Fig. 8.6 was carried out using the substrate configuration shown in Fig. 8.3a. The light beam was transmitted through the glass substrate and absorbed by the silicon substrate. The laser power is 50 W and the bonding

8.6 An optical image of a laser bonded defect-free glass-silicon substrate assembly using the substrate configuration shown in Fig. 8.3a and the top-hat beam profile.

time is 10 s (Wang *et al.*, 2009). A different bonding process for the glass-silicon assembly shown in Fig. 8.6 was also conducted using the reverse arrangement of the substrate configuration shown in Fig. 8.3a. The silicon substrate was arranged to face the laser beam. The thickness of the silicon substrate is ~ 0.5 mm. All of the laser radiation entering the silicon substrate was absorbed and converted into thermal energy. In this configuration it only requires 30 W of laser power to raise the temperature to ~ 300°C for curing the BCB polymer. The bonding time is also 10 s. The significant reduction of the laser power is due to the improvement of the thermal efficiency since the thermal conductivity of the glass is two orders of magnitude lower than that of silicon.

8.4.4 Process monitoring using embedded microsensors

Precise temperature control is important in the laser-assisted polymer bonding process. The laser-induced temperature change depends not only on the laser power and beam profile but also on the dimensions and properties of the substrates as well as the thermal arrangement of the bonding setup. Secondly as shown in Section 8.2, the curing time of the BCB polymer is highly dependent on the curing temperature. It is essential to monitor the laser-induced temperature accurately in order to determine the curing time for achieving proper curing of the BCB polymer for reliable bonding. Excessive curing temperature can cause thermally induced failure of structures and materials in the devices. Although conventional methods based on thermal imaging and thermo-sensitive paints can be used to monitor the temperature of the laser heating effect, these methods cannot provide precise information about the temperature change within the polymer bonding track and the temperature distribution within the packaging cavity since *in situ* monitoring is not possible with these methods.

In order to monitor the packaging temperature accurately and to determine the temperature distribution at the surface of the device substrate, a new method using embedded microsensors has been developed (Liu *et al.*, 2010). Platinum-based sensor arrays are designed and fabricated by plasma etching. The resistance sensors are based on meander designs in order to achieve a small footprint as required for the temperature monitoring work. Figure 8.7 shows a microsensor array on a glass substrate after fabrication and the detail of a peripheral sensor. The widths of the meander lines are 3 µm for all of the peripheral sensors and 10 µm for all of the sensors in the middle of the sensor array. The footprint of the peripheral sensors is only 240 µm × 250 µm. Therefore the peripheral sensors can be covered completely by the BCB bonding ring of track width of 400 µm.

A pre-bonding step is used to attach a cap with a BCB bonding ring to a sensor array on another substrate in order to embed the peripheral sensors

(a)

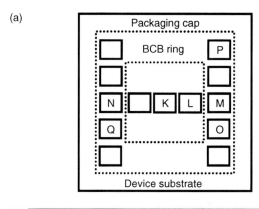

(b)

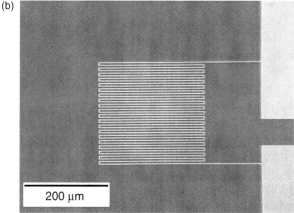

8.7 (a) Schematic layout of a sensor array for temperature monitoring in laser bonding of substrates; (b) optical picture of a peripheral sensor with a track width of 3 μm.

of the sensor array under the BCB bonding ring. The pre-bonding process is carried out on a flip chip bonder at 100°C at which the BCB material is soft and bondable. Figure 8.8 shows an optical image of a sensor array on glass bonded to a silicon substrate. It can be seen that all of the peripheral sensors in the array are well placed under the BCB sealing track. Electrical leads are attached to the contact pads of the sensor array for interfacing with a computer-controlled data acquisition system.

For temperature monitoring the pre-bonded cap and substrate assembly with embedded sensors is placed on the bonding platform. The bonding area is aligned to the beam pattern. The top-hat and frame-shape laser beams are used to investigate the dependence of temperature change on the laser power and the beam profile. Figure 8.9 shows the measured results of temperature profiles for the frame shape beam profile. The cap is glass and

8.8 Optical picture of an embedded sensor array on a glass substrate attached to a silicon substrate after alignment and pre-bonding on a flip chip bonder.

the sensor substrate is silicon. At the onset of the laser radiation, the temperature increases rapidly and then reaches a steady-state value. After the incident laser power is terminated, the temperature decreases to that of the ambient level. It can be seen that the steady-state temperature at the centre of the silicon substrate is lower than that within the bonding track. This is a result of the fact that most of the laser power is directed to the bonding area with the frame-shape beam profile. This is a desirable effect in assembly of temperature sensitive devices since it is possible to use a higher bonding temperature for packaging without causing thermally induced damages to the devices that cannot be achieved using conventional methods based on the global heating approach. The rise and fall of the temperature characteristics both follow an exponential behaviour as expected (Liu *et al.*, 2010). The time constant is about 6 s for both rise and decay of the laser-induced temperature. The time constant has a significant dependence on the heat dissipation arrangement under the substrate assembly. When the ceramic thermal barrier is removed, the thermal constant decreases to a value of about 2 s. There is a corresponding decrease in the steady-state temperature since more laser-generated thermal energy is conducted away from the silicon substrate to the stainless steel platform.

8.5   Laser microwelding for assembly of periodic photonic structures

As discussed in Section 8.2 solid state lasers have been used successfully in assembly of fibre-coupled semiconductor lasers in butterfly packages.

Recently a laser microwelding method has been demonstrated for assembly of polymer-based dielectric microspheres to produce three-dimensional periodic structures for terahertz applications (Takagi *et al.*, 2010). In this work polyethylene-carbon composite spheres of 400 μm of diameter are assembled in sequence to form a diamond lattice-based three-dimensional periodic structure. In this lattice structure each sphere has four neighbours in a tetrahedral configuration (Streetman and Banerjee, 2000). Each polymer-based sphere is joined to the neighbouring spheres simultaneously by laser microwelding at the interfaces between the dielectric spheres.

The assembly process is carried out using a fibre laser-based robotic system and an automated particle (sphere) feeder. The size of each particle is determined before being assembled using an imaging system with a resolution of 0.7 μm. Six continuous wave (CW) fibre lasers operating at 1.1 μm are used to carry out microwelding at the interface between the spheres since the maximum number of contact points is six in the diamond lattice-based periodic structure. Each laser is mounted on a robotic manipulator with 5 degrees of freedom. The laser beam is focused to produce a spot size of 30 μm in diameter using a lens. A computer-based control system is used to carry out the automated imaging and laser welding processes for assembly of the three-dimensional periodic structures.

In fabrication of periodic structures it is essential to control the amount of overlapping between the spheres after laser welding to ensure minimal deviation of a sphere from the expected position in the 3D periodic structure.

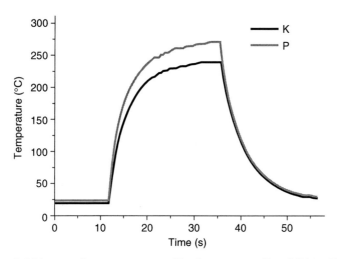

8.9 Measured temperature profiles from sensor P and K (see Fig. 8.7a) for laser bonding of silicon to glass using the arrangement shown in Fig. 8.3b with a frame shape beam profile.

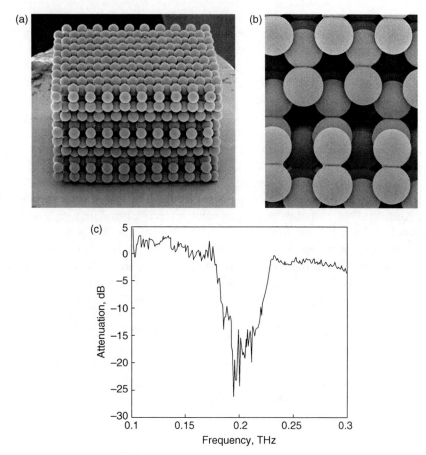

8.10 (a) A picture of a fabricated 3D diamond structure using a laser microwelding method; (b) a close look of the front view of the picture shown in (a); (c) a transmission spectrum of the diamond structure in the <111> crystalline direction (Takagi *et al.*, 2010). (Reprinted with permission from IOP Publishing.)

This is achieved by controlling the amount of overlapping after laser welding using the information from the size measurement for each sphere before assembly. For larger spheres, a higher laser power is used to melt more polymer material at the interface of two adjacent spheres to ensure the same separation between the centres of the spheres as for smaller spheres. The amount of laser power for microwelding in the assembly process for each sphere is predetermined based on the data obtained from a separate study of laser welding of spheres of different sizes. Figure 8.10 shows a laser-assembled diamond structure consisting of 1600 microspheres and the results of the transmission characteristic in the terahertz region. The laser-assembled 3D photonic structure showed a stop band around 0.2 THz of frequency.

This shows that the laser-assembled 3D structure of microspheres is a photonic band gap material and hence demonstrates laser assembly as a method for fabrication of photonic materials.

8.6   Conclusions

Laser-based methods are expected to find many applications in assembly of photonic, micromechanical, microelectronic and sensor devices. Laser welding and soldering methods are already widely used in manufacture of photonic devices to produce fibre coupled semiconductor lasers in butterfly packages. A laser-assisted rapid polymer bonding method for assembly and encapsulation of photonic and other microscale devices has been developed. The method allows application of the mature high efficiency, low cost diode laser systems in photonic and microelectronic manufacturing. An associated temperature monitoring method using an embedded microsensor array has been developed for process control. Due to the localized nature of temperature change it is necessary to use embedded sensors to achieve *in situ* temperature monitoring. Laser assembly of polymer spheres has been demonstrated to produce photonic bandgap materials for terahertz applications. A laser microwelding approach is used to assemble a diamond lattice using polymer spheres. It is expected that laser-based polymer joining approaches will find increasing applications in assembly and packaging of photonic devices and structures.

8.7   References

Bardin, F., Kloss, S., Wang, C. H., Moore, A. J., Jourdain, A., De Wolf, I. and Hand, D. P. (2007), 'Laser bonding of glass to silicon using polymer for microsystems packaging', *IEEE/ASME Journal of Microelectromechanical Systems*, **6**, 571–580.

Brown, D. M., Dickey, F. M. and Weichman, L. S. (2000), 'Multi-aperture beam integration systems'. In Dickey, F. M. and Holswade, S. C. (eds.), *Laser beam shaping: Theory and techniques*. New York: Marcel Dekker, pp. 273–311.

Demir, H. V. Zheng. J. F., Sabnis, V. A., Fidaner, O., Hanberg, J., Harris, J. S. and Miller, D. A. B. (2005), 'Self-aligning planarization and passivation for integration applications in III-V semiconductor devices', *IEEE Transactions on Semiconductor Manufacturing*, **18**, 182–189.

Dow Chemical (2009), 'Processing procedures for CYCLOTENE 4000 series photo BCB resins'. Available from: http://www.microchem.com/PDFs_Dow/cyclotene_4000_immersion_dev.pdf. Accessed 21 December 2011.

Dow Chemical (2011), 'Cyclotene advanced electronics resins'. Available from: http://www.dow.com/cyclotene/solution/index.htm. Accessed 21 December 2011.

Jang, S. (1996), 'Packaging of photonic devices using laser welding', *SPIE*, **2610**, 138–149.

Jang, S. (2000), 'Automation manufacturing systems technology for opto-electronic device packaging', *Electronic Components and Technology Conference*, Las Vegas, 21–24 May 2000, 10–14.

Liu, Y., Zeng, J. and Wang, C. H. (2010), 'Accurate temperature monitoring in laser-assisted polymer bonding for MEMS packaging using an embedded microsensor array', *IEEE/ASME Journal of Microelectromechanical Systems*, **19**, 903–910.

Song, J. H., O'Brien, P. and Peters, F. H. (2009), 'Optimal laser welding assembly sequences for butterfly laser module packages', *Optics Express*, **17**, 16406–16414.

Streetman, B. and Banerjee, S. (2000), *Solid sate electronic devices*. New Jersey: Prentice Hall.

Takagi, T., Masanori, O. and Kawasaki, A. (2010), 'A microsphere assembly method with laser microwelding for fabrication of three-dimensional periodic structures', *Journal of Micromechanics and Microengineering*, **20**, 035032.

Tan, C. W., Chan, Y. C., Leung, B. N. W., Tsun, J. and So, A. C. K. (2005), 'Characterization of Kovar-to-Kovar laser welded joints and its mechanical strength', *Optics and Lasers in Engineering*, **43**, 151–162.

Wiedmann, J., Rai, M. M., Ebihara, K., Matsui, K., Tamura, S. and Arai, S. (2001), 'Deeply etched semiconductor/benzocyclobutene distributed Bragg reflector laser combined with multiple cavities for 1.5 µm wavelength single-mode operation', *Japanese Journal of Applied Physics*, **40**, 4031–4037.

Wang, C. H., Zeng, J. and Liu, Y. (2009), 'Recent advances in laser assisted polymer intermediate layer bonding for MEMS packaging', *Proceedings of 10th International Conference on Electronics Packaging Technology and High Density Packaging (ICEPT-HDP)*, Beijing, 10–13 August 2009, 31–35.

Wong, C. P. (2000), 'Polymers for electronic packaging'. In Harper, C. A. (ed.), *Electronic packaging and interconnection handbook*. New York: McGraw-Hill, pp. 1.1–1.21.

Yagi, H., Onishi, Y., Koyama, K., Ichikawa, H., Yoshinaga, H., Kaida, N., Nomaguchi, T., Hiratsuka, K. and Uesaka, K. (2009), '1.3 µm wavelength AlGaInAs/InP ridge-waveguide lasers utilizing benzocyclobutene planarization process', *SEI Technical Review*, **69**, 92–95.

Part III
Laser fabrication and manipulation of
photonic structures and devices

9

Laser seeding and thermal processing of glass with nanoscale resolution

J. CANNING, University of Sydney, Australia and
S. BANDYOPADHYAY, Central Glass and Ceramic
Research Institute (CGCRI), India

Abstract: The idea of exploiting glass transformation normally associated with thermal heating and quenching by using localized laser processing for various applications has been demonstrated in recent times through the bulk thermal annealing of laser patterned glass. Low temperature stable laser-induced bond breaking, density and stress changes within an optical fibre have since been stabilized to temperatures in excess of 1200°C through the process of regeneration. We review this process in light of recent results.

Key words: regeneration, silica, glass relaxation, annealing, gratings, optical fibre, germanosilicate, fluorine, diffusion, nanophotonics, nanoscale, waveguides.

9.1   Introduction

Silica remains the heart of nearly all modern optical transport systems. Engineering of silica in its various forms ranges from one-dimensional (1D) to three-dimensional (3D) waveguide and periodic structures, including recent interest in 3D photonic crystals. Most of the preparation methods involve complex vapour deposition and various co-dopants, which have an advantage of overcoming the lack of finesse involved with general formation of glass structure through high temperature processing and quenching. Nevertheless, to obtain micron or submicron precision over the processing of glass, post-processing methods are often used, ranging from lithographic etching of systems with dopants to laser processing using UV to mid-IR lasers, through to tapering and drawing of patterned preforms and capillaries into nanostructured fibres and wires. Concrete examples of laser-based micron scale processing of glass include direct written waveguides, Bragg gratings in waveguides and optical fibres and photonic crystals. The drawback with non-thermal post-processing techniques is that, to date, they generally produce glass that is structurally less stable than the starting phase; this includes that produced by femtosecond laser-induced melting

287

and solidification. The relationship between glass stability and the rate of freezing in a structure warrants further investigation.

In contrast, a great deal of control of glass properties is achieved with simple thermal processing, varying both the softening and quenching rates of the glass. By simple processing, a striking range of properties can be achieved with just pure silica alone – controlled quenching of liquid and partial liquid phases can produce a variety of polymorphic phases of varying density and structure. In fact the common distribution of crystalline states of more than a quarter of the Earth's crust is strong evidence that the Earth has cooled down slowly. By comparison, rapid quenching leads to a potential continuum of polyamorphic states – these need not be distinct but can span a significant density range. Rapid quenching is also characteristic of chemical precipitation. Recent discoveries, by one of NASA's rovers, of amorphous silica on the Martian surface would suggest a surprising amount of chemically precipitated silica given the high hydrogen content and amorphous nature (http://www.sciencedaily.com/releases/2007/05/070521154427.htm) (alternatively, rapid thermal quenching of an eruption could have also prevented crystallization, potentially indicative of a distinct thermal history on Mars). This extraordinary range is, within the context of our planet and our brief history, unique to silica and arises principally from the topological rearrangements possible from a simple, but rigid, tetragonal structure of four oxygen atoms and one silicon atom. On a local scale the structure is paradoxically highly ordered whilst on a longer scale that order disappears rapidly into a so-called random network. The rigidity (which tends to lead to large ring collapse within silica) and flexibility of the structure (which is critical for the very low propagation losses enabling the wiring of the internet) can be further expanded by the addition of small dopant quantities, materials which coordinate reasonably well, although the number of defect states rises dramatically. It is no wonder silicon and silica mark the photon epoch we have just entered.

Despite the extraordinary versatility thermal processing offers, very little work has evolved to be able to control it at a dimensional level. There are only a few examples of this, a particularly striking one being the exploitation of two-photon excitation, normally a problem in the semiconductor industry, into the band edge of silica to achieve densification and damage of glass to write Bragg gratings in silica core (Albert *et al.*, 2002) and all-silica (Groothoff *et al.*, 2003) optical fibres. Extending this to higher multiphoton processing, often involving very rapid and localized quenching quickly freezing in large index contrasts, was made possible with the advent of readily available femtosecond lasers. These impressive achievements suggest lasers are one way to go in achieving highly localized, and rapid, glass transformation potentially offering a unique finesse in thermally tuning change compared to conventional radiative or convective heating. Band edge absorption relies heavily on the subsequent thermalization that occurs with localized excitation. A review of these photo-writing processes can be found in Canning (2008). More recently, we have considered the concept of attempting to localize thermal processing on a scale that is unprecedented by combining the localization possible using lasers with thermal processing (Bandyopadhyay *et al.*, 2008; Canning *et al.*, 2008–2010) following earlier work (Canning, 2004; Zhang and Kahriziet, 2007). In effect, the process of 'regeneration', carried out in optical fibres but certainly not restricted to them, involves patterning a seed structure (in any dimension – 1, 2 or 3D) using a laser, perhaps through defect excitation, to effect a difference between one region and another. In so doing, the changes in local structure and stress between regions will lead to a subsequent competitive struggle for change when the whole glass is heated. This competition arises principally from different relaxation rates as a result of changing glass structure by initial irradiation and varying stress fields between such changes. The general concept of patterning the relaxation of glass on a local scale using lasers to achieve localized change was proposed by Canning (2004). The possibility of glass transformation as the template for ultra high temperature gratings and ultra high temperature micro and nano processes was confirmed when we first saw the work of Zhang and Kahriziet (2007), an observation of regeneration in just standard photosensitive fibres. Subsequently, we noted similar work predating this from Juergens *et al.* (2005) – where Zhang and Kahriziet (2007) did not ascribe a mechanism, Juergens *et al.* (2005) assumed a so-called 'chemical composition grating' based on oxygen diffusion, despite some contradictory evidence. Instead, based on the results of Zhang and Kahriziet (2007) we recognized that what we had previously predicted was demonstrated in fibre form: spatial patterning of structural relaxation in glass and that this regeneration approach was one way to achieve this. Hence, we were able to optimize the annealing process to demonstrate strong regenerated gratings (Bandyopadhyay *et al.*, 2008, 2011; Canning *et al.*, 2008, 2009, 2010).

Given the immediate relevance to nanophotonics and the potential of nanoscale processing to produce sophisticated components, together with the growing area of 'extreme' photonics where high temperature operation is regarded as increasingly important, we review that work to show that the regeneration retains fully the seed structure in optical fibres with nanoscale resolution.

9.2   The regeneration process

In reviewing this work and its broader implications, we focus on our work in optical fibres where the picture may appear complicated somewhat by the presence of germanate doping and core–cladding stresses – however, so long as these are exploited correctly the underlying principle is identical. That is the germanate is essential for the fabrication of very strong *seed* gratings through conventional fibre Bragg grating writing; in the presence of hydrogen these seed gratings are proportional to the generation of OH via defect excitation and require no significant contribution to the polarizability change via densification to account for the local induced index change (although clearly this may be present). What is remarkable about hydrogen is that simple physical diffusion of the gas into the fibre alone leads, through hydrogen bonding and OH formation, to significant internal pressures that help reduce tensile stresses produced during optical fibre fabrication which can improve the size of the induced index change. This can significantly alter the distribution of relaxation processes, usually described by the relaxation term, β (Angell, 1995), since the local structural change and the surrounding stresses are altered differently. In contrast, when hydrogen is not used, thermal processing leads to a rollover in index change regenerating gratings that are identical in all aspects to type $1n$ (or type IIa) gratings obtained by continual laser processing (Lindner *et al.*, 2009, 2011a) – this is very strong evidence that the formation of type $1n$ (type IIa) gratings is, in fact, thermally driven. Laser-based annealing becomes a very potent tool if examined in this light. The very slow formation rate along with the much lower annealing temperature suggests' crystallization of the germanate glass rather than silica; it is dopant dependent.

By contrast, an important systematic study was reported using fibres with varying dopants to show that dopants do not directly have a significant contribution to the final regeneration when hydrogen is present (Bandyopadhyay *et al.*, 2011), as was first predicted if the mechanism of regeneration was based primarily on patterning the relaxation kinetics of glass (Bandyopadhyay *et al.*, 2008). Figure 9.1 summarizes a general plot of the obtained regenerated normalized reflectivity versus germanate concentration for seed gratings (all ~ 47 dB) in four types of fibre (SMF 28; two specialty fibres produced at CGCRI in India – NM113 and NM41 – and a UV photosensitive fibre UVS_EPS); in one fibre fluorine is contained in the cladding to see if these regenerated gratings are related to the 'chemical composition gratings' of Fokine (2002–2004) and Trpkovski *et al.* (2005), which are described in terms of chemically assisted diffusive mechanisms.

Interestingly, it would appear that there is an inverse relationship between regenerated grating strength and germanate concentration. This trend immediately rules out germanate as being directly involved with regeneration beyond determining the seed grating strength – in fact, one might mistakenly assume it is deleterious to the process. However, when taking into account the numerical aperture (NA) (V parameter) of the fibre, and based on a model where the structural change takes place in silica and therefore most likely occurs at the core–cladding interface where stress is highest, the

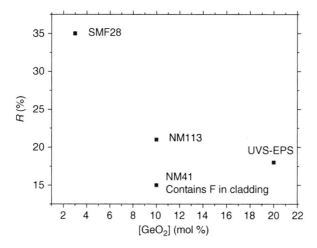

9.1 Regenerated gratings from four different types of fibres: SMF 28, NM113 and NM41 (difference being NM41 contains F in the cladding) and photosensitive UVS_EPS (adapted using data listed in Bandyopadhyay *et al.*, 2009).

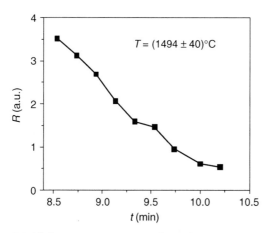

9.2 High temperature annealing of a regenerated grating in boron co-doped germanosilicate optical fibre. Control of the temperature above 1400°C was limited by the available oven – an uncertainty error bar of ±40°C is used in Åslund *et al.* (2010).

dependence of germanate is negligible and can be ruled out. Most interestingly, some support for this interface model is the deleterious effect of fluorine, present in the cladding layers of one fibre (NM41). This has a significant effect in reducing the regeneration and probably can be explained by diffusion of fluorine, which happens readily at these temperatures, leading potentially to a reduced fringe contrast. These results directly contrast with that expected from so-called 'chemical composition gratings' originally thought to arise from fluorine diffusion (Fokine, 2002) and later modified to enhanced fluorine-assisted selective oxygen diffusion (Fokine, 2004, 2009) (after suggestions by others that fluorine could not be responsible but rather oxygen partly based on stoichiometric, diffusion and structural considerations). An oxygen diffusion model was also taken at face value by Trpkovski et al. (2005) and Juergens et al. (2005), although the latter did point out conflicting results. It appears likely that the origin of these diffusive approaches can be traced to the work on valid thermally induced diffusion gratings (varying 'chemical composition gratings') demonstrated by Dianov et al. (1997) using localized heat sources (> 1000°C) such as CO_2 lasers to fabricate long period gratings. Although it is unclear how uniform heating of a standard grating, chemically activated by periodic irradiation of fluorine or OH formation, can generate such distinct and stable periodic structures based on chemically assisted diffusion alone at lower temperatures, the logical extension is that any dopant can be used to create regenerated gratings. Our results are inconsistent with such a mechanism as the primary cause of change – in contrast, our mechanism which is generic and not dependent on the local chemical details but rather glass structure, does explain the results and is based on a simple extension of well-known glass kinetics and relaxation (Canning, 2004) so it almost certainly must be able to occur.

It seems clear that regenerated gratings do not follow the type of behaviour expected solely based on diffusion. Further, whilst diffusion obviously can occur at these temperatures, such diffusive models would lead to fringe decay, often uneven, and therefore changes in optical phase as the index also changes in between the periodically laser processed regions amplifying non-uniformities. It is difficult to rationalize the results obtained in high NA germanosilicate optical fibre annealed at 1295°C (Canning et al., 2009), where the fibre core is soft, and at even higher temperatures until 1400°C was exceeded (Åslund et al., 2010) and whether it would be possible if the changes were dependent on diffusing dopants (since there is no reason why diffusion should stop at these higher temperatures). The evidence would indicate that dopant diffusion, whilst probably present, is not critically involved and *that the major changes are associated with silica structural phase changes* probably at the core–cladding interface. Since silica can operate at these temperatures, the exclusion of dopants in the first instance makes the system a very simple one to understand in terms of conventional glass processing as we have proposed (making the optical fibre potentially a wonderful laboratory for studying glass and its changes optically). From a fundamental science perspective, the broader picture we arrived at was that it is possible to tailor and pattern any amorphous material, with either laser or thermal systems or indeed any alternative means, by patterning its local relaxation kinetics and thermodynamics, either 1D, 2D or indeed 3D (enabled by multiphoton absorption processing such as those possible with near-IR femtosecond lasers). Practically speaking this is also extremely exciting – the higher temperature resistance of silica, in combination with laser processing conditions using dopants, means that potentially high resolution, nanoscale and thermally robust patterning of glass is possible, as originally predicted (Bandyopadhyay *et al.*, 2008). Interestingly, after our original work, so-called chemical composition gratings were redescribed in terms of changing the local structure (through diffusion) to achieve differing relaxation profiles between processed and unprocessed glass (Fokine, 2009). Although limited to a single relaxation process, the description in terms of relaxation is identical to what we have been saying (Bandyopadhyay *et al.*, 2008; Canning, 2004) and is an acknowledgement that nanoscale processing of glass will be critically dependent on the physical picture we have been espousing. Whether chemical composition gratings are indeed identical to high temperature regenerated gratings, but perhaps described in terms of details that are independent of the larger picture of glass transformation and consequently may not be optimized appropriately, may be an interesting topic for some readers to pursue – given some inconsistencies with our experimental results, we have continued to assume the two gratings to be distinct. More importantly, the general consensus is that a model based on glass transformation of the type we first suggested is probably correct. Undoubtedly, the opportunity for exploring new routes to such transformations will herald significant results not only in optical fibres but in two- and three-dimensional photonic waveguides and devices, for potential use in extreme environments.

In summary, the dopants are therefore important in inscribing the initial seed grating since there is a direct correlation with regeneration and the seed grating strength, not surprising given that it should also correlate with the variation in local relaxation rates. Further, the retention of phase information (Canning *et al.*, 2009, 2010) throughout what would otherwise be regarded as substantially harsh thermal processing, confirms nanoscale processing is possible. More immediately, we may conclude that if this is the case, then gratings that are regenerated should be able to retain all the complex phase information of the seed grating if thermally processed uniformly. This could have extremely important ramifications since in principle ultra high temperature stable complex patterning in glass can be achieved, something that is increasingly important in a number of industries, including semiconductor optical lithography and ultra high power fibre laser filters and reflectors. Further evidence decoupling the role of hydrogen in regeneration completely from the initial seed grating is the recent demonstration of post-hydrogen loading regeneration of seed gratings written with no hydrogen (Canning *et al.*, 2011; Lindner *et al.*, 2011b): the presence of hydrogen leads to regeneration that appears identical to that so far reported, far exceeding the thermal stability from type 1*n* (type IIa) – like regeneration (Lindner *et al.*, 2009, 2011a). In addition to its scientific importance, this decoupling allows for a significant expansion of the regeneration process to gratings that are produced without hydrogen, including on-line draw tower gratings (Lindner *et al.*, 2011b).

9.2.1 Upper temperature limits

Much more extensive work is required to explore the limits of both thermal and mechanical stability of these gratings if they are to enhance the value of such gratings. More generally, for example, Juergens *et al.* (2005) do report on surface micro cracks from long-term exposure of optical fibre gratings at very high temperatures. Such effects, attributed to potential water ingress and resulting expansion of the glass (similar to that underlying the stress role of hydrogen we suggested), can reduce the loading strain of an optical fibre let alone any grating structure. It can also affect relaxation of the glass overall leading to Bragg wavelength shifts that are not reversible.

In recent work (Åslund *et al.*, 2010) rapid annealing of regenerated gratings above 1400°C was reported, signalling that depending on the fibre core softening temperature an upper limit around this value is expected (summarized in Fig. 9.2). It also suggests that much more robust results can be obtained using better cores, either free of dopants or with compositions that have higher softening points, such as aluminosilicate cores. Another significant issue when processing at these very high temperatures is clearly packaging. This grating had to be fused within a large silica capillary to prevent breaking from increased fragility – this form of packaging turns out to be quite effective in preventing the normal moisture ingress associated with fibre surfaces at high temperatures, although for some sensing work it can reduce access to the grating. Alternatively, using dry inert atmospheres or in vacuum can provide better resistance (though these are unlikely to exist in many commercial applications such as high temperature smelting operations but may do so in niche areas). Packaging requirements will be determined by applications ranging from the oil and gas industries to structural health monitoring of spacecraft and equipment.

9.2.2 Characterizing seed and regenerated optical fibre gratings

The first step in precipitating structural change associated with regenerated gratings is a seed grating. We have observed that the stronger the seed grating, the stronger the final regenerated grating, typically 10%–15% in grating index modulation to date, though this will vary somewhat depending on uniformity, fringe contrast and duty cycle (and assuming the V parameter is considered). Given the importance of evaluating the nanoscale resolution possible with regeneration, we reproduce earlier work here (Canning *et al.*, 2009, 2010).

A conventional Bragg grating ($L = 50$ mm) was inscribed into a relatively highly Ge doped step index fibre with no boron ($r_{core} \sim 2$ μm, [$GeO_2 \sim 10.5$ mol%], $\Delta n_{co/cl} = 0.012$) using the 244 nm output from a frequency doubled Ar^+ laser ($P \sim 50$ W/cm², $f_{cumulative} \sim (6{-}12)$ kJ/cm²). The use of a small core fibre was based on the possibility that the changes may be occurring at the stressed core–cladding interface (Bandyopadhyay *et al.*, 2008). Ignoring the slight quadratic chirp in the transmission profile in Fig. 9.3a, the simulation spectra for a uniform grating, based on transfer matrix solution of the coupled mode equations, was fitted to the bandwidth to estimate the index modulation achieved: $\Delta n_{mod} \sim 1.6 \times 10^{-3}$, consistent with a grating > 120 dB in strength.

Using a processing procedure identical to that optimized in earlier work, ultra strong seed gratings were thermally processed. At 950°C the onset of regeneration is observed, and as the seed grating disappears completely, the regenerated grating appears (Fig. 9.3b). Numerical simulation indicates an index modulation of $\Delta n_{mod} \sim 1.55 \times 10^{-4}$. This is consistent with a general observation that the regenerated grating index modulation is typically $\Delta n_{reg}/\Delta n_{seed} \sim (10\%{-}15\%)$ of the seed grating modulation, although this will be sensitive to the level of fringe contrast involved with seed grating fabrication.

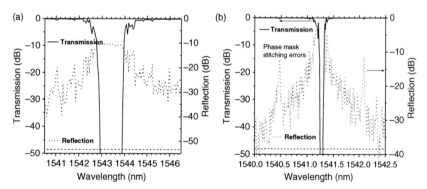

9.3 (a) T & R spectra of the seed grating. Noticeably, the large side lobes of this structure obscure the stitching errors expected from the phase mask used. The dashed line represents the noise floor; (b) T & R spectra of the regenerated grating. Noticeably, the stitching errors of the phase mask are clearly visible indicating all the relative phase information has been retained. The dashed line represents the noise floor. (Reproduced from Canning *et al.*, 2009a.)

This parameter will become increasingly important as a potential means of comparing regenerated gratings produced by different means in different fibres. The overlap integral may be added for completeness.

In Canning *et al.* (2010) we reported using a second regenerated grating made from a weaker seed grating, written with a cumulative fluence ~ 30% less than that of the first grating, so that the full transmission spectrum can be observed within the noise floor of the tunable laser and power meter setup. The second regenerated grating was used to confirm the longer term performance at 1000°C and 1100°C.

9.3   Estimating the retention of nanoscale information in regenerated grating structures

In order to determine whether this process can be applied beyond simple Bragg grating writing as a realistic approach to the production of complex gratings and patterns and structures that can operate at high temperature whilst retaining the complexity of a nanoscaled device, the impact of the regeneration process on two complex grating structures was explored (Canning *et al.*, 2010):

(1) Structure consisting of two superposed gratings with λ_1 ~ 1548.73 nm and λ_2 ~ 1553.56 nm, that is, with $\Delta\lambda$ ~ 4.8 nm; and
(2) A dual channel grating produced by writing a Moiré grating. In a Moiré grating, the refractive index variation along the length of the grating is also different where a uniform period, Λ_B, is modulated by a low spatial frequency sinusoidal envelope of period, Λ_e, that produce two sidebands (essentially a phase shifted structure built up from a periodic distribution of identical phase shifts). Given the sensitivity of the Moire grating to any perturbation in phase, the preservation of the transmission notch and overall profile will be indicative of nanoscale resolution in the regenerated structure.

For the superposed gratings (L ~ 5 mm) were inscribed into a H_2 loaded (24 h, $P = 100$ atm, $T = 100°C$) GeO_2 doped core silica fibre ([GeO_2] ~ 10%, fabricated at CGCRI) using a pulsed KrF exciplex laser (248 nm, pulse duration = 20 ns, f_{pulse} ~ 70 mJ/cm², repetition rate = 200 Hz). The Moiré grating was written into a fibre which was similar to that used for the superposed grating but also had boron to increase the seed photosensitivity. Regeneration is carried out with an identical recipe to that described earlier but inside a short fibre micro-heater. The hot zone of this heater is supposedly uniform over 5 mm only (the exact variation along this length is not known but we suspect a Gaussian profile), and this dictates the grating length.

9.3.1 Superposed gratings

Sample #1 was prepared by superposing two seed gratings with Bragg wavelengths $\lambda_1 \sim 1548.73$ nm and $\lambda_2 \sim 1553.56$ nm, that is, with $\Delta\lambda \sim 4.8$ nm. Each of the seed gratings was of moderate strength with transmission loss at $\lambda \sim -20$ dB (grating with λ_1 being slightly stronger than that at λ_2). The superposition of two gratings leads to a compound form of the local index modulation described as (Bao *et al.*, 2001):

$$\Delta n(z) = 2\Delta n_0 Cos\left(\frac{\pi(2\Lambda_{B1} + \Delta\Lambda)}{\Lambda_{B1}(\Lambda_{B1} + \Delta\Lambda)}z + \frac{\Delta\Phi}{2}\right)Cos\left(\frac{\pi\Delta\Lambda z}{\Lambda^2} - \frac{\Delta\Phi}{2}\right) \qquad [9.1]$$

Λ_{B1} and Λ_{B2} are the periods of the gratings with $\Lambda_{B2} = \Lambda_{B1} + \Delta\Lambda$ and $\Delta\Phi$ is the initial phase difference of the gratings. It is clear from this expression any non-uniformity introduced by the thermal annealing process will result in a spread of $\Delta\Phi$ and broadening of the peaks. The structure was then thermally processed as described earlier until regeneration was complete. The results are summarized in Fig. 9.4. Within experimental uncertainty, the Bragg wavelength separation remains the same (~ 4.8 nm) although, as expected, the annealing has led to a decrease in average index so that the Bragg wavelengths are blue-shifted. This reduction leads to a change in the phase distribution and the regenerated gratings have a more asymmetric profile, shown in the inset of Fig. 9.4c. This is consistent with a very weak Gaussian, or quadratic, chirp on the grating. The origin for this chirp almost certainly arises from the hot zone temperature distribution of the microheater rather than any intrinsic grating property.

9.3.2 Moiré gratings

In a Moiré grating, a uniform period, Λ_B, is modulated by a low spatial frequency sinusoidal envelope of period, Λ_e (Fig. 9.5) that produces two sidebands. The structure is equivalent to two gratings with stopgaps that overlap sufficiently to produce a resonant phase shift-like structure in the stop gap of the superstructure. A similar profile is obtained by placing phase shifts with a low frequency period along a uniform grating.

The position dependent index amplitude modulation profile can be described as (Ibsen *et al.*, 1998):

$$\Delta n(z) = 2n\Delta n_0 F(z)\cos\left(\frac{2\pi Nz}{\Lambda_B}\right)\cos\left(\frac{2\pi Mz}{\Lambda_e}\right) \qquad [9.2]$$

where N and M are integer and $2n\Delta n_0$ is the UV induced index change, $F(z)$ is the apodization profile. On simplifying, Eq. [9.2] directly leads to the

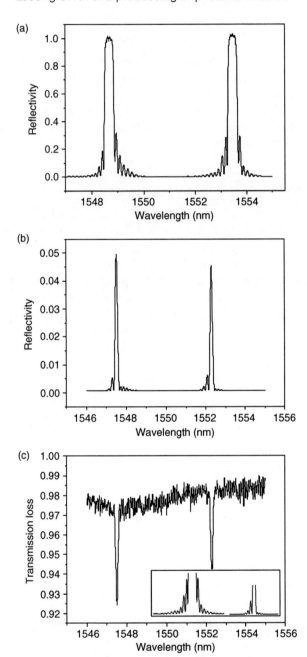

9.4 Spectrum of dual over-written gratings. (a) Normalized reflection spectrum of the seed; (b) and (c) reflection and transmission spectrum of the regenerated grating respectively represented in absolute scale. Inset: close-up of side lobe structure of right hand peaks of seed and regenerated grating for comparison. (Modified from Bandyopadhyay *et al.*, 2009; Canning *et al.*, 2009.)

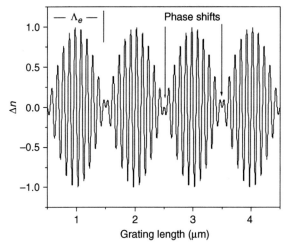

9.5 Index modulation introduced into the seed Moiré grating (from Bandyopadhyay *et al.*, 2009; Canning *et al.*, 2009).

resultant spatial frequencies at the sum and difference frequencies where two Bragg reflections will occur and may be represented as:

$$\Delta n(z)=n\Delta n_0 F(z)\left\{\cos\left(\frac{2\pi N}{\Lambda_B}\left[1+\frac{M\Lambda_B}{N\Lambda_e}\right]z\right)+\cos\left(\frac{2\pi N}{\Lambda_B}\left[1-\frac{M\Lambda_B}{N\Lambda_e}\right]z\right)\right\} \qquad [9.3]$$

The new reflection has two effective bands separated in wavelength, $\Delta\lambda$, as:

$$\Delta\lambda=\frac{\lambda_B^2}{2n_{eff}\Lambda_e} \qquad [9.4]$$

Based on the principle mentioned above a dual seed grating with a 100 GHz separation, that is, $\Delta\lambda \sim 0.8$ nm, was written. Selected $\Lambda_B = 533.17$ nm produces a grating with $\lambda_B \sim 1554$ nm. Modulating Λ_B with $\Lambda_e = 1028$ μm we could generate two channels Bragg wavelengths, $\lambda_1 = 1553.51$ nm and $\lambda_2 = 1554.34$ nm respectively. The effective refractive index of the fibre is $n_{eff} = 1.4573$. A precisely controlled scanning beam writing setup was used to produce π-phase shifts at specific locations of the grating to generate the required low frequency sinusoidal modulation of the index profile. A summary of the induced profile is shown in Fig. 9.6. The seed grating reflection profile is shown in Fig. 9.6a and the regenerated grating reflection and transmission profiles are shown in Fig. 9.6b and 9.6c. Unlike the superposed gratings, where the sidebands of the grating are a result of the interference between end reflections of the grating and therefore susceptible to temperature

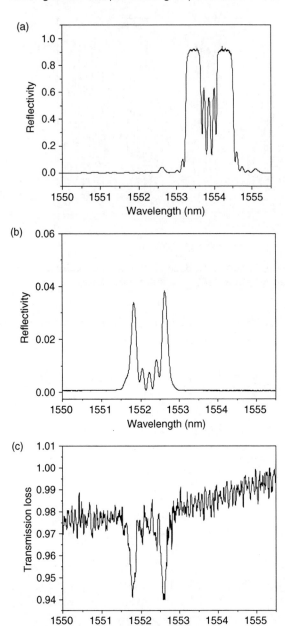

9.6 Spectrum of Moiré grating. (a) Normalized reflection spectrum of the seed, (b) and (c) reflection and transmission spectrum of the regenerated grating respectively (from Bandyopadhyay *et al.*, 2009; Canning *et al.*, 2009).

gradients in the micro-heater hot zone, the interference in the phase shift region is a result of the distributed interference between the grating and super period of the phase shifts. This means the structure is less sensitive to overall gradients on a macroscale. Importantly, the interference in the phase shift region is preserved after regeneration indicating that despite the very large macro-heating process involved in creating the regenerated grating, the structure retains full memory of the seed grating, indicating that there is no evidence of a diffusive process that would alter the phase relationship anywhere over the grating length. Full preservation on a nanoscale is maintained through regeneration – this is a remarkable result.

9.4   Conclusions

The preservation of optical phase information within a grating is evidence that the resolution of features within a laser-written seed grating can be preserved on a nanoscale after annealing and subsequent regeneration. Clearly information is retained; almost certainly the memory is stored through differences in glass structural properties between original seed processed and unprocessed regions. Hydrogen preferentially selects out changes in silica; without hydrogen it would seem that lower temperature effects dominate. The exact role of hydrogen is unclear, but above 500°C we know Si-OH is readily formed (Sørenson et al., 2005); laser irradiation may likely increase the defect sites where OH can form leading to a higher concentration and therefore greater stress changes (swelling) which can relieve core–cladding tensile stresses more within these regions, for example. This explains the original observed sensitivity of regeneration to fibre tension (Bandyopadhyay et al., 2008). However, more studies are necessary, including potential simulations such as that previously used to study radiation bond breaking of Si-O (Wootten et al., 2001).

The recent work decoupling the role of hydrogen from seed grating fabrication (Canning et al., 2011; Lindner et al., 2011b) points quite clearly to a common mechanism between both with and without hydrogen type I gratings in retaining memory, despite one being associated with mostly OH formation and the other with density change and defect polarizability. In spite of some differences, both lead to periodic stress along the grating, although where one involves compaction the other may involve dilation or net reduced compaction. This, together with the original observation of Bragg wavelength mismatch between regenerated and seed gratings as a function of applied tension, suggests stress and structural changes at the interface are critical. Separating out the contributions of dopants, especially those in the core, strongly supports the hypothesis that important change involves silica transformation most likely near the core–cladding interface (consistent with differences in regenerated Bragg wavelength shift from the original seed grating as a function of tension, as originally first observed). The nature of this transformation remains an area of important investigation. These structures can withstand temperatures up to and beyond 1400°C for short periods before rapid annealing is observed, although more detailed work is required to better quantify the conditions of these upper limits. At this temperature the majority of core dopants will lead to core glass softening, making it less likely that the important change is occurring in the core. Core–cladding interface changes are strongly suggestive of stress as a significant contribution to the type of glass polymorph or polyamorph that is induced. Thus further work with applied stresses is warranted.

The retained complex functionality of a laser-written seed structure after very high temperature processing, tantalizingly hints at the prospect of ultra high temperature patterning of glass, something vital for many industries, including semiconductor lithography where more robust masks are necessary and also in applications that require robust holographic filters or reflectors such as high power fibre lasers and a new generation of high power holograms (including a potentially lower cost approach to fabricating optical phase masks for fibre grating writing).

Given that much of the general process appears to involve glass re-quenching under a different thermal history, both the thermal stabilization and the regenerative processes described here are unlikely to be confined to silica fibres, or waveguides, loaded with or without hydrogen (as recent results indicate). Lower temperature regeneration without hydrogen suggests changes in structure of the germanate glass, which is less stable than silica. The correlation with laser-written regeneration such as type I*n* (type IIa) gratings suggests that high multiphoton lasers can be fine-tuned to achieve similar laser regeneration in pure silica through continued exposure, another potential subject area of investigation.

This work indicates potentially superior performance may still be extracted by optimizing the thermal robustness (or indeed melting point) of the core glass relative to the cladding. Exploring other systems such as aluminosilicate, phosphosilicate and indeed other glasses (oxide or otherwise) to explore the predicted dependency, or in some cases limitations, is another area of important work.

There is no obvious reason why these changes could not be reproduced under different material systems and indeed different laser preparations given the foundation of glass transformation under appropriate cooling and quenching kinetics – all that is required is an amorphous network (and even some possible discrete transitions within crystals could be prepared analogously). As such, the first suggestion for controlling relaxation kinetics and thermodynamics was by laser (Canning, 2004) – here, we have reviewed a hybrid approach where a seed is patterned with a laser and thermal macro annealing in the presence of hydrogen completes the transformation. There is no reason why the entire process could not be done using laser or other variations. Anyone skilled in the area of amorphous materials can see the broader potential of such controlled transformations in their particular niches.

9.5   Acknowledgements

We would like to acknowledge various colleagues who have worked on several aspects of regenerated gratings at different stages. Some of the more recent results warrant special mention of the following: Palas Biswas from CGCRI, Michael Stevenson, Kevin Cook, Mattias Aslund from iPL, and Hongyan Fu and Hwayaw Tam from Hong Kong Polytechnic.

Funding from the Australian Research Council (ARC) and an International Science Linkage Grant from the Department of Industry, Innovation, Science and Research (DIISR), Australia and the Council of Scientific and Industrial Research (CSIR), India, under the 11th five-year plan is acknowledged.

9.6   References

Albert, J., Fokine, M. and Margulis, W. (2002), 'Grating formation in pure silica-core fibers', *Optics Letters*, **27**, 809–811.

Angell, C. A. (1995), 'Formation of glasses from liquids and biopolymers', *Science*, **267**, 1924–1935.

Åslund, M. L., Canning, J., Fu, H. and Tam, H. (2010), 'Rapid disappearance of regenerated fibre Bragg gratings at temperatures approaching 1500°C in Boron-codoped Germano silicate optical fibre', *4th European Workshop on Optical Fiber Sensors* (EWOFS), Porto Portugal.

Bandyopadhyay, S., Canning, J., Stevenson, M. and Cook, K. (2008), 'Ultrahigh-temperature regenerated gratings in boron-codoped germanosilicate optical fiber using 193 nm', *Optics Letters*, **33**(16), 1917–1919.

Bandyopadhyay S., Canning J., Biswas P., Chakraborty, R., Dasgupta, K. (2009), 'Regeneration of complex Bragg gratings', *Proceedings of SPIE*, **7503**, 750371.

Bandyopadhyay, S., Canning, J., Biswas, P., Stevenson, M. and Dasgupta, K. (2011), 'A study of regenerated gratings produced in germanosilicate fibers by high temperature annealing', *Optics Express*, **19**(2), 1198–1206.

Bao, J., Zhang, X., Chen, K. and Zhou, W. (2001), 'Spectra of dual overwritten Bragg grating', *Optics Communications*, **188**, 31–39.

Canning, J. (2004), 'The characteristic curve and site-selective laser excitation of local relaxation in glass', *Journal of Chemical Physics*, **120**(20), 9715–9719.

Canning, J. (2008), 'Fibre gratings and devices for sensors and lasers', *Lasers and Photonics Reviews*, **2**(4), 275–289.

Canning, J., Stevenson, M., Bandyopadhyay, S. and Cook, K. (2008), 'Extreme silica optical fibre gratings', *Sensors*, **8**, 6448–6452.

Canning, J., Bandyopadhyay, S., Stevenson, M., Biswas, P., Fenton, J. and Aslund, M. (2009), 'Regenerated gratings', *Journal of the European Optical Society*, **4**, 09052.

Canning, J., Bandyopadhyay, S., Stevenson, M., Biswas, P., Fenton, J., Chakraborty, R., and Aslund, M. (2009), 'Thermal preparation of highly stable glass periodic changes with nano-scale resolution using a laser-inscribed hydrogen loaded seed template', *IEEE Laser and Electro Optics Society (LEOS) Conf.*, 30–31, (2009)

Canning, J., Bandyopadhyay, S., Biswas, P., Aslund, M., Stevenson, M. and Cook, K. (2010), 'Regenerated fibre Bragg gratings'. In Pal, B. (ed.), *Frontiers in guided wave optics and optoelectronics*. Vienna: In-Tech.

Canning, J., Lindner, E., Cook, K., Chojetzki, C., Brückner, S., Becker, M., Rothhardt, M. and Bartelt, H. (2011), 'Regeneration of gratings by post-hydrogen loading', *The International Quantum Electronics Conference (IQEC)/The Conference on Lasers and Electro-Optics (CLEO) Pacific Rim*, (IQEC/CLEO-Pacific Rim 2011), Sydney, Australia.

Dianov, E. M., Karpov, V. I., Grekov, M. V., Golant, K. M., Vasiliev, S. A., Medvedkov, O. I. and Khrapko, R. R. (1997), 'Thermo-induced long period fibre gratings', *IOOC-ECOC'97*, Edinburgh, UK, **2**, 53–56.

Fokine, M. (2002a), 'Formation of thermally stable chemical composition gratings in optical fibers', *Journal of the Optical Society of America B*, **19**, 1759–1765.

Fokine, M. (2002b), 'Thermal stability of chemical composition gratings in fluorine-germanium-doped silica fibers', *Optics Letters*, **27**, 1016–1018.

Fokine, M. (2004), 'Thermal stability of oxygen-modulated chemical-composition gratings in standard telecommunication fiber', *Optics Letters*, **29**, 1185–1188.

Fokine, M. (2009), 'Manipulating glass for photonics', *Physica Status Solidi A*, **206**(5), 880–884.

Groothoff, N., Canning, J., Buckley, E., Lyttikainen, K. and Zagari, J. (2003), 'Bragg gratings in air-silica structured fibers', *Optics Letters*, **28**, 233–235.

Ibsen, M., Durkin, M. K. and Laming, R. I. (1998), 'Chirped Moiré fiber gratings operating on two wavelength channels for use as dual-channel dispersion compensators', *IEEE Photonics Technology Letters*, **10**, 84–86.

Juergens, J., Adamovsky, G., Bhatt, R., Morscher, G. and Floyd, B. (2005), 'Thermal evaluation of fiber Brgag gratings at extreme temperatures', *43rd AIAA Aerospace Sciences Meeting and Exhibit*, paper 1214 (AIAA – American Institute of Aeronautics and Astronautics).

Lindner, E., Chojetzki, C., Brückner, S., Becker, M., Rothhardt, M. and Bartel, H. (2009), 'Thermal regeneration of fiber Bragg gratings in photosensitive fibers', *Optics Express*, **17**, 12523–12531.

Lindner, E., Chojetzki, C., Canning, J., Brückner, S., Becker, M., Rothhardt, M. and Bartel, H. (2011a), 'Thermal regenerated type IIa fiber Bragg gratings for ultra-high temperature operation', *Optics Communications*, **284**(1), 183–185.

Lindner, E., Canning, J., Chojetzki, C., Brückner, S., Becker, M., Rothhardt, M. and Bartelt, H. (2011b), 'Post-hydrogen-loaded draw tower fiber Bragg gratings and their thermal regeneration', *Applied Optics*, **50**(17), 2519–2522.

Sørenson, H. R., Canning, J. and Kristensen, M. (2005), 'Thermal hypersensitisation and grating evolution in Ge-doped optical fibre', *Optics Express*, **13**(7), 2276–2281.

Trpkovski, S., Kitcher, D. J., Baxter, G. W., Collins, S. F. and Wade, S. A. (2005), 'High-temperature-resistant chemical composition Bragg gratings in Er3+-doped optical fiber', *Optics Letters*, **30**, 607–609.

Wootten, A., Thomas, B. and Harrowell, P. (2001), 'Radiation-induced densification in amorphous silica: a computer simulation study', *Journal of Chemical Physics*, 115, 3336–3341.

Zhang, B. and Kahriziet, M. (2007), 'High temperature resistance fiber Bragg grating temperature sensor fabrication', *IEEE Sensor Journal*, **7**, 586–590.

Femtosecond-laser-induced refractive index modifications for photonic device processing

M. AMS, D. J. LITTLE and M. J. WITHFORD,
Macquarie University, Australia

Abstract: The femtosecond laser direct-write technique has emerged as a powerful tool for fabricating high-quality integrated photonic components. The direct-write platform combines flexible, rapid prototype with genuine 3D capability, compatibility with fibre-optics and the ability to process almost any transparent material. In this chapter, the physics that underpin the femtosecond laser-induced modification and the mechanisms that result in the emergence of a local refractive index change are reviewed. Experimental methods for leveraging this process to make photonic components and the current capabilities of the femtosecond laser direct-write platform are also discussed.

Key words: femtosecond laser, direct-write, waveguide, refractive index, photonic.

10.1 Introduction

The fabrication of photonic devices via laser systems is widely recognized as an enabling technology for the innovation of advanced components for application in fields such as communications, quantum information science, biophysics and medicine, to name but a few. Since the development of affordable and commercially available ultrashort pulsed lasers there has been an explosion in this research with groups all around the world reporting breakthrough science in these fields. This is because ultrashort pulsed lasers are capable of producing high irradiances which can drive a controllable nonlinear optical breakdown in transparent materials, extending the range of materials that can be processed in order to realize photonic devices.

Initially, research groups concentrated on fabricating photonic devices on the surface of their chosen material. Ultrafast laser surface processing of metals, polymers, glasses and crystalline materials became commonplace. As technologies and fabrication techniques advanced, a diverse range of opportunities and activity in subsurface laser processing of materials emerged. An example of this can be seen in the fibre sensing and fibre laser fields of research where ultrafast laser-written gratings inscribed into the core of optical fibres exhibit similar spectral properties with superior

305

thermal stability compared to those produced using conventional methods (Martinez *et al.*, 2005). In this chapter we will focus on subsurface ultra-fast processing of bulk transparent materials for the fabrication of photonic devices and associated opportunities.

In 1996, two distinct research groups demonstrated optical breakdown in transparent bulk glass using focused femtosecond laser pulses. Both groups used amplified Ti:Sapphire laser pulses centred near 800 nm to modify micrometre-sized volumes inside a variety of dielectrics. Glezer *et al.* (1996) created damage points inside a variety of transparent materials whereas Davis *et al.* (1996) induced a localized point of positive refractive index change in numerous glass samples. This latter result led to a new field in photonics in which guided wave devices, functioning in the same way as optical fibres, could be directly written into either passive or active bulk materials simply by moving the sample through the focus of the femtosec-ond laser beam and creating a pathway of modified material. The material surrounding the focal volume remains largely unaffected by the propagat-ing writing beam due to the highly nonlinear nature of the laser absorption by the material, allowing structures to be written at arbitrary depths and in a three-dimensional fashion. Further advantages of the technique include integration of a suite of photonic devices, rapid prototyping, compatibility with existing fibre systems, and freedom from lithographic masks and clean room fabrication environments. Ultrafast laser-written photonic devices are also tolerant of changing environmental conditions by virtue of being embedded inside a bulk material.

Ultrafast direct writing has been used to fabricate photonic devices using a variety of transparent substrates, including glasses, crystals and polymers. Owing to their high purity and large transparency window, glasses and crys-tals are commonly used as base materials. The dominant material change in most of these materials is a positive refractive index change, however when applied to common materials such as ZBLAN (Lancaster *et al.*, 2011), YAG (Okhrimchuk *et al.*, 2005; Siebenmorgen *et al.*, 2009) and Ti:Sapphire (Apostolopoulos *et al.*, 2004) for example, a decrease in the refractive index can result. The use of directly written suppressed cladding arrangements or induced stress fields can still allow waveguiding regions to be realized in these cases.

A wide variety of stand-alone femtosecond laser-written devices have already been demonstrated including waveguides, couplers, filters and resonators. Additionally, ultrafast laser-written devices are being integrated into commercially available devices in the form of optical interconnects and sensors. In fact, integrated devices are fast becoming the new direction/trend for the ultrafast laser direct-write technique. Astronomers, biologists and quantum physicists are now harnessing the three-dimensional capability of the direct-write technique to integrate, and in some cases replace, conventional apparatus from their fields of research with laser-written devices. Furthermore, researchers are beginning to engineer optical materials and tailor their properties to better match the direct-write technique. Several recent review articles on the femtosecond laser direct-write technique highlight the considerable attention this field of research is already generating (Ams *et al.*, 2009b; Della Valle *et al.*, 2009; Gattass and Mazur, 2008; Szameit and Nolte, 2010; Thomas *et al.*, 2011).

This chapter presents the femtosecond laser direct-write technique as it stands today. In Section 10.2 the underlying processes of subsurface material modification of dielectrics by ultrafast laser beams are outlined. In particular, the nonlinear ionization mechanisms underpinning light–matter interactions are discussed. Section 10.3 introduces the mechanisms responsible for creating refractive index modifications in transparent dielectrics. The roles of densification, rarefaction, phase changes and colour centres are reviewed. Section 10.4 presents a discussion of the typical fabrication and characterization methods used in the field of ultrafast laser direct-write photonics. Example experimental components and layouts are also detailed. The characteristics of various photonic devices fabricated in both passive and active glasses are discussed in Section 10.5 with particular attention being directed to all-optical waveguide amplifiers and monolithic laser oscillators. The chapter is completed with a summary of its contents in Section 10.6.

10.2 Ultrafast laser interactions with dielectric materials

10.2.1 Transparency in glasses and crystals

A dielectric is said to be transparent when light can be transmitted through the medium with minimal loss. Optical glasses and crystals are transparent over a large range of wavelengths, often spanning from the ultraviolet (UV) to the near-infrared. These properties arise partially as a result of the large energy bandgap that separates the valence and conduction bands of these materials. Photons propagating through a transparent medium possess insufficient energy to excite an electronic transition, and thus continue to propagate without being absorbed. Glasses and crystals tend to exhibit a distinct threshold wavelength, sometimes called the UV cut-off wavelength, corresponding to the point at which the photon energy is equal to the energy bandgap. For wavelengths smaller than the UV cut-off the photon energy is no longer smaller than the energy bandgap of the material, making the medium opaque. The finite size of the energy bandgap in a glass or crystal is generally what limits transmission in the UV.

10.2.2 Multiphoton absorption

At sufficiently high irradiances, the nonlinear response of a medium gives rise to multiphoton absorption (MPA) (Göppert-Mayer, 1931). MPA is the simultaneous absorption of two or more photons, where the combined energy of the absorbed photons exceeds the energy bandgap of the medium. The number of photons that are simultaneously absorbed depends on the photon energy and the energy bandgap of the medium and is sometimes referred to as the multiphoton order. MPA is often seen as an undesirable loss mechanism; however, for applications where energy transfer between optical field and medium is sought after such as laser machining, MPA offers some unique benefits. Since MPA is a nonlinear interaction, it only occurs in a highly localized volume, at the focus of a laser beam for example, and can occur deep within the bulk of a transparent material. Furthermore, since MPA is a non-resonant process there are virtually no restrictions on the range of materials that can be laser-processed, as opposed to schemes that require chemical photosensitivity.

10.2.3 Photo-ionization

Photo-ionization is the generation of free (charge) carriers through direct interaction with an optical field. One of the ways free carriers can be generated is through MPA, where the combined energy of the absorbed photons is sufficient to promote an electron from the valence band, across the energy bandgap and into the conduction band, resulting in the formation of a mobile electron–hole pair. A contributing mechanism for generating free carriers is the spatial distortion of energy bands that arises in the presence of a strong electric field. If the energy bands are sufficiently distorted, electrons can tunnel through the potential barrier that separates the valence and conduction bands. Three distinct regimes are normally associated with photo-ionization events: the multiphoton ionization (MPI) regime, where a free carrier is predominantly generated through MPA, the tunnelling ionization (TI) regime, where a free carrier is predominantly generated through electron tunnelling, and the intermediate regime where a free carrier is generated through significant contributions from both MPA and electron tunnelling (Fig. 10.1).

The degree to which MPA and electron tunnelling contributes to a photo-ionization event can be quantified using the Keldysh parameter (Keldysh, 1965):

$$\gamma^2 = \frac{\varepsilon_0 c}{2e^2} \frac{n\omega^2 m^* E_g}{I},$$ [10.1]

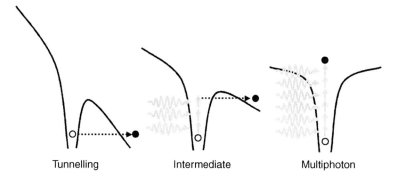

Tunnelling   Intermediate   Multiphoton

10.1 Nonlinear photo-ionization regimes underlying femtosecond-laser material modification.

where n is the refractive index, m^* is the effective mass of the electron–hole pair, E_g is the energy band gap of the material and I and ω are the irradiance and angular frequency of the optical field respectively. The regime is said to be MPI if $\gamma > 2$, TI if $\gamma < 0.5$ and intermediate otherwise, although these delineation points are somewhat arbitrary and vary between authors. Inspection of Eq. [10.1] reveals that MPI is more predominant at lower irradiances, while TI is more predominant at higher irradiances. For a wide bandgap material (~ 5 eV) at a wavelength of 800 nm, the MPI regime corresponds to irradiances around 10 TW/cm^2 or less, while the TI regime corresponds to irradiances around 100 TW/cm^2 or greater.

10.2.4   Avalanche ionization

Free carriers generated through photo-ionization will continue to interact with optical fields, acquiring energy through inverse Bremsstrahlung absorption. Once a free carrier becomes energized it can transfer this energy to other electrons through collisions, liberating them and resulting in the formation of more free carriers (Wright, 1964). This form of ionization is termed impact ionization, also referred to as avalanche ionization as the growth of free carriers becomes exponential with time when this ionization mechanism dominates. Avalanche ionization is termed a secondary ionization process as it requires a population of free carriers (generated via photo-ionization for example) to seed the exponential growth. The large amount of absorbed energy associated with the rapid buildup of free carriers usually results in destruction of the substrate, thus avalanche ionization is typically associated with material damage. Avalanche ionization dominates when laser pulses are of the order of 1 ps or more. Ultrafast lasers with pulse durations of 250 fs or less are employed to achieve energy deposition within a bulk substrate without the destructive buildup of energy that occurs for longer pulse durations. Modifying a substrate without causing wholesale damage is the key to fabricating photonic components using the laser direct-write platform.

10.2.5 Self-focusing, plasma defocusing and filamentation

As a result of high incident irradiance, a host of other nonlinear mechanisms accompany ionization processes in a transparent medium. The most significant of these, from a laser machining point of view, are self-focusing and plasma defocusing, as these mechanisms result in significant distortion of the laser beam.

Self-focusing

The refractive index in an intense optical field is given as

$$n = n_0 + n_2 I(r,t),$$
[10.2]

where the function $I(r, t)$ describes the spatiotemporal irradiance profile of the optical field (Chiao et al., 1964). The coefficient n_2 is the nonlinear Kerr index and is related to the third-order nonlinear susceptibility,

$$\chi^{(3)} = \frac{4}{3}\varepsilon_0 c n_2 n_0^2.$$
[10.3]

Irradiance profiles that are higher at the centre of the beam and fall away off axis induce a curvature in the wavefront similar to that induced by a lens. If n_2 is positive (which is usually the case) this effect mimics that of a focusing lens. If the self-focusing effect is sufficiently strong, then the beam profile becomes increasingly more focused until the beam collapses in on itself. This power, beyond which occurs beam collapse, is called the critical power,

$$P_{cr} = \frac{\alpha\lambda_0^2}{8\pi n_0 n_2},$$
[10.4]

where α is a constant related to the beam profile ($\alpha = 3.77$ for a Gaussian beam for example). In practice, beam collapse is arrested by other nonlinear mechanisms, including plasma defocusing.

Plasma defocusing

The presence of free carriers results in a local reduction in the refractive index,

$$n \approx n_0 - \frac{e^2}{2\varepsilon_0 m^* \omega^2} \rho(r,t),$$ [10.5]

where ρ is the plasma density (Feit and Fleck, 1974). By the same argument used above, this results in a curvature of the wavefront that has a defocusing effect on the beam. The interplay between self-focusing and plasma defocusing leads to laser pulses exhibiting complex behaviour as they propagate. This complex behaviour tends to be more prevalent when low (< 0.4) numerical aperture (NA) lenses are used to focus the laser beam.

Filamentation

One example of such complex behaviour is filamentation (Couairon and Mysyrowicz, 2007). Under certain conditions, the beam can repeatedly cycle between collapse due to self-focusing and plasma defocusing arresting the collapse. This continuing cycle of focusing and defocusing causes the beam to remain confined over long distances, a 'filament', rather than a beam that spreads out as it propagates. Filamentation cannot be sustained indefinitely as the energy of the beam is continually being depleted through plasma generation. Eventually the irradiance of the beam is not sufficient for self-focusing to overcome diffraction, at which point the beam ceases to propagate as a filament. It is possible for a beam to break up into multiple filaments, this can occur deterministically due to some form of anisotropy being present in either the beam or substrate, or stochastically due to fluctuations (noise) in the irradiance profile.

10.2.6 Ultrafast laser-induced modification

The formation of a plasma inside a dielectric leads to a cascade of physical processes that can result in a permanent local modification of the solid-state structure (Gattass and Mazur, 2008). Free carriers have a lifetime of the order of 1–10 ps, after which they begin to decay. When the plasma decays the energy of the free carriers is transferred to the phonon modes of the substrate, causing it to heat. If there is sufficient energy, the substrate undergoes localized softening and melting. Shock waves form during heating, dissipating energy away from the heated volume. The remaining energy is eventually lost through thermal diffusion. Since the heated volume is typically of the order of several μm^3, the subsequent cooling is rapid, of the order of

1 μs or so. Rapid cooling means that structural changes that occur during heating become 'frozen in' which can result in a host of structural modifications including densification, rarefaction, devitrification, amorphization and colour centre formation. The change in refractive index that occurs in conjunction with these structural changes is particularly useful in photonics.

10.2.7   Cumulative heating

For ultrafast lasers with a sufficiently high pulse repetition rate (> 500 kHz), the deposited energy cannot fully dissipate away from the heated volume before the arrival of the next pulse. This results in thermal energy accumulating over many pulses, until thermal equilibrium is reached (Eaton *et al.*, 2005). In this 'cumulative heating' regime, the nature of the ultrafast laser-induced modification depends on inter-pulse characteristics as well as properties of individual pulses. The advantage of the cumulative heating regime is the natural symmetry that modifications possess, as well as the rapid speed with which they can be written. There are some drawbacks to using this regime; most notably not all materials are suited for it (e.g. fused silica).

10.3   Refractive index modification mechanisms

On an atomic level, the refractive index of a medium is directly related to its polarizability, the dipole moment obtained per unit of applied electric field strength, per unit volume. Polarizability can be related to the refractive index via the Lorentz–Lorenz equation,

$$\frac{4\pi}{3}N_A\alpha = R\frac{\rho}{M} = \frac{n^2-1}{n^2+2},$$ [10.6]

where α is the mean polarizability, ρ is the density, M is the molar mass and R is the molar refractivity. Changes to the refractive index of a material can therefore be related to a change in either ρ, R or M.

10.3.1   Densification and rarefaction

Densification and rarefaction can occur as a result of thermally induced volumetric changes that are 'frozen in' as a result of rapid cooling or mechanically induced volumetric changes due to the formation of shock waves or voids. From the Lorentz–Lorenz equation (Eq. [10.6]) an increase in density will result in an increase in the refractive index and vice versa. For small refractive index changes ($< 1 \times 10^{-2}$) the density is approximately proportional to the refractive index

$$\frac{d\rho}{dn} \simeq \frac{M}{R} \frac{6n_0}{\left(n_0^2 + 2\right)^2},$$ [10.7]

where n_0 is the unmodified refractive index. Conservation of mass dictates that densification must be accompanied by rarefaction, thus regions of positive refractive index change must accompany regions of negative refractive change where density change is the only refractive index change mechanism present.

10.3.2   Phase changes

Phase changes in this context refer to a material switching between crystalline states, or from a crystalline to amorphous state (or vice versa). The bonding arrangement of a given composition of atoms varies substantially between amorphous and various crystalline forms, for example, fused silica, α-quartz and β-quartz. This usually brings about a change in density, which will contribute to a modification in the refractive index; however, there is an additional contribution brought about by the change in the bond structure of the material (Little *et al.*, 2010a). The effect of bond structure in the Lorentz–Lorenz equation is encapsulated by a change in the molar refractivity, R. Birefringence can also be induced or removed via phase changes through transitions between anisotropic and isotropic states.

10.3.3   Colour centres

Colour centres are defects within the bond structure of a material. They can manifest as atoms with incomplete octets (holes), electrons 'stuck' in excited states (excitons), missing atoms (vacancies), impurities (substitutions) or a combination of these. Colour centres are localized points in the material where the bond structure has been altered. The formation of colour centres therefore changes the molar refractivity of a material, independent of density change.

   Colour centres have been shown to accompany ultrafast laser-induced modifications in the non-cumulative heating regime, and are thought to arise as a result of free electrons decaying into trapped excited (defect) states rather than the valence band. More recently, it has been suggested that colour centres also arise due to the breaking of bonds within the material (Fig. 10.2), or alternatively, the inability for bonds to reform after

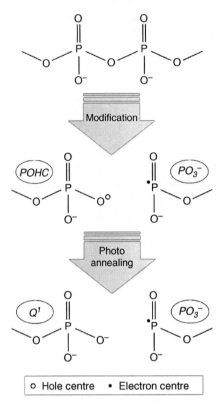

10.2 Colour centre formation/removal in phosphate glass under kHz repetition rates. Phosphorous–oxygen hole centres (POHCs) and PO_3^- ions form as a result of P–O bonds being broken during the modification process, and the subsequent removal of POHCs give rise to the increased proportion of P-tetrahedra bonded to a single bridging oxygen (Q^1 P-tetrahedra).

cooling (Little *et al.*, 2010a). Colour centres are typically not photo-stable, meaning exposure to UV/visible wavelengths can result in the colour centres being removed (or photo-bleached) and the accompanying refractive index change washed out.

When colour centres are removed in glass, it has been observed that only part of the accompanying refractive index change is erased (Dekker *et al.*, 2010). This indicates that the glass does not revert back to its unmodified form after colour centres have been removed. By extension, changes in the bond structure of the glass that do not consist of colour centres must contribute to the refractive index change. The conversion of bridging oxygen atoms to non-bridging oxygen atoms was recently identified as a contributor of this type (Little *et al.*, 2010b).

10.3.4 Stress

The application of compressive stress in a material can also result in a change in the refractive index. There is an isotropic contribution due to density changes that accompany mechanical compression, and an anisotropic component due to the stress itself. In crystalline media, it is generally difficult to induce a positive change in the refractive index directly, instead lines of negative refractive index change or even voids are written parallel to one another (Thomas *et al.*, 2007). The space in between the voids is subject to compressive stress, which can act as the core of a waveguide. As the stress is anisotropic, it tends to induce birefringence, so waveguides of this type are often polarization-selective.

10.3.5 Voids

If the density of a plasma generated within a medium becomes sufficiently high then the Coulombic repulsion between particles results in a micro-explosion, forming a void (Glezer and Mazur, 1997). Voids are localized points of material damage consisting of vacuum pockets surrounded by a densified shell. The refractive index contrast that occurs as a result of void formation is $(n - 1) \approx 0.5$, which is very large in comparison to other refractive index change mechanisms. Voids can be written with a great degree of repeatability and precision, making them useful for applications that make use of strong scattering centres in a precise formation, such as optical data storage and recovery (Glezer *et al.*, 1996), and photonic crystals (Juodkazis *et al.*, 2002).

Void formation is also the basis for writing point-by-point Bragg gratings in optical fibres with femtosecond laser pulses. This method has proven advantageous when writing Bragg gratings in active fibres as they can be inscribed in non-photosensitive media and do not degrade under high-power operation. In addition, voids can be anisotropic which allows the polarization of a fibre laser output to be manipulated (Jovanovic *et al.*, 2009).

10.3.6 Nanostructure in fused silica

Focusing femtosecond pulses into fused silica can result in a quasi-periodic modification where the orientation of periods depends on the polarization of the femtosecond laser pulses. These types of modifications are termed *nanostructures*, or *nanogratings*, as the period of the modulation can be as small as 20 nm. It is not yet understood why this modification appears to be unique to fused silica and the mechanism by which these structures appear is still under debate. Pattathil *et al.* (2005) proposed that the quasi-periodic modulation arises as a result of the field enhancement around an initial 'seed' defect which in turn causes nanogratings to grow in a self-organized fashion. Shimotsuma *et al.* (2005) proposed an alternative explanation where the quasi-periodic modulation arises as a result of interference between waves excited within the electron plasma and the laser pulse.

Nanostructure formation is thought to be the underlying reason why modified glass regions experience an enhanced etch rate when exposed to hydrofluoric (HF) acid, and why the etch rate can be increased by a factor of 100 by varying the polarization of the femtosecond pulses (Hnatovsky *et al.*, 2005). Modifying fused silica in this fashion is advantageous for fabricating microfluidic channels.

10.4 Photonic device processing

10.4.1 Device fabrication: experimental components and procedures

Figure 10.3 illustrates a typical femtosecond laser direct-write setup used for fabricating photonic devices inside bulk materials. The most common femtosecond laser systems used in this field are:

(i) regeneratively amplified Ti:Sapphire laser systems that provide high pulse energies (µJ–mJ) at kHz repetition rates
(ii) oscillator-only Ti:Sapphire systems with low energy (nJ) and high repetition rates (MHz)

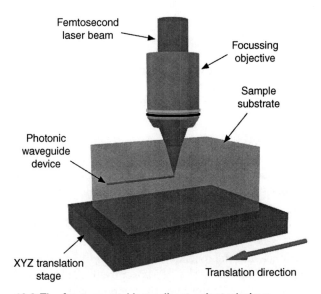

10.3 The femtosecond laser direct-write technique.

(iii)   high pulse energy (500 nJ–μJ) ytterbium or erbium-doped fibre lasers at high repetition rates (100 kHz–MHz)
(iv)   low energy (nJ) cavity dumped ytterbium or Ti:Sapphire laser oscillators operating at high repetition rates (500 kHz–MHz).

As mentioned in the previous section, whilst all of these systems are effective at modifying transparent dielectrics, significant differences exist between the mechanism underlying the modification, and therefore, also the strength of the modification, level of damage (if any), and most importantly in terms of waveguides, whether the index change is positive or negative.

Femtosecond laser writing is achieved by focusing the laser beam from one of these systems (which can range from 780 nm up to 1560 nm, or include one of its harmonics) into a bulk sample using either a fixed microscope objective or lens. The sample to be processed is typically mounted on high precision motion controlled stages which are translated (by computer-linked drive control units) with respect to the laser beam to create (directly write) continuous structures in the bulk of the sample.

Optical head considerations

Most fabricators of photonic devices use microscope objectives rather than lenses given that they are well corrected with higher numerical aperture (NA). A variety of objectives with different NA and working distances have been used by researchers in the field allowing the size and shape of the fabricated structures to be tailored to a certain degree. Usually high numerical aperture (NA) objectives (typically > 0.6) are used in conjunction with high repetition rate systems as a tight focus is required in order to achieve sufficient intensities to modify the sample substrate. Such a tight focus is not required when using low repetition rate systems and lower NA objectives are typically used in those cases.

In practice, the higher the magnification the higher the fluence at focus and the smaller the working distance. Hence there is a trade-off that must be considered when selecting an objective to fabricate a specific device. The shorter the objective's working distance the closer the focal point is to the surface of the glass sample. If the focal point gets too close to the surface, damage to the sample can arise and the capability to fabricate true 3D devices is reduced. The use of lower NA objectives requires higher writing powers to induce modification, which causes issues with self-focusing distorting the beam. Furthermore, employing low NA objectives produces challenges in achieving circularly symmetric local area index changes in bulk materials because the depth of field of the focal spot in these cases is larger than the lateral dimensions. In these cases simple beam-shaping techniques including asymmetric focusing using a slit aperture (Ams *et al.*, 2005) or two-dimensional deformable mirror (Thomson *et al.*, 2008), astigmatic cylindrical telescope arrangements (Cerullo *et al.*, 2002), the multiscan method (Liu *et al.*, 2004) or adaptive wavefront control using spatial light modulators (Mauclair *et al.*, 2008) can recover this symmetry. Such techniques thus enable photonic devices with circular symmetry to be written using low magnification and long working distance objectives.

In the case of directly written photonic devices created below the surface of a substrate it is important to also consider the effect of spherical aberration that subsurface focusing causes. Spherical aberration and its effect on waveguide cross-sections can be controlled using objectives that are corrected for focusing through a fixed depth of material (for example, a cover slip corrected objective); however, this limits the 3D capabilities of the writing platform. A more suitable solution is to use oil-immersion focusing objectives that are not sensitive to the depth of focus in the material, since all optical path lengths to the focus remain constant, or to use adaptive optics techniques that compensate for the change in phase a wavefront experiences inside the material before the focal point (Jesacher and Booth, 2010).

Material and laser fabrication parameters

The material's interaction processes at play within the laser focus are strongly dependent on both the material and the laser parameters, and it is common to observe both positive and negative changes in the material's refractive index under different laser processing conditions or even within the same interaction region.

In most optical materials, the maximum refractive index change that can be induced using the direct-write technique is limited by material damage. At the other extreme, the minimum intensity that results in refractive index change depends on the material's modification threshold, which in turn depends on nonlinear absorption coefficients and ionization cross-sections. The pulse energy required will depend on both the material and the chosen application. Different pulse energies will result in different peak index changes and associated variations in the guided mode field diameter (MFD) of resulting photonic devices. For example, certain phosphate glasses can display an interesting property in that not just the magnitude but also the sign of the net refractive index change induced by the writing laser is a function of pulse energy (Ams *et al.*, 2008).

When using high repetition rate femtosecond laser systems, hundreds of pulses accumulate to heat the focal volume constituting an approximate point source of heat within the bulk of the material. Longer exposure of the material to this heat source gives rise to higher temperatures resulting in a larger affected region. Due to symmetric thermal diffusion outside of the focal volume, a spherically shaped modified region is produced. In contrast, when using a low repetition rate femtosecond laser system, the focal volume returns to room temperature before the arrival of the next pulse resulting in the same region of the material being heated and cooled many times by successive pulses. This repetitive type of machining means that the structural modification of the material is confined to the focal volume alone. It has also been shown that the refractive index contrast of a modified region can be increased by overwriting a waveguide with more than one pass of a low repetition rate femtosecond laser beam in a multiple fabrication scan fashion (Hirao and Miura, 1998).

It has been demonstrated that, for non-cumulative heating regimes, the material change (and accompanying refractive index changes) induced by circularly polarized pulses is different to that induced by linearly polarized pulses (Little *et al.*, 2008). It has been suggested that this effect is due to the photo-ionization cross-sections being different for circularly polarized and linearly polarized pulses and the scaling of the multiphoton order as the power of the writing beam is increased (Little *et al.*, 2011).

Other parameters which affect the writing properties, and thus the resulting device, include the sample translation speed and direction, focused beam shape, M^2 value, wavelength and pulse duration. Resulting devices are not only dependent on these fabrication parameters but are also heavily influenced by the properties of the actual material in which the device is to be created: for example, bandgap energy, whether the sample is crystalline or amorphous, thermal characteristics, variations of impurities and dopants, and fracture strength. Typically, the parameter windows are relatively small for high-quality results in any given application (Ams *et al.*, 2008).

10.4.2   Characterization of fabricated devices

After fabrication, the input and output facets of the material are ground and polished to expose the photonic device because the direct-write process cannot easily access the final few microns of glass near the edge of the target substrate. Photonic waveguide devices are generally characterized in terms of their transmission and reflection data, near- and far-field mode distributions, insertion, coupling, propagation and polarization dependent losses, induced refractive index contrasts, and if applicable, device gain. General experimental setups used to take such measurements from a device under test (DUT) are shown in Fig. 10.4. Various light sources (free space, fibre coupled, swept or tunable) at a variety of wavelengths (application dependent) are used to probe fabricated devices. Optical spectrum analysers (OSAs), power meters and charge-coupled device (CCD) cameras are used to analyse device properties. If characterization fibres are used to either

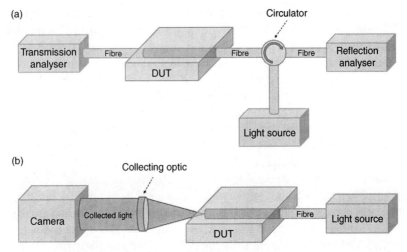

10.4 Photonic device characterization. Typical experimental layouts used to take (a) transmission and reflection measurements and (b) near-field distributions. DUT – device under test.

pump light into or collect light from a DUT, index matching gel/oil is generally inserted at the fibre–DUT interface to reduce losses.

Mode field diameter

Near-field distributions exiting a fabricated device (Fig. 10.4) give an indication of the type of guided modes that are present in the device plus provide a means to measure the MFD of a guided mode. The MFD is defined as the distance between the $1/e$ points of the amplitude profile or $1/e^2$ points of the power density profile. The MFD is not necessarily the same dimension as the core. This can clearly be seen in Fig. 10.5, for example, where the MFD is plotted with respect to a step-index waveguide's core diameter for a number of refractive index contrasts. The NA and the MFD of a single mode waveguide are inversely proportional to each other. The smaller the MFD, the greater the maximum exit radiation angle. Therefore, the MFD has a strong influence on a waveguide's bending sensitivity and coupling efficiency with other optical devices.

Refractive index measurements

Variants of Fig. 10.5 (modified for the wavelength and type of glass of interest) can be used to estimate the peak refractive index change between the bulk material and the waveguide structures written using the ultrafast laser direct-write technique. Another computational method which uses the near-field distribution directly can also be used (Mansour and Caccavale, 1996).

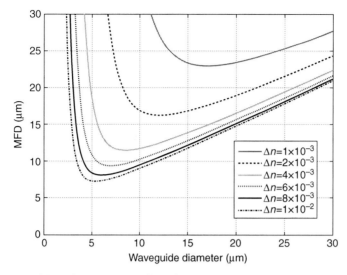

10.5 Mode field diameter (MFD) at 1535 nm with respect to a step-index waveguide's core diameter for a number of refractive index contrasts. The waveguide's host in this example is a fused silica sample where $n = 1.444$.

Refractive index profilometers and QPM software packages are the most common methods of generating refractive index profiles of directly written photonic devices.

Photonic devices fabricated using the femtosecond laser direct-write technique are very small, of the order of microns. In addition to these dimensions being impossible to visualize via the naked eye, the human eye also cannot resolve the small refractive index contrasts that constitute an optical waveguide device. Transverse and end-on images of fabricated devices are therefore taken with a phase contrast or differential interference contrast (DIC) microscope. From these images approximate waveguide dimensions are measured. DIC microscopy is a beam-shearing interference technique in which a reference beam is sheared by a minuscule amount resulting in the interference of two distinct wavefronts that reach the final image plane slightly out of phase with each other. Essentially, the technique produces an image that displays the gradient of optical paths for both high and low spatial frequencies present in the material, that is, refractive index contrasts.

Photonic device losses

The insertion loss (IL) of a fabricated device is taken to be the ratio of the measured transmitted powers with and without the DUT in the setup shown in Fig. 10.4a and includes coupling, propagation, Fresnel and absorption losses. Coupling losses are usually estimated using the mode matching coupling efficiency formula outlined by Marcuse (1972) while absorption losses are material-specific and can be measured using a spectrometer. Propagation loss can then be determined by either subtracting the coupling, Fresnel and absorption losses from the IL, using the standard cutback method, or by taking the difference between the IL in reflection when the collecting fibre in Fig. 10.4a is replaced with a highly reflecting mirror aligned square to the device's output and the IL in transmission without the mirror.

Device gain

The setup shown in Fig. 10.4a is slightly modified for active waveguide characterization in that wavelength division multiplexers (WDMs) are inserted at the device's outputs so that both a signal source and a pump source can co-propagate along the device in a bidirectional configuration as outlined by Ams *et al.* (2009b).

10.5   Photonic devices

The femtosecond laser direct-write technique could potentially provide reliable and cheap devices with the best mode match to optical fibres, and thus find increasing applications in astronomy, medicine, biophotonics and quantum information. In fact, using this technique, researchers have shown that it is possible to fabricate a range of photonic devices of arbitrary dimension and design inside many different kinds of transparent glasses, polymers and crystals. The most common device created is of course the fundamental optical waveguide. Other devices, most of which are also based on the principle of guided light waves, have been fabricated by many international research groups.

10.5.1   Passive devices

The femtosecond laser direct-write technique has already been used to fabricate waveguides, waveguide arrays, power splitters, couplers, Mach Zehnder interferometers, gratings, Bragg gratings, computer-generated holograms, optical storage devices, interconnects, sensors (temperature, vibration and refractive index) and quantum simulation circuits inside a large host of bulk materials (details pertaining to these devices can be found in the reference list of this chapter). Although this field is still in its infancy, compared to other well-established techniques (silicon, ion-based and photosensitivity technologies), researchers are now benchmarking ultrafast laser direct-written devices against those made by the standard methodologies and are demonstrating comparable performance.

As mentioned previously, either a positive or negative refractive index change can be induced in a sample by the ultrafast direct-write technique. Investigations into increasing this contrast have been explored with a maximum positive index change of 2.2×10^{-2} being reported (Eaton *et al.*, 2010). The fabrication of single mode devices is possible with propagation losses typically smaller than 0.4 dB/cm. These characteristics make ultrafast laser-written waveguides ideal for use in applications where standard single mode fibres are used. The optical vibration sensor fabricated by Kamata *et al.* (2005) is an example of this. The sensor consists of a single straight waveguide written across a series of three pieces of glass with the central glass piece mounted on a suspended beam. Displacement of the central piece is detected by optical fibres that are used to measure the change in optical transmission through the device making it sensitive to mechanical vibration and acceleration. The sensor has a linear response over the frequency range 20 Hz–2 kHz and can detect accelerations as small as $0.01 \ \text{m/s}^2$.

The technique is not limited to 1D linear structures. For example, 2D arrays of evanescently coupled waveguides have been used to tune the dispersion and diffraction of propagating light (Szameit and Nolte, 2010) and 2D directional couplers have been used in quantum simulation circuits (Marshall *et al.*, 2009) and as wavelength dependent splitters (Chen *et al.*, 2008). Other wavelength dependent structures have also been reported. High quality waveguide Bragg-gratings (WBGs) (Marshall *et al.*, 2006; Zhang *et al.*, 2007), cascaded WBGs (Zhang *et al.*, 2006), chirped WBGs (Zhang and Herman, 2009) and coupled WBGs (Ha *et al.*, 2011) have all been demonstrated (for an example see Fig. 10.6). Typical WBGs have a coupling coefficient of $\kappa = 2.5 \ \text{cm}^{-1}$ with 3 dB bandwidths less than 300 pm. One of the real potentials of the ultrafast laser direct-write technique was first demonstrated when Nolte *et al.* (2003) reported a 3D splitter with almost equal splitting ratios at 1.05 μm. This paved the way for further 3D devices to be realized. For example, a 3D electro-optic subsurface modulator coupled to surface electrodes resulting in an integrated device with an effective electro-optic coefficient of 0.17 pm/V (Li *et al.*, 2006) and the demonstration of 3D interconnections of silica-based planar lightwave circuit (PLC) waveguides using ultrafast laser-written waveguides (Nasu *et al.*, 2009).

10.5.2   Active devices

Active waveguide devices fabricated using the femtosecond laser direct-write technique have been reported in various materials: Nd-doped silicate glass, Er/Yb co-doped phosphate glass, LiF crystal, Nd-doped and Yb-doped YAG crystal, Er-doped bismuthate glass, Er-doped and Er/Yb co-doped oxyfluoride silicate glass, and Er-doped and Er/Yb co-doped phospho-tellurite glass. These samples were specifically chosen due to their target application, region

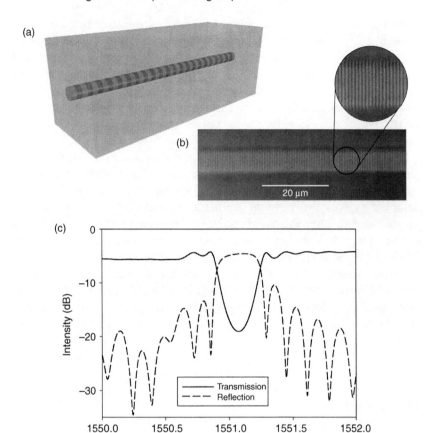

10.6 Example waveguide Bragg-grating (WBG) demonstrating the control of the ultrafast laser direct-write technique. (a) Schematic of a WBG fabricated using the modulated writing beam method. (b) Differential interference contrast (DIC) top-view micrograph of a WBG structure fabricated in a borosilicate glass sample clearly showing the 500 nm grating period. (c) C-band transmission and reflection spectra from the WBG.

of wavelength emission and ease of doping with rare-earth ions. Of these materials, the phosphate glass hosts are best suited to active device fabrication for use in telecommunications as tens of percent by weight of rare-earth ions can be held in solution offering the highest gain-per-unit length value (4 dB/cm) in the C-band without detrimental effects such as ion-clustering.

Waveguide amplifiers

Small signal gain characteristics of a waveguide written in a doped material are measured at the wavelength corresponding to the peak of the material's gain curve. To date, internal gains of 3.18 dB/cm in the C-band have been reported for Er/Yb doped phosphate systems (Osellame *et al.*, 2008) and a gain of 1.5 dB/cm at 1054 nm for a femtosecond laser-written waveguide amplifier fabricated in Nd-doped silicate glass (Sikorski *et al.*, 2000). There is still potential for further improvement of the gain of these devices by optimizing the doping concentrations of the active ions, optimizing the waveguide amplifier length and optimizing pump power configurations. Furthermore, there is scope to develop glasses that are engineered for the direct-write process, as the doped glasses typically used are those developed for use in fibre preforms and ion exchange processes. An example of why this may be necessary is demonstrated by Dekker *et al.* (2010), who show that annealing of colour centres during characterization causes device degradation over time. It is worth noting, however, that the waveguide amplifiers reported to date created using the ultrafast laser direct-write technique (Ams *et al.*, 2009b; Della Valle *et al.*, 2009) have reached a performance level (net gain of 2.7 dB/cm) comparable to those demonstrated with conventional fabrication techniques (net gain of 3 dB/cm) in similar glasses with the same dopant concentrations (Jaouen *et al.*, 1999, for example).

To turn these amplifiers into laser oscillators, appropriate feedback of the particular amplified frequency is required. Positive feedback may be obtained by placing the gain material between a pair of suitable mirrors which form an optical cavity or resonator. With guided wave devices, the cavity usually takes the form of a Distributed Bragg Reflector (DBR) geometry or a Distributed Feedback (DFB) design, although external dielectric mirrors or dichroic coatings on the amplifier's end facets can also be used.

DBR waveguide oscillators

The laser cavity employed by some research groups in the field is formed by butt-coupling fibre Bragg gratings (FBGs) (which act as cavity mirrors) to the waveguide amplifier (Osellame *et al.*, 2008; Psaila *et al.*, 2008). The FBGs are centred at the same wavelength but have different reflectivities and bandwidth. This enables a laser cavity to be created with a high reflector at one end and an output coupler at the other. Such a cavity led to the demonstration of ultrafast laser-written waveguide lasers (WGLs) able to provide wavelength coverage over the entire C-band, single longitudinal mode and stable mode-locking operation. In particular, recent reports have indicated that using a cavity length of 5.5 cm and a 57% output coupler, a WGL with threshold pump power of 124 mW and slope efficiency of 21% is achievable. Furthermore, a mode-locked WGL source based on carbon nanotubes technology (used to create a saturable absorber) has been reported (Osellame *et al.*, 2008). This WGL had a repetition rate of 16.7 MHz and pulse duration of 1.6 ps.

Integrated DFB waveguide oscillators

A serious handicap of the feedback provided by mirror resonators is that many longitudinal modes may fit under the system's gain bandwidth thus impeding single longitudinal mode operation. In addition, the device is not completely monolithic. Several experimental techniques have been reported that enable the realization of Bragg grating structures inside femtosecond laser-written waveguides to create a cavity. Kawamura *et al.* (2004) developed a hologram technique to encode grating structures inside an LiF crystal by a single interfered femtosecond laser pulse. When side-pumping the grating with 450 nm light, a DFB laser oscillator at 707 nm was created. The grating structure, however, was weak and not permanent as the colour centres creating it were eliminated by annealing.

More recently, by square-wave modulating a low repetition rate femtosecond laser's output, in a fashion reported by Zhang *et al.* (2007), first order permanent waveguide Bragg gratings (WBGs) were directly written inside Er/Yb co-doped and Yb-doped phosphate glasses in a single fabrication step (Ams *et al.*, 2009a; Marshall *et al.*, 2008). A waveguide variant of a distributed feedback (DFB) laser was hence demonstrated (Fig. 10.7). The total output power of these lasers measured 0.37 mW at 1535 nm and 102 mW at 1032 nm respectively with linewidths less than 4 pm and efficiencies up to 17%. Although it is known that the WBG structure contributes to an increase in the propagation loss of the amplifier device (Zhang *et al.*, 2007), clearly the WBG is of a high enough quality that the internal gain in the system still exceeds this increase.

Other waveguide oscillators

A near Gaussian mode WGL at 1064 nm was realized in an Nd-doped YAG crystal by writing parallel damage tracks in the crystalline lattice (Siebenmorgen *et al.*, 2009). Waveguiding was observed in the unmodified region between the pair of tracks due to stress-induced birefringence of the material. Laser operation was achieved without the use of external mirrors but by using only the Fresnel reflection of about 9% at the device end facets. Pumping the device with a continuous wave 808 nm Ti:Sapphire laser resulted in an output power of 25.5 mW and a power conversion efficiency of 23%. By using the same techniques, Siebenmorgen *et al.* (2010) also demonstrated waveguide laser action at 1030 nm in a Yb-doped YAG crystal. An output power of 0.8 W with 1.2 W of launched pump power was achieved, resulting in a record WGL slope efficiency of 75%.

Very recently a Tm^{3+} doped ZBLAN glass WGL was reported which produced 47 mW at 1880 nm (Lancaster *et al.*, 2011). This is the first report of a mid-IR WGL fabricated using the femtosecond laser direct-write technique. The waveguide cladding was defined by two overlapped rings made

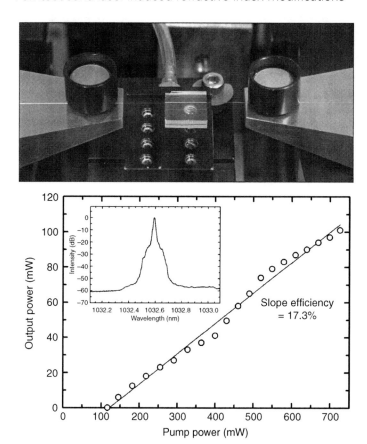

10.7 Example distributed feedback (DFB) waveguide laser (WGL) at 1032 nm fabricated in a 9% ytterbium (Yb) doped phosphate glass sample. The top image shows the cooperative luminescence along the device while pumping at 976 nm. The bottom spectrum plots the laser output power of the 9.5 mm long WGL as a function of pump power. The inset shows the output spectrum of the WGL at 102 mW output power with the ordinate axis scale referenced to the peak.

up of directly written modified glass tubes, resulting in a depressed-index cladding structure. The laser resonator was created using external dielectric mirrors resulting in a 50% internal slope efficiency and a measured M^2 value of 1.7.

10.5.3   Hybrid devices

Microfluidics is the science and technology of systems that process or manipulate small amounts of liquids using channels with dimensions of tens to hundreds of micrometres. Introduction of photonic technologies into these systems created the field of optofluidics and developed an analytical tool capable of gleaning information such as the size, shape, concentration, chemical composition and structural organization of liquid-borne particles. Many comprehensive articles are available on the applications and current state-of-the-art of the optofluidics discipline (Hunt and Wilkinson, 2008; Monat *et al.*, 2007).

A key advantage of the femtosecond laser direct-write technique is that the laser exposed regions of the material may be preferentially removed using an acid etchant (typically hydrofluoric [HF] acid) to create hollow channels and thus microfluidic pathways (Marcinkevicius *et al.*, 2001). Hence, the femtosecond laser direct-write technique is a powerful tool offering a single laser system which can be used to create both microphotonic and microfluidic devices inside the same glass sample with the opportunity to also integrate these devices in 3D. A handful of groups in the world have this capability and are beginning to make simple integrated optofluidic modules (Osellame *et al.*, 2007). In addition, there is commercial interest in this technology with Translume Inc. now offering microfluidic devices fabricated in fused silica glass for purchase.

10.6   Conclusions

The femtosecond laser direct-write technique is a technology capable of producing high-quality integrated photonic devices inside bulk transparent materials without the need for lithography, etching, a controlled environment or much sample preparation. The nonlinear ionization processes that underpin light–matter interactions and the means by which local refractive index changes are resultant were reviewed. Densification and rarefaction (of which void formation is an extreme example) were identified as primary contributors. Phase changes and colour centres could also contribute, due to the change in local bond structure (molar refractivity) and the consequent change in the polarizability they impart; even in the absence of density changes. The typical experimental methods for utilizing these processes to fabricate and characterize subsurface photonic waveguide devices were presented. Photonic devices fabricated in both passive and active glasses were briefly discussed with particular attention being directed to the fabrication of all-optical waveguide amplifiers and monolithic laser oscillators. The comparable performance of ultrafast laser-written devices with devices fabricated using traditional methods has generated a lot of interest in a variety of research fields. Biologists, astronomers, quantum physicists and engineers alike are already harnessing the capabilities of the femtosecond laser direct-write technique, and with breakthrough scientific fields continuing to emerge it appears that this trend will continue.

10.7   References

Ams, M., Dekker, P., Marshall, G. D. and Withford, M. J. (2009a), 'Monolithic 100 mW Yb waveguide laser fabricated using the femtosecond-laser direct-write technique', *Optics Letters*, **34**, 247–249.

Ams, M., Marshall, G. D., Dekker, P., Dubov, M., Mezentsev, V. K., Bennion, I. and Withford, M. J. (2008), 'Investigation of ultrafast laser–photonic material interactions: Challenges for directly written glass photonics', *IEEE Journal of Selected Topics in Quantum Electronics*, **14**, 1370–1381.

Ams, M., Marshall, G. D., Dekker, P., Piper, J. A. and Withford, M. J. (2009b), 'Ultrafast laser written active devices', *Laser & Photonics Review*, **3**, 535–544.

Ams, M., Marshall, G. D., Spence, D. J. and Withford, M. J. (2005), 'Slit beam shaping method for femtosecond laser direct-write fabrication of symmetric waveguides in bulk glasses', *Optics Express*, **13**, 5676–5681.

Apostolopoulos, V., Laversenne, L., Colomb, T., Depeursinge, C., Salathe, R. P., Pollnau, M., Osellame, R., Cerullo, G. and Laporta, P. (2004), 'Femtosecond-irradiation-induced refractive-index changes and channel waveguiding in bulk Ti^{3+}: Sapphire', *Applied Physics Letters*, **85**, 1122–1124.

Cerullo, G., Osellame, R., Taccheo, S., Marangoni, M., Polli, D., Ramponi, R., Laporta, P. and De Silvestri, S. (2002), 'Femtosecond micromachining of symmetric waveguides at 1.5 μm by astigmatic beam focusing', *Optics Letters*, **27**, 1938–1940.

Chen, W.-J., Eaton, S. M., Zhang, H. and Herman, P. R. (2008), 'Broadband directional couplers fabricated in bulk glass with high repetition rate femtosecond laser pulses', *Optics Express*, **16**, 11470–11480.

Chiao, R. Y., Garmire, E. and Townes, C. H. (1964), 'Self-trapping of optical beams', *Physical Review Letters*, **13**, 479–482.

Couairon, A. and Mysyrowicz, A. (2007), 'Femtosecond filamentation in transparent media', *Physics Reports: Review Section of Physics Letters*, **441**, 47–189.

Davis, K. M., Miura, K., Sugimoto, N. and Hirao, K. (1996), 'Writing waveguides in glass with a femtosecond laser', *Optics Letters*, **21**, 1729–1731.

Dekker, P., Ams, M., Marshall, G. D., Little, D. J. and Withford, M. J. (2010), 'Annealing dynamics of waveguide Bragg gratings: evidence of femtosecond laser induced colour centres', *Optics Express*, **18**, 3274–3283.

Della Valle, G., Osellame, R. and Laporta, P. (2009), 'Micromachining of photonic devices by femtosecond laser pulses', *Journal of Optics A: Pure and Applied Optics*, **11**, 013001.

Eaton, S. M., Ng, M. L., Osellame, R. and Herman, P. R. (2010), 'High refractive index contrast in fused silica waveguides by tightly focused, high-repetition rate femtosecond laser', *Journal of Non-Crystalline Solids*, **357**, 2387–2391.

Eaton, S. M., Zhang, H. B. and Herman, P. R. (2005), 'Heat accumulation effects in femtosecond laser-written waveguides with variable repetition rate', *Optics Express*, **13**, 4708–4716.

Feit, M. D. and Fleck, J. A. (1974), 'Effect of refraction on spot-size dependence of laser-induced breakdown', *Applied Physics Letters*, **24**, 169–172.

Gattass, R. R. and Mazur, E. (2008), 'Femtosecond laser micromachining in transparent materials', *Nature Photonics*, **2**, 219–225.

Glezer, E. N. and Mazur, E. (1997), 'Ultrafast-laser driven micro-explosions in transparent materials', *Applied Physics Letters*, **71**, 882–884.

Glezer, E. N., Milosavljevic, M., Huang, L., Finlay, R. J., Her, T. H., Callan, J. P. and Mazur, E. (1996), 'Three-dimensional optical storage inside transparent materials', *Optics Letters*, **21**, 2023–2025.

Göppert-Mayer, M. (1931), 'Über Elementarakte mit zwei Quantensprüngen', *Annalen der Physik Berlin*, **401**, 273–294.

Ha, S., Ams, M., Marshall, G. D., Neshev, D. N., Sukhorukov, A. A., Kivshar, Y. S. and Withford, M. J. (2011), 'Control of light transmission in laser-written phase-shifted Bragg grating couplers', *Optics Letters*, **36**, 1380–1382.

Hirao, K. and Miura, K. (1998), 'Writing waveguides and gratings in silica and related materials by a femtosecond laser', *Journal of Non-Crystalline Solids*, **239**, 91–95.

Hnatovsky, C., Taylor, R. S., Simova, E., Bhardwaj, V. R., Rayner, D. M. and Corkum, P. B. (2005), 'Polarization-selective etching in femtosecond laser-assisted microfluidic channel fabrication in fused silica', *Optics Letters*, **30**, 1867–1869.

Hunt, H. C. and Wilkinson, J. S. (2008), 'Optofluidic integration for microanalysis', *Microfluidics and Nanofluidics*, **4**, 53–79.

Jaouen, Y., Du Mouza, L., Barbier, D., Delavaux, J. M. and Bruno, P. (1999), 'Eight-wavelength Er-Yb doped amplifier: Combiner/splitter planar integrated module', *Photonics Technology Letters, IEEE*, **11**, 1105–1107.

Jesacher, A. and Booth, M. J. (2010), 'Parallel direct laser writing in three dimensions with spatially dependent aberration correction', *Optics Express*, **18**, 21090–21099.

Jovanovic, N., Thomas, J., Williams, R. J., Steel, M. J., Marshall, G. D., Fuerbach, A., Nolte, S., Tunnermann, A. and Withford, M. J. (2009), 'Polarization-dependent effects in point-by-point fiber Bragg gratings enable simple, linearly polarized fiber lasers', *Optics Express*, **17**, 6082–6095.

Juodkazis, S., Matsuo, S., Misawa, H., Mizeikis, V., Marcinkevicius, A., Sun, H. B., Tokuda, Y., Takahashi, M., Yoko, T. and Nishii, J. (2002), 'Application of femtosecond laser pulses for microfabrication of transparent media', *Applied Surface Science*, **197**, 705–709.

Kamata, M., Obara, M., Gattass, R. R., Cerami, L. R. and Mazur, E. (2005), 'Optical vibration sensor fabricated by femtosecond laser micromachining', *Applied Physics Letters*, **87**, 051106.

Kawamura, K., Hirano, M., Kurobori, T., Takamizu, D., Kamiya, T. and Hosono, H. (2004), 'Femtosecond-laser-encoded distributed-feedback color center laser in lithium fluoride single crystals', *Applied Physics Letters*, **84**, 311–313.

Keldysh, L. V. (1965), 'Ionization in the field of a strong electromagnetic wave', *Soviet Physics JETP*, **20**, 1307–1314.

Lancaster, D. G., Gross, S., Ebendorff-Heidepriem, H., Kuan, K., Monro, T. M., Ams, M., Fuerbach, A. and Withford, M. J. (2011), 'Fifty percent internal slope efficiency femtosecond direct-written Tm³⁺:ZBLAN waveguide laser', *Optics Letters*, **36**, 1587–1589.

Li, G. Y., Winick, K. A., Said, A. A., Dugan, M. and Bado, P. (2006), 'Waveguide electro-optic modulator in fused silica fabricated by femtosecond laser direct writing and thermal poling', *Optics Letters*, **31**, 739–741.

Little, D. J., Ams, M., Dekker, P., Marshall, G. D., Dawes, J. M. and Withford, M. J. (2008), 'Femtosecond laser modification of fused silica: The effect of writing polarization on Si-O ring structure', *Optics Express*, **16**, 20029–20037.

Little, D. J., Ams, M., Dekker, P., Marshall, G. D. and Withford, M. J. (2010a), 'Mechanism of femtosecond-laser induced refractive index change in phosphate glass under a low repetition-rate regime', *Journal of Applied Physics*, **108**, 033110–033115.

Little, D. J., Ams, M., Gross, S., Dekker, P., Miese, C. T., Fuerbach, A. and Withford, M. J. (2010b), 'Structural changes in BK7 glass upon exposure to femtosecond laser pulses', *Journal of Raman Spectroscopy*, **42**, 715–718.

Little, D. J., Ams, M. and Withford, M. J. (2011), 'Influence of bandgap and polarization on photo-ionization: Guidelines for ultrafast laser inscription [Invited]', *Optical Materials Express*, **1**, 670–677.

Liu, J. R., Zhang, Z. Y., Flueraru, C., Liu, X. P., Chang, S. D. and Grover, C. P. (2004), 'Waveguide shaping and writing in fused silica using a femtosecond laser', *IEEE Journal of Selected Topics in Quantum Electronics*, **10**, 169–173.

Mansour, I. and Caccavale, F. (1996), 'An improved procedure to calculate the refractive index profile from the measured near-field intensity', *Journal of Lightwave Technology*, **14**, 423–428.

Marcinkevicius, A., Juodkazis, S., Watanabe, M., Miwa, M., Matsuo, S., Misawa, H. and Nishii, J. (2001), 'Femtosecond laser-assisted three-dimensional microfabrication in silica', *Optics Letters*, **26**, 277–279.

Marcuse, D. (1972), *Light transmission optics*. New York: Van Nostrand Reinhold Company.

Marshall, G. D., Ams, M. and Withford, M. J. (2006), 'Direct laser written waveguide-Bragg gratings in bulk fused silica', *Optics Letters*, **31**, 2690–2691.

Marshall, G. D., Dekker, P., Ams, M., Piper, J. A. and Withford, M. J. (2008), 'Directly written monolithic waveguide laser incorporating a distributed feedback waveguide-Bragg grating', *Optics Letters*, **33**, 956–958.

Marshall, G. D., Politi, A., Matthews, J. C. F., Dekker, P., Ams, M., Withford, M. J. and O'Brien, J. L. (2009), 'Laser written waveguide photonic quantum circuits', *Optics Express*, **17**, 12546–12554.

Martinez, A., Khrushchev, I. Y. and Bennion, I. (2005), 'Thermal properties of fibre Bragg gratings inscribed point-by-point by infrared femtosecond laser', *Electronics Letters*, **41**, 176–178.

Mauclair, C., Mermillod-Blondin, A., Huot, N., Audouard, E. and Stoian, R. (2008), 'Ultrafast laser writing of homogeneous longitudinal waveguides in glasses using dynamic wavefront correction', *Optics Express*, **16**, 5481–5492.

Monat, C., Domachuk, P. and Eggleton, B. J. (2007), 'Integrated optofluidics: A new river of light', *Nature Photonics*, **1**, 106–114.

Nasu, Y., Kohtoku, M., Hibino, Y. and Inoue, Y. (2009), 'Waveguide interconnection in silica-based planar lightwave circuit using femtosecond laser', *Journal of Lightwave Technology*, **27**, 4033–4039.

Nolte, S., Will, M., Burghoff, J. and Tunnermann, A. (2003), 'Femtosecond waveguide writing: a new avenue to three-dimensional integrated optics', *Applied Physics A: Materials Science and Processing*, **77**, 109–111.

Okhrimchuk, A. G., Shestakov, A. V., Khrushchev, I. and Mitchell, J. (2005), 'Depressed cladding, buried waveguide laser formed in a YAG:Nd3+ crystal by femtosecond laser writing', *Optics Letters*, **30**, 2248–2250.

Osellame, R., Della Valle, G., Chiodo, N., Taccheo, S., Laporta, P., Svelto, O. and Cerullo, G. (2008), 'Lasing in femtosecond laser written optical waveguides', *Applied Physics A: Materials Science and Processing*, **93**, 17–26.

Osellame, R., Maselli, V., Vazquez, R. M., Ramponi, R. and Cerullo, G. (2007), 'Integration of optical waveguides and microfluidic channels both fabricated by femtosecond laser irradiation', *Applied Physics Letters*, **90**. 231118.

Pattathil, R. P., Hnatovsky, C., Bjardwaj, V. R., Simova, E., Taylor, R. S., Rayner, D. M. and Corkum, P. B. (2005), 'Femtosecond laser-induced nanostructures in fused silica', *Proceedings of SPIE*, **5971**, 59711D.

Psaila, N. D., Thomson, R. R., Bookey, H. T., Chiodo, N., Shen, S., Osellame, R., Cerullo, G., Jha, A. and Kar, A. (2008), 'Er:Yb-doped oxyfluoride silicate glass waveguide laser fabricated using ultrafast laser inscription', *IEEE Photonics Technology Letters*, **20**, 126–128.

Shimotsuma, Y., Hirao, K., Qiu, J. R. and Kazansky, P. G. (2005), 'Nano-modification inside transparent materials by femtosecond laser single beam', *Modern Physics Letters B*, **19**, 225–238.

Siebenmorgen, J., Calmano, T., Petermann, K. and Huber, G. (2010), 'Highly efficient Yb:YAG channel waveguide laser written with a femtosecond-laser', *Optics Express*, **18**, 16035–16041.

Siebenmorgen, J., Petermann, K., Huber, G., Rademaker, K., Nolte, S. and Tünnermann, A. (2009), 'Femtosecond laser written stress-induced Nd:Y3Al5O12 (Nd:YAG) channel waveguide laser', *Applied Physics B: Lasers and Optics*, **97**, 251–255.

Sikorski, Y., Said, A. A., Bado, P., Maynard, R., Florea, C. and Winick, K. A. (2000), 'Optical waveguide amplifier in Nd-doped glass written with near-IR femtosecond laser pulses', *Electronics Letters*, **36**, 226–227.

Szameit, A. and Nolte, S. (2010), 'Discrete optics in femtosecond-laser-written photonic structures', *Journal of Physics B: Atomic, Molecular and Optical Physics*, **43**, 163001.

Thomas, J., Heinrich, M., Burghoff, J., Nolte, S., Ancona, A. and Tunnermann, A. (2007), 'Femtosecond laser-written quasi-phase-matched waveguides in lithium niobate', *Applied Physics Letters*, **91**, 151108.

Thomas, J., Heinrich, M., Zeil, P., Hilbert, V., Rademaker, K., Riedel, R., Ringleb, S., Dubs, C., Ruske, J.-P., Nolte, S. and Tünnermann, A. (2011), 'Laser direct writing: Enabling monolithic and hybrid integrated solutions on the lithium niobate platform', *Physica Status Solidi (A)*, **208**, 276–283.

Thomson, R. R., Bockelt, A. S., Ramsay, E., Beecher, S., Greenaway, A. H., Kar, A. K. and Reid, D. T. (2008), 'Shaping ultrafast laser inscribed optical waveguides using a deformable mirror', *Optics Express*, **16**, 12786–12793.

Wright, J. K. (1964), 'Theory of electrical breakdown of gases by intense pulses of light', *Proceedings of the Physical Society*, **84**, 41–46.

Zhang, H. and Herman, P. R. (2009), 'Chirped Bragg grating waveguides directly written inside fused silica glass with an externally modulated ultrashort fiber laser', *Photonics Technology Letters, IEEE*, **21**, 277–279.

Zhang, H. B., Eaton, S. M. and Herman, P. R. (2007), 'Single-step writing of Bragg grating waveguides in fused silica with an externally modulated femtosecond fiber laser', *Optics Letters*, **32**, 2559–2561.

Zhang, H. B., Eaton, S. M., Li, J. Z. and Herman, P. R. (2006), 'Femtosecond laser direct writing of multiwavelength Bragg grating waveguides in glass', *Optics Letters*, **31**, 3495–3497.

11

Thermal writing of photonic devices in glass and polymers by femtosecond lasers

S. M. EATON, Istituto di Fotonica e Nanotecnologie (IFN)-CNR, Italy, G. CERULLO, Politecnico di Milano, Italy and R. OSELLAME, Istituto di Fotonica e Nanotecnologie (IFN)-CNR, Italy

Abstract: Femtosecond laser microprocessing is a direct, maskless technique capable of inducing a permanent refractive index increase buried beneath the surface of transparent glasses and polymers, enabling photonic circuit fabrication in 3D geometries. We describe how the repetition rate influences the heat accumulation between laser pulses, which determines the regime of modification and the resulting morphological change in glasses in polymers. In most silicate and phosphate glasses, higher repetition rates are shown to be beneficial for driving increased heat accumulation, leading to rapid fabrication of low-loss optical waveguides. In pure silica which has low absorption due to its high bandgap, heat accumulation effects are reduced and higher fluences provided by the second harmonic visible wavelength from Yb-based femtosecond lasers are required to form highly confining optical waveguides. In polymers, femtosecond laser waveguide writing generally leads to depressed refractive index changes and a time decaying behavior.

Key words: femtosecond lasers, micromachining, optical waveguides, glasses, polymers.

11.1 Introduction

Femtosecond laser microprocessing offers the possibility to tailor the refractive index in the bulk of many transparent materials, including glasses and polymers. Due the nonlinear laser–material interaction, the modification is confined to a micrometer-sized region within the focal volume and can be written along three-dimensional pathways, unlike traditional photolithographic techniques. Chapter 10 describes the fundamentals of the interaction of focused femtosecond laser pulses with transparent materials and gives an overview of the many novel integrated devices fabricated by this technology. In this chapter, we focus on the key exposure variables which influence the resulting morphological changes when femtosecond laser pulses are focused in the bulk of transparent dielectrics.

333

By carefully tuning the laser exposure conditions, these bulk modifications can result in a smooth refractive index contrast yielding optical waveguides, the building blocks for more complex photonic circuits such as passive optical splitters (Eaton *et al.*, 2009) and active lasers (Della Vallè *et al.*, 2009) for use in fiber-to-the-home telecommunications, optofluidic fluorescence sensors (Osellame *et al.*, 2007) and temperature/strain (Zhang, 2007) and refractive index sensors (Crespi *et al.*, 2010). We classify the laser–material interaction regime according to the laser repetition rate, since it controls the relative strength of thermal diffusion and heat accumulation. In glasses, high repetition rates are crucial for improving fabrication speeds, increasing refractive index contrast, lowering propagation losses and enabling wide processing windows with waveguide size and mode size easily tuned by the laser dwell time. In polymers, high repetition rates instead are shown to be detrimental to forming high-quality waveguides.

In Section 11.2, we describe the exposure variables of importance in optical waveguide writing. In Section 11.3, we describe waveguide writing in fused silica, borosilicate, phosphate and exotic glasses at both low and high repetition rates. In Section 11.4, we review the current literature on waveguide formation in polymers. We provide a summary of the chapter in Section 11.5 and give insight into future research directions in Section 11.6.

11.2   Femtosecond laser–material interaction in waveguide writing

Peak intensities on the order of 10 TW/cm^2 can be readily produced by focused femtosecond laser pulses from today's commercial laser systems. Such intensities result in strong nonlinear absorption, allowing for localized energy deposition in the bulk of dielectrics. After several picoseconds, the laser-excited electrons transfer their energy to the lattice, leading to a permanent material modification. While a complete physical model of the laser–material interaction has thus far eluded researchers, the process can be simplified by subdivision into three main steps: the initial generation of a free electron plasma followed by energy relaxation and modification of the material. In Chapter 10, an excellent overview of the excitation processes are given. Here we further discuss the relaxation and modification processes, in terms of both single pulse and cumulative pulse interactions at high repetition rates. In this section, we focus our attention on glasses since most progress has been made in this important class of dielectrics. Further insight into the role of the numerous exposures parameters on the modification of polymeric materials is given in Section 11.4.

11.2.1 Modification mechanisms and the influence of pulse energy

It is well accepted that nonlinear photo-ionization and avalanche ionization from absorbed femtosecond laser pulses are responsible for the creation of a free electron plasma. However, once the electrons have transferred their energy to the lattice, the physical mechanisms for material modification are not fully understood. Of the hundreds of published articles citing the pioneering work by Davis *et al.* (1996), the observed morphological changes can be generally classified into three types of structural changes: a smooth refractive index change (Miura *et al.*, 1997), a form birefringent refractive index modification (Hnatovsky *et al.*, 2005a; Shimotsuma *et al.*, 2003; Sudrie *et al.*, 1999) and microexplosions leading to empty voids. The regime of modification and resulting morphological change depend on many exposure parameters (energy, pulse duration, repetition rate, wavelength, polarization, focal length, scan speed and others) but also material properties (bandgap, thermal conductivity and others). However, in pure fused silica which is the most commonly processed material for waveguide writing, these three different morphologies can be observed by simply changing the incident laser energy (Itoh *et al.*, 2006).

Smooth refractive index change

An isotropic regime of modification is useful for optical waveguides, where smooth and uniform refractive index modification is required for low propagation loss. At low pulse energies just above the modification threshold (~ 100 nJ for 0.6-NA focusing of 800-nm, 100-fs pulses), a smooth refractive index modification has been observed in fused silica (Itoh *et al.*, 2006), which is attributed to densification from rapid quenching of the melted glass in the focal volume (Chan *et al.*, 2001). In fused silica, the density (refractive index) increases when glass is rapidly cooled from a higher temperature (Bruckner, 1970). Micro-Raman spectroscopy has confirmed an increase in the concentration of 3 and 4 member rings in the silica structure in the laser-exposed region, indicating a densification of the glass (Chan *et al.*, 2001). Shock waves generated by focused femtosecond laser pulses giving rise to stress have been shown to play a role in causing densification under certain conditions (Sakakura *et al.*, 2008).

It has been argued that laser-induced color centers may be responsible for the laser-induced refractive index change through a Kramers-Kronig mechanism (a change in absorption leads to a change in refractive index) (Hirao and Miura, 1998). Although induced color centers have been observed in glasses exposed to femtosecond laser radiation (Chan *et al.*, 2003b; Streltsov and Borrelli, 2002), to date only a weak link between color center formation and the induced refractive index change has been demonstrated in the literature. Waveguides formed in fused silica with an infrared femtosecond laser (Saliminia *et al.*, 2005) were found to exhibit photo-induced absorption peaks at 213 and 260 nm corresponding to SiE' (positively charged oxygen vacancies) and non-bridging oxygen hole centers (NBOHC) defects, respectively. However, both color centers were completely erased after annealing at 400°C, although waveguide behavior was still observed up to 900°C. Therefore, it is unlikely that color centers played a significant role in the refractive index change (Saliminia *et al.*, 2005). Other research found that the thermal stability of color centers produced in borosilicate and fused silica glasses by femtosecond laser irradiation is not consistent with that of the induced refractive index change (Streltsov and Borrelli, 2002).

Recently, Withford's group has shown for Yb-doped phosphate glasses used in waveguide lasers, femtosecond laser-induced color centers contribute approximately 15% to the observed refractive index increase (Dekker *et al.*, 2010). Using integrated waveguide Bragg gratings, the authors were able to accurately study the photobleaching and thermal annealing of the induced color centers. The color centers were stable for temperatures below 70°C, which is below the operating point during lasing. However, the green luminescence generated by the Yb ions results in a photobleaching of the color centers during laser operation, resulting in reduced lifetime which must be corrected by pre-aging techniques.

Although a complete understanding of the femtosecond laser material interaction in forming optical waveguides has presently eluded researchers, it is evident that densification and color centers play a role. However, their contributions will vary depending on the glass composition and the femtosecond laser exposure conditions, further adding to the complexities in modeling femtosecond laser waveguide writing. In glasses with structures that are more complex than fused silica, further contributions must be considered. For example, in multicomponent crown glass, the authors concluded that the ring-shaped refractive index profile during femtosecond laser irradiation was the result of ion exchange between network formers and network modifiers (Kanehira *et al.*, 2008).

Birefringent refractive index change

For higher pulse energies (~150–500 nJ for 0.6-NA focusing of 800-nm, 100-fs pulses), birefringent refractive index changes have been observed in the bulk of fused silica glass (Itoh *et al.*, 2006), as first reported by Sudrie *et al.* (1999). Kazansaky *et al.* argued that the birefringence was due to periodic nanostructures that were caused by interference of the laser field and the induced electron plasma wave (Shimotsuma *et al.*, 2003). In similarly exposed fused silica samples, Taylor's group observed periodic layers of alternating refractive index with sub-wavelength period that were clearly visualized after etching the laser-written tracks with HF acid (Fig. 11.1a and 11.1b). The orientation of the nanogratings was perpendicular to the writing laser polarization in all cases. The period of the nanostructures was found to be approximately $\lambda/2n$, regardless of scan speed, which implies a self-replicating formation mechanism (Hnatovsky *et al.*, 2006). However, new research suggests a slight variation of the nanograting period with exposure parameters (Ramirez *et al.*, 2010). Taylor's group proposed that inhomogeneous dielectric breakdown results in the formation of a nanoplasma resulting in the growth and self-organization of nanoplanes (Hnatovsky *et al.*, 2006). The model was found to accurately predict the experimentally measured nanograting period, but further development is needed to explain why nanostructures have not been observed in borosilicate glasses, and why they only form in a small window of pulse energy and pulse duration in fused silica (Hnatovsky *et al.*, 2006). Preferential HF etching of laser-written tracks was observed when the nanogratings were parallel to the writing direction (polarization perpendicular to scan direction) allowing the HF acid to diffuse more easily in the track, as shown in Fig. 11.1c. This effect can be exploited to fabricate buried microchannels for microfluidic applications (Osellame *et al.*, 2011).

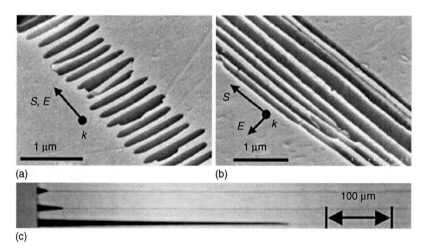

(a)   (b)   (c)

11.1 Scanned electron microscope image of nanogratings formed at 65-μm depth (sample cleaved and polished at writing depth) with polarization parallel (a) and perpendicular (b) to the scan direction. Overhead view (c) of etched microchannels demonstrating polarization selective etching with parallel (top), 45° (middle) and perpendicular (bottom) linear polarizations (Hnatovsky *et al.*, 2006).

Void formation

At high pulse energies (>500 nJ for 0.6-NA focusing of 800-nm, 100-fs pulses) giving peak intensities greater than ~ 10^{14} W/cm², pressures greater than Young's modulus are generated in the focal volume, creating a shock wave after the electrons have transferred their energy to the ions (~ 10 ps) (Itoh *et al.*, 2006). The shock wave leaves behind a less dense or hollow core (void), depending on the laser and material properties (Juodkazis *et al.*, 2006). By conservation of mass, this core is surrounded by a shell of higher refractive index. Such voids may be exploited for 3D memory storage (Glezer *et al.*, 1996) or photonic bandgap materials (Juodkazis *et al.*, 2002), but are not suitable for optical waveguides.

Cumulative pulse interaction

The above interpretations for the structural changes induced by focused femtosecond lasers were based on single pulse interactions, but can likely be extended to the explain modification from multiple pulses within the same laser spot, assuming the repetition rate is low enough that thermal diffusion has carried the heat away from the focus before the next pulse arrives (Itoh *et al.*, 2006). In this situation, the ensuing pulses may add to the overall modification, but still act independently of one another.

For high repetition rates (>100 kHz), the time between laser pulses is less than the time for heat to diffuse away, resulting in an accumulation of heat in the focal volume. If the pulse energy is sufficient, the glass near the focus is melted and as more laser pulses are absorbed, this melted volume increases in size until the laser is removed, at which point the melt rapidly cools into a structure with altered refractive index. For a scanned waveguide structure, the size of the melted volume can be controlled by the effective number of pulses in the laser spot size, $N = 2w_0R/v$, where $2w_0$ is the spot size diameter ($1/e^2$), R is the repetition rate and v is the scan speed.

Figure 11.2 shows optical microscope images of borosilicate glass modified by static laser exposure of 450-nJ pulse energy with varied repetition rate and number of pulses. Spherical laser modified zones were observed for all static exposures tested, and arise from the three-dimensional symmetry of heat diffusion from a small laser absorption volume of ~ 2-µm diameter. These refractive index structures are the result of localized melting within a cumulative heating zone that is built up over many laser pulses, which then cools rapidly to resolidify after the laser exposure. Evidence of cumulative heating is noted at repetition rates above 200 kHz, where the diameter of the modified volume significantly exceeds the ~ 2-µm laser spot size. Within each row (constant repetition rate) in Fig. 11.2, one notes a modest increase in the diameter of the heat-affected zone despite a four order-of-magnitude increase in exposure. More dramatic is the ~ 10-fold increase

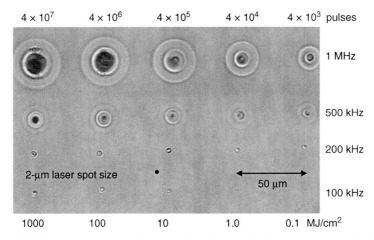

11.2 Optical microscope images showing heat-affected zones created in borosilicate glass with 450-nJ pulse energy from a 1045-nm femtosecond laser. Total pulse (top) and fluence accumulation (bottom) is shown for each column and the laser repetition rate is indicated for each row. The laser direction is normal to page and the approximate laser spot size is shown by the black circle.

in modified zone radius as noted when the repetition rate is increased from 0.1 to 1 MHz in each column. Since the total laser exposure is identical within any column, 200-kHz repetition rate defines the onset for cumulative heating effects above which thermal diffusion controls the properties of optical circuits formed by the femtosecond laser. One also notes that the size of the modification zone grows more quickly with the number of pulses when in the cumulative heat regime.

11.2.2   Influence of focusing

Linear optical effects such as dispersion, diffraction, aberration and nonlinear effects such as self-focusing, plasma defocusing and energy depletion influence the propagation of focused femtosecond laser pulses in dielectrics, thereby altering the energy distribution at the focus and the resulting refractive index modification.

Linear propagation

Incident femtosecond laser pulses are focused with an external lens to achieve a small micrometer-sized focal spot and drive nonlinear absorption. Neglecting spherical aberration and nonlinear effects, the spatial intensity profile of a femtosecond laser beam can be well represented by the paraxial wave equation and Gaussian optics. The diffraction-limited minimum waist radius w_0 (1/2 the spot size) for a collimated Gaussian beam focused in a dielectric is given by:

$$w_0 = \frac{M^2 \lambda}{\pi NA} \qquad [11.1]$$

where M^2 is the Gaussian beam propagation factor (beam quality) (Johnston, 1998), NA is the numerical aperture of the focusing objective and λ is the free space wavelength. The Rayleigh range z_0 (1/2 the depth of focus) inside a transparent material of refractive index n is given by:

$$z_0 = \frac{M^2 n \lambda}{\pi NA^2} \qquad [11.2]$$

Chromatic and spherical aberration cause a deviation in the intensity distribution near the focus so that Eq. [11.1] and Eq. [11.2] are no longer valid approximations. Chromatic aberration as the result of dispersion in the lens is corrected by employing chromatic aberration-corrected micro-scope objectives for the wavelength spectrum of interest. For lenses made with easily formed spherical shapes, light rays which are parallel to the optic axis but at different distances from the optic axis fail to converge to the same point, resulting in spherical aberration. This issue can be addressed by using multiple lenses such as those found in microscope objectives or employing an aspheric focusing lens. In waveguide writing where light is focused inside glass, the index mismatch at the air–glass interface introduces additional spherical aberration. As a result, there is a strong depth dependence for femtosecond laser-written buried structures (Eaton *et al.*, 2008b; Hnatovsky *et al.*, 2005b; Marcinkevicius *et al.*, 2003). This depth dependence is more pronounced for higher NA objectives (Schaffer *et al.*, 2001), except in the case of oil-immersion lenses (Osellame *et al.*, 2006) or dry objectives with collars that enable spherical aberration correction at different focusing depths (Hnatovsky *et al.*, 2005b).

Dispersion from mirror reflection and transmission through materials can broaden the pulse width (Osellame *et al.*, 2005) which can reduce the peak intensity and alter the energy dissipation at the focus. For a typical Yb-based amplified femtosecond laser with 1-μm wavelength and 10-nm bandwidth, the dispersion in glass is −50 ps/km/nm and the pulse duration increase per length is 5 fs/cm. Since these sources have pulse durations >200 fs, the dispersion is negligible for most laser micromachining beam delivery systems which have less than 1 cm of transmission through glass. It is only for short pulse <40-fs oscillators with large bandwidths that dispersion becomes an issue. In this case, pre-compensation of the dispersion through the microscope objective is required to obtain the shortest pulse at the focus (Osellame *et al.*, 2005).

Nonlinear propagation

The spatially varying intensity of a Gaussian laser beam can create a spatially varying refractive index in dielectrics. Because the nonlinear refractive index n_2 is positive in most materials, the refractive index is higher at the center of the beam compared to the wings. This variation in refractive index acts as a positive lens and focuses the beam inside a dielectric with a strength dependent on the peak power. If the peak power of the femtosecond laser pulses exceeds the critical power for self-focusing (Schaffer *et al.*, 2001):

$$P_c = \frac{3.77\lambda^2}{8\pi n_0 n_2} \qquad [11.3]$$

the collapse of the pulse to a focal point is predicted. However, as the beam self-focuses, the increased intensity is sufficient to nonlinearly ionize the material to produce a free electron plasma, which acts as a diverging lens that counters the Kerr lens self-focusing. A balance between self-focusing and plasma defocusing leads to filamentary propagation, which results in axially elongated refractive index structures, which are undesirable for transversely written waveguide structures described in the next section. Self-focusing can be suppressed in waveguide fabrication by tightly focusing the laser beam with a microscope objective to reach the intensity for optical breakdown without exceeding the critical power for self-focusing.

In fused silica, $n_0 = 1.45$ and $n_2 = 3.5 \times 10^{20}$ m²/W (Sudrie *et al.*, 2002) so that for $\lambda = 800$ nm, the critical power is ~ 1.8 MW. From Eq. [11.3], the critical power is proportional to the square of the laser wavelength, therefore, lower critical powers result when working with the second and third harmonic frequencies of femtosecond lasers. Also, the critical power is inversely related to the nonlinear (and linear) refractive index, presenting a challenge in forming waveguides in nonlinear materials such as heavy metal oxide ($n_0 \sim 2, n_2 \sim 10^{-18}$ m²/W; Siegel *et al.*, 2005), chalcogenide glasses ($n_0 \sim 2.5, n_2 \sim 10^{-17}$ m²/W; Ta'eed *et al.*, 2007) and polymers ($n_0 \sim 1.5, n_2 \sim 10^{-18}$ m²/W).

11.2.3   Influence of writing geometry

The standard configurations for laser writing of optical waveguides are shown in Fig. 11.3. In longitudinal writing, the sample is scanned parallel, either towards or away from the incident laser. In this configuration, the resulting waveguide structures have cylindrical symmetry, owing to the

Longitudinal   Transverse

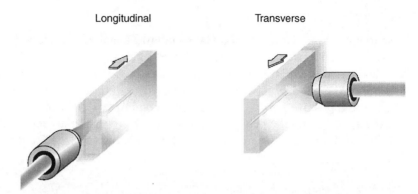

11.3 Longitudinal and transverse writing geometries for femtosecond laser waveguide fabrication in the bulk of transparent materials. In transverse (longitudinal) writing, the sample is scanned transversely (parallel) with respect to the incident femtosecond laser (Gattass and Mazur, 2008).

transverse symmetry in the Gaussian intensity profile of the laser beam. The main disadvantage of the longitudinal writing geometry is that the waveguide length is limited by the working distance of the lens, which for a typical focusing objective with NA = 0.4, is approximately 5 mm. To overcome this issue, researchers have employed looser focusing lenses (NA = 0.2) (Yamada *et al.*, 2001), requiring higher laser power to reach the intensity required for optical breakdown. At such peak powers (~ 1 MW), the optical Kerr effect results in self-focusing, producing filaments which yield refractive index change structures elongated in the axial direction by up to several hundred microns (Yamada *et al.*, 2001). Despite the long length of the filaments, fabrication speeds are still relatively slow at ~1 µm/s to build up enough refractive index increase to efficiently guide light (Yamada *et al.*, 2001).

In the transverse writing scheme of Fig. 11.3, the sample is scanned orthogonally relative to the incoming laser. The working distance no longer restricts the waveguide length and structures may be formed over a depth range of several millimeters, which is sufficient flexibility for many applications to provide 3D optical circuits. The disadvantage of the transverse geometry is that the waveguide cross-section is asymmetric due to the ratio between depth of focus and spot size $2z_0/2w_0 = n/NA$, where n is the refractive index and NA is the numerical aperture. For waveguides formed in glasses with $n = 1.5$ with typical NA of 0.25–0.85, the focal volume asymmetry n/NA varies from 6.0 to 1.8. This asymmetry results in elliptical waveguide cross-sections with elliptical guided modes, which couple poorly to optical fibers. Methods for overcoming this waveguide asymmetry are discussed below.

11.3   Femtosecond laser waveguide writing in glasses

The properties of femtosecond laser-written waveguides in bulk glass depend strongly on laser exposure conditions such as scan speed, average power and numerical aperture. These parameters influence the total laser dosage, or net fluence, NF:

$$\text{NF} = \frac{2w_0 R F_p}{v} \qquad\qquad [11.6]$$

where w_0 is the waist radius, R is the repetition rate, F_p is the per-pulse fluence (pulse energy per area) and v is the scan speed. Related to the dwell time ($t_{\text{dwell}} = 2w_0 / v$) the effective number of pulses per spot size diameter ($2w_0$) during scanning is given by $N = 2w_0 R / v$. Refractive index modification in femtosecond laser waveguide writing occurs above a bulk modification threshold intensity, typically $\sim 10^{13}$ W/cm^2 in dielectrics (Schaffer et al., 2001). Above this threshold, increasing net fluence via decreased scan speed or increased fluence often leads to increased refractive index, and at sufficiently high net fluence, damaged and irregular modification tracks.

11.3.1   Low repetition rate fabrication of optical waveguides in glasses

At low repetition rates, the time between pulses is long enough so that thermal diffusion has carried the heat away from the focus before the next pulse arrives. The threshold repetition rate for heat accumulation depends on several factors including the glass properties (heat capacity, thermal diffusivity, absorption), the laser pulse energy, the focusing NA and the scanning speed, which will be discussed in further detail in the next section. In the low repetition rate waveguide writing regime, the ensuing pulses may add to the overall modification, but still act independently of one another. Most results in the field of femtosecond laser microfabrication have been carried out at low repetition rates, and usually at 1 kHz, due to the common availability of 800-nm regeneratively amplified Ti:Sapphire femtosecond lasers at this repetition rate. One limitation of waveguide writing in the single-pulse interaction regime is that waveguide cross-sections take on a shape similar to the asymmetric focal volume since the depth of focus is larger than the transverse spot size. The resulting waveguides written with the transverse writing scheme, where the sample is translated transversely relative to the incident laser, are elliptical-like, giving modes that couple poorly to optical fiber. Several methods have been proposed to produce a more symmetric focal volume including astigmatic focusing with a cylindrical lens telescope

(Osellame *et al.*, 2003), slit reshaping (Ams *et al.*, 2005; Cheng *et al.*, 2003), multiscan writing (Nasu *et al.*, 2005) and two-dimensional deformable mirrors (Thomson *et al.*, 2008).

Silicate and phosphate glasses

Although pure fused silica is the most common glass for photonic applications due to its high transmission, excellent temperature resistance and compatibility with biomaterials, silicate and phosphate glasses offer similar characteristics at a reduced cost. Further, silicate and phosphates may be doped with active ions for amplification or lasing applications.

The most common silicate glasses are boro-aluminosilicate glasses, which in addition to silica (SiO_2) contain significant concentrations of boron trioxide (B_2O_3), aluminum oxide (Al_2O_3) and sodium oxide (Na_2O). Waveguides have been successfully fabricated by low repetition rate femtosecond lasers in several borosilicates including Schott Duran (Ehrt *et al.*, 2004), Corning 7890 (Streltsov and Borrelli, 2002) and Corning 1737 (Low *et al.*, 2005). In Corning EAGLE2000, a common glass used primarily in displays owing to its low density, Zhang *et al.* explored a wide range of processing conditions with a 1-kHz Ti:Sapphire femtosecond laser. By tuning the compressor alignment, pulse durations of 50 fs–2 ps were studied, revealing promising windows for waveguide writing at both 100 fs and 1 ps (Zhang *et al.*, 2007), disproving the widely held belief that sub-200 fs pulses were needed to form optical waveguides in glass. The most important consequence of this work was a serendipitous discovery from scanning the sample at 0.5 mm/s so that successive pulses only partially overlapped, resulting in a periodic refractive index modulation along the waveguide. The partially overlapped refractive index voxels resulted in segmented waveguides that showed strong and narrowband Bragg reflection while maintaining high optical confinement and low-loss waveguiding at 1550-nm wavelength. For more discussion of waveguide Bragg gratings and related sensing and lasing devices, the reader is referred to Chapter 10.

In Schott BK7, the most common borosilicate glass used in commercial optics, initial reports of waveguide writing with 1-kHz Ti:Sapphire lasers suggested that only negative refractive index modification was possible (Bhardwaj *et al.*, 2005; Ehrt *et al.*, 2004; Mermillod-Blondin *et al.*, 2006). However, it was later shown that by using temporally shaped pulses of ~ 1 ps duration, positive index changes are indeed possible in BK7. It was also recently demonstrated that optical waveguides may be formed at low repetition rates in BK7 using a slit beam reshaping technique (Dharmadhikari *et al.*, 2011). Waveguides may be more easily formed in BK7 without correction techniques by applying higher repetition rates, with demonstrations of low-loss (~ 0.2 dB/cm) waveguides at both 2 MHz (Eaton *et al.*, 2008a) and 11 MHz (Allsop *et al.*, 2010).

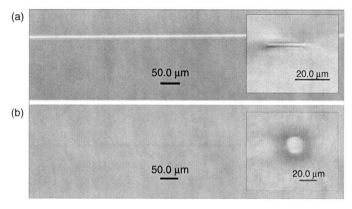

11.4 Microscope images of waveguides fabricated in phosphate glass (a) without and (b) with a 500 μm slit (Ams *et al.*, 2005).

Good quality optical waveguides have been demonstrated in phosphate glass, which is easily doped with Er and Yb ions for active waveguide applications. Using the astigmatic writing method with a cylindrical lens telescope, waveguides with low damping loss (0.25 dB/cm) at 1550-nm wavelength were written at 20 μm/s with a 1-kHz, 150-fs Ti:Sapphire laser with a 0.3-NA microscope objective and 5-μJ pulse energy (Osellame *et al.*, 2003). This work is significant since it was the first time a beam-shaping method was applied to correct for the asymmetric intensity distribution in the transverse waveguide writing. The slit shaping method was later applied by Withford and coworkers to produce symmetric waveguides (Fig. 11.4) with similar propagation loss in the same active glass (Ams *et al.*, 2005).

Pure fused silica glass

Many groups have applied the femtosecond laser-writing method to pure fused silica glass, but few have shown good quality waveguides with operation at both visible and near-infrared wavelengths, for use in biophotonics and telecom devices, respectively. The best result in fused silica to date was obtained with a multiscan writing procedure to form waveguides with nearly square cross-sections (7.4 μm × 8.2 μm) with a refractive index change of 4×10^{-3}, suitable for low-loss coupling to single mode fiber (SMF) (Nasu *et al.*, 2005). The refractive index profile obtained by refracted near field (RNF) profilometry for a waveguide written in Ge-doped silica glass for planar lightwave circuits (PLCs) is shown in Fig. 11.5. The waveguides written in PLC glass were nearly identical to the waveguides written in pure silica. A 775-nm, 150-fs Ti:Sapphire laser with 1-kHz was applied in the transverse writing geometry with a 0.4-NA objective, 182-nJ pulse energy and 10-μm/s writing speed. The waveguides were fabricated with 20 scans

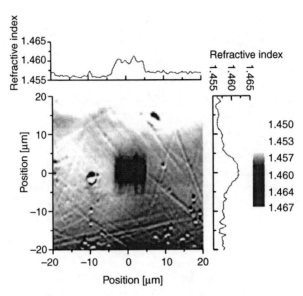

11.5 Refractive index profile of a waveguide written in Ge-doped silica PLC glass by multiple scans (Nasu *et al.*, 2005).

separated transversely by 0.4 μm. A propagation loss of 0.12 dB/cm at 1550 nm was reported, which is the lowest reported in the field, and is attributed to the gentle refractive index modification enabled by the novel low-fluence, multiscan fabrication method.

The effect of writing speed on waveguide properties was evidenced in a study in pure silica with a 120-fs 1-kHz Ti:Sapphire laser (Will *et al.*, 2002). With 3-μJ energy pulses focused 0.5 mm below the sample surface with a 0.25-NA lens, the initially single mode waveguide at 1 mm/s scan speed (Fig. 11.6a) showed higher confinement at a decreased scan speed of 0.5 mm/s (Fig. 11.6b). As the scan speed was further reduced, increasing the net fluence, the waveguide became multimode as the effective index was further increased with the *V*-number exceeding 2.4, the single-mode cut-off value (Agrawal, 1997).

The effect of writing laser polarization on waveguide transmission properties was studied using a 1-kHz, 120-fs Ti:Sapphire laser with a 0.5-mm slit placed before the 0.46-NA focusing objective (Little *et al.*, 2008) to obtain symmetric waveguide cross-sections. With 3-μJ pulse energy measured after the slit and 25-μm/s writing speed, a refractive index change of 2.3×10^{-3} was measured for circular polarization, which was about twice as high as that obtained with linear polarizations. This enhancement was attributed to the higher photo-ionization rate for circular polarization compared to linear polarization in the range of intensities studied (42–50 TW/cm²).

It is probable that nanogratings also influence the transmission properties of femtosecond laser-written waveguides in fused silica. For more details on nanogratings and their application to post-etching of buried microchannels, the reader is referred to a review by Taylor *et al.* (2008).

Exotic glasses

Waveguides were written in a highly nonlinear heavy-metal oxide (HMO) glass with a 1-kHz, 800-nm, 100-fs Ti:Sapphire laser (Siegel *et al.*, 2005). HMO glasses are attractive due to their high optical nonlinearity ($n_2 \approx 10^{-18}$ m^2/W), but this presents significant challenges in femtosecond laser writing because of strong self-focusing, resulting in a delocalized spatial distribution of the laser energy which is difficult to control. By focusing 1.8-µJ femtosecond laser pulses with a 0.42-NA objective and scanning the sample transversely at 60 µm/s, elongated damage structures of 65-µm vertical length were observed when the sample was viewed from the end facets. These elongated structures were the result of filamentation when self-focusing balances against plasma defocusing. The waveguiding regions were found to be adjacent to the filament-induced damage zone, with propagation losses below 0.7 dB/cm demonstrated at 633-nm wavelength. The regions of refractive index increase adjacent to the filament were attributed to compressive stress induced outside the laser-damaged zone, similar to observations during waveguide writing in crystalline materials.

11.3.2 High repetition rate fabrication of optical waveguides in glasses

The recent development of high repetition rate, high power femtosecond lasers opens new avenues for manipulating thermal relaxation effects that control the properties of optical waveguides formed when ultrashort laser pulses are focused inside glasses. At low to moderate repetition rates (1–100 kHz), an increase in laser pulse energy leads to formation of larger modification structures as thermal diffusion extends the laser-heated region far outside the focal volume. As repetition rate increases, the time between laser pulses becomes shorter than the time for the absorbed laser radiation to diffuse out of the focal volume and heat builds up in the focal volume. Schaffer *et al.* (2003) first reported heat accumulation in the bulk of glass using a 25-MHz femtosecond laser oscillator. With increased dwell time, a dramatic increase in the size of laser-modified structures was observed compared to structures formed with single-pulse interactions, where no variation in modification size with dwell time was observed. The combination of high repetition rate and heat accumulation offers fast writing speeds and cylindrically symmetric waveguides together with benefits of annealing and

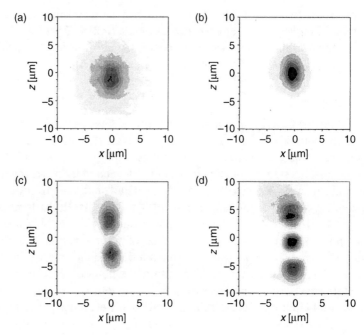

11.6 Influence of the writing speed on waveguide properties at a wavelength of 514 nm. Only the highest-order guided modes are shown for a writing speed of (a) 1 mm/s, (b) 0.5 mm/s, (c) 80 μm/s and (d) 25 μm/s (Will *et al.*, 2002).

decreased thermal cycling that are associated with low propagation and coupling loss to standard optical fiber.

Silicate and phosphate glasses

Following the early work by Schaffer *et al.* (2003), further insight into heat accumulation effects was provided by Eaton *et al.* (2008b). Using a finite difference thermal diffusion model, the temperature in the focal volume was calculated as a function of pulse number (dwell time) for repetition rates of 0.1, 0.5 and 1 MHz in Corning EAGLE2000 borosilicate glass. Typical writing conditions of 200 nJ of absorbed energy, 0.55-NA focusing and a melting point of 985°C were assumed. As shown in Fig. 11.7, at 100-kHz repetition rate, the temperature relaxes to below the softening point before the next pulse arrives, resulting in minimal heat accumulation and significant temperature cycling during waveguide writing. At 0.5 and 1 MHz repetition rates, heat accumulation is strongly evident, leading to a larger melted volume which increases with pulse number and repetition rate. Decreased thermal cycling with increased repetition rate is anticipated to lead to smoother waveguides with less propagation loss and birefringent stress.

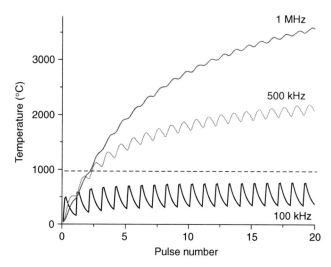

11.7 Model of glass temperature versus exposure at repetition rates of 100 kHz, 500 kHz and 1 MHz, at a radial position of 3 μm from the center of the focal volume. The absorbed energy of 200 nJ was identical at each repetition rate.

To unravel the contributions of thermal diffusion and heat accumulation to the resulting waveguide morphology, a variable repetition rate (0.1–5 MHz) 300-fs, 1045-nm Yb-doped fiber laser was applied to waveguide writing in Corning EAGLE2000 borosilicate glass (Eaton *et al.*, 2008b) with a 0.55-NA focusing lens. Structures were formed at 150 μm below the surface unless specified otherwise. The onset for heat accumulation was determined by comparing waveguide diameters produced by scanned exposures with that of single-pulse exposures, where the only contribution is from thermal diffusion. The threshold was defined as the minimum pulse energy to increase the waveguide diameter 2-fold over the diameter produced by diffusion in a single-pulse interaction. Single-pulse diffusion diameters were found to vary from 2 to 6 μm for pulse energies of 0.1 to 1 μJ (Eaton *et al.*, 2008b).

The threshold pulse energy for driving heat accumulation is shown in Fig. 11.8 as a function of repetition rate (200 kHz–2 MHz) and for scan speeds of 2, 10 and 40 mm/s. The single pulse modification threshold was 50 nJ (2.5 J/cm^2) and invariant with repetition rate. In contrast, the energy threshold for heat accumulation decreased sharply from 900 nJ at 200 kHz to 80 nJ at 2 MHz and was only weakly dependent on scan speed. In this 200 kHz–2 MHz range of repetition rates, the thermal diffusion scale length $\sqrt{D/R}$ decreases from 1.6 to 0.5 μm, where D is the thermal diffusivity and R is the repetition rate. This decrease in the characteristic thermal diffusion

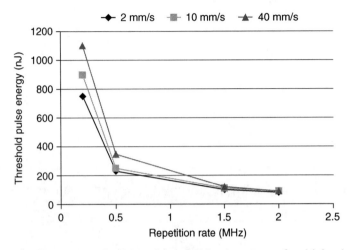

11.8 Experimental values of threshold pulse energy for driving heat accumulation in borosilicate glass as a function of laser repetition rate for scan speeds of 2, 10 and 40 mm/s and 0.55-NA focusing (Eaton *et al.*, 2008b).

length indicates that the effective laser heating volume decreases dramatically with increasing repetition rate, thereby reducing the threshold pulse energy for heat accumulation. For $R > 1$ MHz, the thermal diffusion scale length of 0.7 μm falls inside the laser waist radius of $w_0 = 0.8$ μm, resulting in an asymptotic limit of the heat accumulation threshold energy to a minimum value of 80 nJ at 2-MHz repetition rate in Fig. 11.8. Beyond 2 MHz, the available laser pulse energy was below this value, preventing the observation of heat accumulation effects in EAGLE2000 glass. Similarly, when the laser was operated at the lowest 100-kHz repetition rate, the 2-μJ maximum pulse energy available was insufficient to drive heat accumulation effects beyond a larger thermal diffusion diameter of 8 μm.

In the same study, a new laser processing window was discovered for producing low-loss waveguides across the large 200 kHz–2 MHz range of repetition rates (Eaton *et al.*, 2008b). By holding the average power constant and delivering the same net fluence exposure at each repetition rate, strong thermal diffusion from high energy pulses at 200-kHz repetition rate balanced the strong heat accumulation with low pulse energy delivered at high repetition rates to produce waveguides with similar diameter and strong guiding at 1550-nm wavelength. Figure 11.9 shows cross-sectional refractive index profiles measured by RNF for waveguides written with 200-mW average power at repetition rates of 0.2–2 MHz and 25-mm/s scan speed. This average power was found to give the lowest insertion loss (IL) at each repetition rate as described later. The core of the waveguides shown in Fig. 11.9

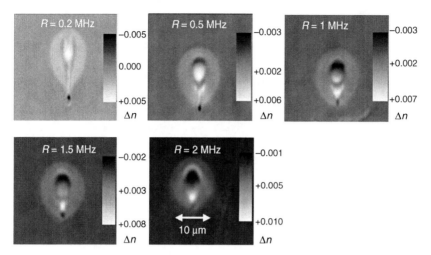

11.9 Cross-sectional refractive index profiles of waveguides written with 200-mW average power, 25-mm/s scan speed and repetition rates of 0.2, 0.5, 1, 1.5 and 2 MHz (Eaton *et al.*, 2008b).

is attributed to the high temperature spikes induced within the laser spot size by each laser pulse, while the outer lower-contrast cladding is formed by a more slowly evolving near-Gaussian temperature distribution, with the overall size determined by the maximum diameter where the temperature exceeds the melting point. Because of the variable temperature across the modified zones, the cooling rates are highly non-uniform, and therefore are expected to lead to a non-uniform distribution of the final glass density (Chan *et al.*, 2003a).

The small, dark spot at the bottom of the index profiles in Fig. 11.9 is attributed to the focus plane location since its depth was constant with varying exposure conditions. The images show that most of the laser energy was deposited upstream of the focus. At 200-kHz repetition rate, the waveguides showed a vertically elongated central core guiding region with peak Δn = 0.005. The waveguide is elliptical due to thermal diffusion from a laser heating volume extended vertically at this high pulse energy of 1 μJ and also from minimal cumulative heating between laser pulses. At 500 kHz, owing to increased heat accumulation, a more circular guiding region was found with a maximum Δn = 0.006 in the guiding region. A small region of increased refractive index is also observed below the main guiding region. At 1-MHz repetition rate, this region below the core has increased in size and magnitude to a peak Δn = 0.007. This region is now responsible for guiding of 1550-nm light but the mode also extends into the weaker central core to yield two transverse modes when formed with a higher average power exposure (> 250 mW). At 1.5 MHz, the guiding region has clearly transitioned below

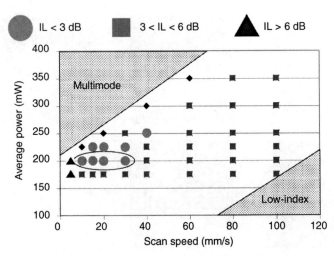

11.10 Processing window map: waveguide properties as a function of average power and scan speed for 1.5-MHz repetition rate.

the central core to form a strong guiding region with peak $\Delta n = 0.008$. At 2-MHz repetition rate, a similar profile is observed but beyond this repetition rate, the pulse energy dropped below the heat accumulation threshold of 90 nJ (Fig. 11.8) and formed weakly guiding structures. The highly non-linear laser interactions lead to vertical shifts of waveguide position and differing waveguide profiles that must be monitored and accounted for during fabrication with varying exposure conditions and focal depths in the glass.

Figure 11.10 aids in visualizing waveguide properties as a function of average power and scan speed, shown for 1.5-MHz repetition rate. Insertion loss is classified by circles, squares and triangles representing low (< 3 dB), medium (3–6 dB) and high (>6 dB) IL, respectively, for the 2.5-cm long waveguides. Waveguides exhibiting multiple transverse modes (diamonds) and typically damaged morphology, written with the highest net fluence are found at the top-left. Conversely, the bottom-right corner indicates under-exposed waveguides where the index change was too low to efficiently guide 1550-nm light. At 1.5-MHz repetition rate, the lowest IL of ~ 1.2 dB for fiber–waveguide–fiber coupling was observed over a large 10–25 mm/s range of scan speeds, but in a narrow 200 mW ± 10 mW average power range (encircled data in Fig. 11.10).

Similar analysis was carried out for waveguides formed with 0.2, 0.5, 1 and 2-MHz repetition rates and in all cases, revealed a similar processing window of 200 mW and 10–25 mm/s for the lowest IL. The minimum IL and MFD at each repetition rate are presented in Fig. 11.11. This constant 200-mW average power exposure window appears consistent with the optimum 250-mW power for generating low-loss waveguides in phosphate glass at repetition rates of 505–885 kHz with a similar femtosecond laser (Osellame *et al.*, 2006). The decreasing IL with increasing repetition rate in Fig. 11.11 is associated with increasingly stronger heat accumulation that results in higher refractive change and smaller MFD for best coupling to optical fibers at 1.5 MHz. The increased IL and MFD from 1.5 to 2 MHz is attributed to inadequate laser pulse energy (100 nJ) at 2 MHz for driving sufficient laser heating above the ~ 90 nJ threshold for heat accumulation shown in Fig. 11.8. Beyond 2-MHz repetition rate, only narrow ~2 μm diameter waveguides were formed that were barely guiding and showed no evidence of heat accumulation.

The refractive index profiles of waveguides written with 0.55-NA lens, 200-mW power, 150-μm depth, 1.5-MHz repetition rate and scan speeds of 5, 15 and 30 mm/s are shown in Fig. 11.12. Due to the decreased net fluence, the cladding and guiding region diameters and also the peak refractive index

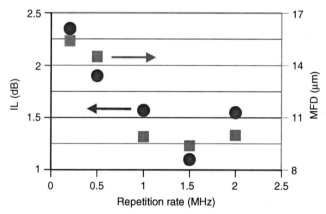

11.11 IL and MFD versus repetition rate for waveguides formed with 200-mW average power and 15-mm/s scan speed (Eaton *et al.*, 2008b).

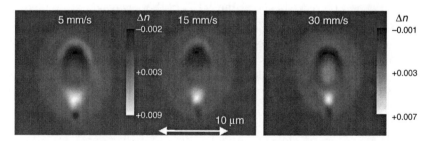

11.12 Refractive near-field measurements of cross-sectional refractive index profiles for waveguides written with 0.55-NA lens, 200-mW power, 150-μm depth, 1.5-MHz repetition rate and scan speeds of 5, 15 and 30 mm/s. The writing laser was incident from the top (Eaton *et al.*, 2008b).

change decrease with increasing scan speed. As shown in Fig. 11.11, a scan speed of 15 mm/s provided the lowest insertion loss of 1.2 dB at 1550-nm wavelength with ~10-μm MFD allowing efficient coupling to single-mode fiber. The highest exposure at 5-mm/s scan speed led to high loss (Fig. 11.10), while 30-mm/s speed resulted in weakly confined modes of ~15-μm MFD, coupling poorly to fiber. At the maximum scan speed of 100 mm/s, the MFD increased to ~20 μm.

The refractive index profile for the optimum waveguide at 15-mm/s scan speed (1.2-dB insertion loss, ~ 0.3-dB/cm propagation loss) was used to confirm the accuracy of the RNF measurements, as shown in Fig. 11.13. The RNF data were imported into a numerical mode solving routine (Lumerical MODE Solutions 2.0) and the resulting mode profile shows excellent agreement with the experimentally measured mode profile, confirming the accuracy of the RNF measurements. The arrow indicates the relative position of mode and waveguide cross-section.

To take advantage of femtosecond laser writing of waveguides in all three dimensions, one must carefully address the problem of spherical aberration at the air–glass interface, which varies dramatically with the focusing depth. The effect of spherical aberration is reduced with oil-immersion lenses (Osellame et al., 2006), objectives with collars for variable depth correction (Hnatovsky et al., 2005b) and asymmetric focusing with slit reshaping (Ferrer et al., 2007). It well known that spherical aberration is less pronounced with lower NA focusing (Schaffer et al., 2001). Further, one can take advantage of strong heat accumulation effects to drive spherically symmetric heat flow which compensates for an axially elongated focal volume. A combination of heat accumulation effect and low numerical aperture of the focusing optic (NA < 0.55) was shown to enable depth-independent, low-loss waveguides (Eaton et al., 2008b).

For the cumulative heating regime of 0.55-NA focusing applied above (200 mW, 1.5 MHz, 15 mm/s), waveguides with similar low loss and mode size could only be obtained in a narrow depth range of $d = 50$–200 μm (Eaton et al., 2008b). Figure 11.14 shows the MFD of these waveguides to increase 60% from 10 μm at 50-μm depth to 16 μm at 300-μm depth, spherical aberration precluding a deeper waveguide writing range. Much deeper waveguide writing was possible with the 0.25-NA lens, but 5-fold lower peak intensity at the maximum available laser power (400 mW) yielded only small diameter waveguides and weak refractive index change. A better balance was found with the 0.4 NA lens, providing only a small increase in MFD from ~11.0 to 13.5 μm as the depth was increased from $d = 50$ to 520 μm. The measured propagation loss of ~ 0.35 dB/cm was nearly independent of focal depth. The ability to write waveguides to depths of 520 μm is a substantial improvement over the maximum depth of ~ 200 μm reported by other groups employing MHz repetition rate femtosecond lasers with higher NA focusing objectives (Osellame et al., 2005).

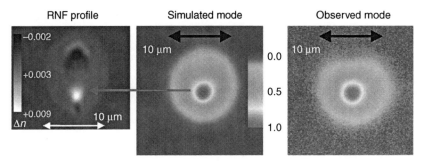

11.13 Waveguide fabricated with 1.5-MHz repetition rate, 200-mW power, 0.55-NA lens, 150-mm depth and 15-mm/s scan speed: cross-sectional refractive index profile (left), simulated mode profile (middle) and measured mode profile (right). A gray arrow indicates position of mode relative to waveguide cross-section (Eaton *et al.*, 2008b).

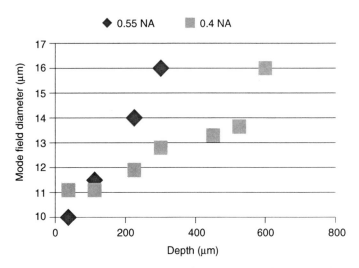

11.14 Mode field diameter versus focusing depth for waveguides formed with 1.5-MHz repetition rate, 15-mm/s scan speed with 0.55 NA (230 mW) and 0.4 NA (200 mW) (Eaton *et al.*, 2008b).

Thermal annealing experiments were performed on the above borosilicate waveguides to test their thermal stability (Eaton *et al.*, 2008b). Waveguides were written at 230-mW power, 1.5-MHz repetition rate, 0.4-NA, 150-μm depth and 8–20-mm/s scan speed. After waveguide characterization, the EAGLE2000 sample was baked for 1 h in a tube furnace in repeated heating and testing cycles of increased temperature in 100°C steps. A peak temperature of 800°C was tested that exceeds the annealing point of 722°C, the temperature at which stresses are relieved after several minutes, but remains below the softening point of 985°C at which the glass deforms under its own

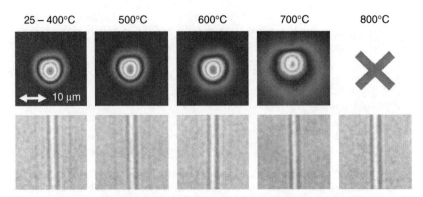

11.15 1550-nm mode profile (top) and overhead microscope (bottom) image at different annealing temperatures for waveguide fabricated in EAGLE2000 with 230-mW power, 1.5-MHz repetition rate, 0.4-NA, 150-μm depth and 15-mm/s scan speed.

weight. For reference, the strain point of 666°C is the temperature below which glass can be rapidly cooled without introducing stresses.

Figure 11.15 shows the mode profile (top) and overhead morphology (bottom) of the waveguide written with intermediate 15-mm/s scan speed for increasing annealing temperature. There is little change in the mode size and waveguide morphology at temperatures up to 500°C. At 600°C, the mode diameter increased ~ 10% and there is less contrast in the cladding. At 700°C, the mode diameter increased ~ 40% compared with the unheated sample (25°C), and the cladding is barely visible. At the last heating step of 800°C, the mode was undetectable due to high losses and the cladding has completely disappeared.

It was found that above 500°C, the MFD for all waveguides tested increased with temperature (Eaton *et al.*, 2008b), with the degree of degradation being smallest for the waveguides written with the lowest speed or highest net fluence. Glass has a frozen-in structure that depends on the cooling rate which corresponds to a fictive temperature of the equilibrium melt (Bruckner, 1970). Waveguides fabricated with the highest net fluence are expected to cool fastest from the highest temperatures, creating modification structures with the highest fictive temperatures. Therefore, waveguides written at the highest exposure required annealing at higher temperatures to undo the thermal history of the laser-modified glass and restore its properties to that of the unmodified bulk. Further, the disappearance of the cladding before the central core in Fig. 11.15 is attributed to a lower fictive temperature in the outer cladding due to lower temperatures and slower cooling rates (Eaton *et al.*, 2008b).

By comparison, waveguides written in EAGLE2000 borosilicate glass with 1-kHz repetition rate were less stable than the present 1.5 MHz results, undergoing an 80% increase in MFD after annealing at 500°C, and resulting in undetectable guiding at 1550 nm after annealing at 750°C (Zhang *et al.*, 2007). The higher temperature stability of waveguides written with 1.5-MHz repetition rate is attributed to the higher fictive temperatures driven by the cumulative heating regime.

Fused silica glass

As demonstrated by Shah *et al.* (2005), the fundamental 1045-nm wavelength led to weak refractive index contrast and irregular morphology in high repetition rate waveguide writing in pure fused silica. However, processing with the second harmonic wavelength of 522-nm wavelength enabled relatively low loss (1 dB/cm) waveguides and moderately high refractive index change ($\Delta n = 0.01$). The benefits of thermal diffusion acting with heat buildup to form circular waveguide cross-sections were not observed in fused silica as reported in lower bandgap silicate glasses (Eaton *et al.*, 2008b; Schaffer *et al.*, 2003). The lack of cumulative heating in fused silica was previously attributed to less absorption in high bandgap fused silica (Shah *et al.*, 2005) and later confirmed experimentally with a 2-fold lower absorption in fused silica measured compared to borosilicate glass for the same laser fluence (Eaton *et al.*, 2011). Also, the working point temperature of fused silica is about 1.5-fold higher than that of borosilicate glasses making it more difficult to melt fused silica and drive heat accumulation. As demonstrated by Osellame *et al.* (2005) with a tightly focused 26-MHz repetition rate femtosecond laser, heat accumulation is possible in fused silica with a combination of higher repetition rate and fluence. However, unlike other glasses (Eaton *et al.*, 2005, 2008b; Minoshima *et al.*, 2001; Osellame *et al.*, 2006; Schaffer *et al.*, 2003) where low propagation loss waveguides were reported, the structures defined by heat accumulation in fused silica were non-uniform and unable to guide light (Osellame *et al.*, 2005). By exploring repetition rates from 0.25 to 2 MHz, a more comprehensive study of waveguide optimization with the second harmonic wavelength was recently performed in fused silica (Eaton *et al.*, 2011) compared to the initial study (Shah *et al.*, 2005).

A repetition rate of 1 MHz yielded the absolute minimum IL of 1.0 dB (sample length 2.5 cm) with 0.55-NA focusing. Cross-sectional profiles taken by optical microscopy and RNF for the optimum waveguide written at 1 MHz, 175 nJ and 0.2 mm/s are shown in Fig. 11.16a and 11.16b, respectively. The modified structures for 0.55-NA focusing were vertically elongated significantly beyond the 2.2-µm depth of focus by self-focusing and plasma defocusing because the peak power was equal to the 0.8-MW critical power. Good qualitative agreement between the microscope and RNF images was found as shown in Fig. 11.16, with the RNF profile showing a guiding region with peak $\Delta n = 0.016$ formed below an irregular damaged region.

To improve the symmetry of the guiding structures in fused silica, a high 1.25-NA oil-immersion lens was applied in the same study (Eaton *et al.*, 2011) to enable increased laser absorption from higher peak intensity and a more symmetric focal volume. With tighter focusing by the 1.25-NA lens, the best IL of 1.2 dB ($L = 1.25$ cm) was obtained for 500-kHz repetition rate, 0.2-mm/s speed and 133-nJ energy, with the corresponding microscope and RNF images shown in Fig. 11.16c and 11.16d, respectively. Weaker waveguides were produced at 1-MHz repetition rate, limited by the maximum on-target energy of 100 nJ. Compared to 0.55-NA, the waveguide formed by 1.25-NA focusing was significantly less elongated, which was attributed to a more symmetric focal volume, and reduced self-focusing (0.6-MW peak power). The peak $\Delta n = 0.022$ represented the highest refractive index increase ever reported for a femtosecond laser-written waveguide in fused silica.

The mode profiles at 1550 nm for the optimum waveguides written by 0.55- and 1.25-NA focusing are shown in Fig. 11.17b and 11.17c, respectively, along with the mode for SMF in Fig. 11.17a with 10.5-μm MFD. The waveguide mode produced by 0.55-NA focusing (MFD = 9.7 μm × 11.9 μm) becomes significantly smaller and more symmetric (MFD = 7.1 μm × 7.4 μm) with 1.25-NA focusing, defining the smallest reported mode to date for a laser-written waveguide in fused silica with simulations predicting that

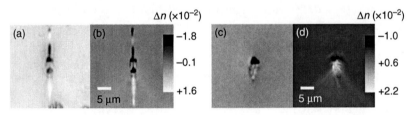

11.16 Cross-sectional microscope images (a, c) and RNF profiles (b, d) for waveguides fabricated with 0.55 NA (a, b) and 1.25 NA (c, d) lenses.

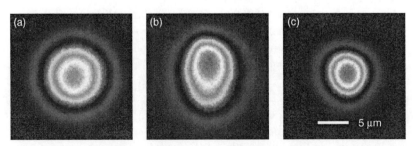

11.17 Mode profile at 1550-nm wavelength for (a) SMF, waveguide writing with 0.55-NA (b) and 1.25-NA (c) objectives.

small $R = 15$ mm bends are now feasible (Eaton *et al.*, 2011), opening the door for much higher density integrated optical circuits.

It is not immediately evident why processing with the green wavelength offers much better quality waveguides than the fundamental wavelength in pure silica glass. It was previously thought the benefit of the green wavelength was due to an enhanced absorption from a lower order nonlinear process, thus providing stronger index contrast waveguides (Shah *et al.*, 2005). However, a similar 40% absorption for 522- and 1045-nm wavelengths for the same laser fluence was found in a recent study (Eaton *et al.*, 2011). Schaffer *et al.* (2001) have shown that multiphoton ionization plays a smaller role in large bandgap materials like fused silica. Instead, avalanche ionization dominates, and since this process is relatively independent of wavelength, the second harmonic wavelength does not offer an advantage in terms of increased absorption. Instead, the benefit of the green wavelength may be due to its higher on-target fluence (Eaton *et al.*, 2011). After accounting for the 50% conversion efficiency of the SHG process and the 4-fold smaller focal area, a 2-fold higher maximum fluence is possible with 522-nm wavelength. In previous studies, the maximum fluence was required to fabricate the optimum waveguide structures at 522-nm wavelength. It is expected that if higher power lasers were available providing fluences at the fundamental wavelength similar to the optimal value at green wavelength, stronger guiding structures could be formed. This is supported by the recent demonstration by Pospiech *et al.* (2009) of waveguides with 1.2-dB/cm loss in fused silica using a 1-MHz repetition rate femtosecond laser with more energetic pulses (500 nJ) at the 1030-nm fundamental wavelength. However, the reported waveguides showed more irregular morphology compared to the structures demonstrated here with the green wavelength. It is likely that other factors which have yet to be identified may contribute to the green wavelength's advantage in femtosecond laser waveguide formation in fused silica.

Exotic glasses

High-quality waveguides were written in an exotic bismuth borate glass (Nippon BZH7) with a 150-fs, 250-kHz regeneratively amplified Ti:Sapphire laser (Yang *et al.*, 2008a). At this moderate 250-kHz repetition rate, the regime of waveguide fabrication is likely the result of both thermal diffusion and heat accumulation. Bismuth borate has similar nonlinear properties as the HMO glass described in the section titled 'Exotic glasses' and is therefore a good candidate for nonlinear optics applications but because of its low critical power, presents a challenge for writing symmetric waveguide structures. To overcome the problem of an axially elongated distribution of laser energy at the focus, the slit reshaping method was applied. With

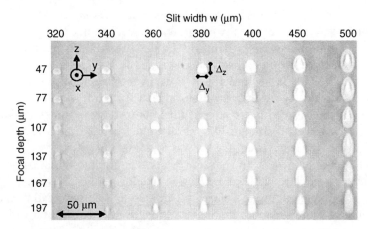

11.18 Cross-sectional microscope image of buried waveguides for different slit widths and depths (Yang *et al.*, 2008a).

200-nJ pulses focused with a 0.55-NA objective and with a writing speed of 200 µm/s, a symmetric waveguide cross-section was produced with a slit width of 380 µm, as shown in Fig. 11.18. The symmetric aspect ratio was found to be preserved up to focal depths of about 100 µm but beyond this value, spherical aberration due to the large refractive index of the glass ($n = 2$) led to an increase in the aspect ratio. When probed with 1550-nm light, the waveguide exhibited a circular mode with MFD of 11 µm, well-matched to SMF. Using the Fabry–Perot method, a propagation loss of 0.2 dB/cm was measured, which is the lowest loss achieved in a high index glass waveguide fabricated by femtosecond laser writing.

11.3.3   Influence of other exposure variables within low and high repetition rate regimes

In addition to pulse energy, scan speed, and focusing, several other exposure parameters have been found to influence the resulting properties of femtosecond laser-written waveguides. These factors include pulse duration (Fukuda *et al.*, 2004; Zhang *et al.*, 2007), polarization (Eaton *et al.*, 2008b; Little *et al.*, 2008), direction (Yang *et al.*, 2008b) and wavelength (Eaton *et al.*, 2011).

Buried structures formed in fused silica with moderate fluence show no evidence of heat accumulation even as the repetition rate is raised from 1 kHz (Little *et al.*, 2008) to 1 MHz (Eaton *et al.*, 2008b) and the waveguide properties were found to be strongly dependent on the incident writing polarization. In contrast, no detectable difference in insertion loss or mode size was found when waveguides were formed with different polarizations in borosilicate glass within the heat accumulation regime at MHz repetition rates (Eaton *et al.*, 2008b). In addition, the waveguide properties in borosilicate glass were invariant to pulse duration when varied 300–700 fs, which is in contrast to results in fused silica, where pulse duration was observed to strongly affect waveguide mode size and loss (Little *et al.*, 2008). The sensitivity to pulse duration and polarization in fused silica is associated with form birefringence arising from nanogratings formed within the laser-modified volume (Hnatovsky *et al.*, 2005a). In borosilicate glass, nanogratings have not been observed (Hnatovsky *et al.*, 2006), and such polarization and pulse duration dependence may possibly be erased by the strong thermal annealing (Hnatovsky *et al.*, 2006) within the heat accumulation regime.

Due to energy depletion, self-focusing and plasma defocusing, pulse duration influences the spatial distribution of the energy density in the focal volume (Rayner *et al.*, 2005). At 1-kHz repetition rate, where heat accumulation is not present, the dependence of waveguide properties on pulse duration in lithium niobate (Burghoff *et al.*, 2007) and fused silica glass (Zhang *et al.*, 2006) was attributed to nonlinear pulse propagation. However, in the heat accumulation regime, nearly spherically symmetric thermal diffusion washes out the elliptical distribution of energy in the focal volume to yield waveguides with cross-sections that are relatively circular. Therefore, one would expect pulse duration, despite its effect on the energy distribution at the focus, to play a lesser role on the properties of waveguides fabricated in the heat accumulation regime.

Kazansky *et al.* recently discovered the quill effect (Yang *et al.*, 2008b), in which laser material modification is influenced by the writing direction, even in amorphous glass with a symmetric laser intensity distribution. The researchers conclusively showed that the cause of the directional writing dependence is due to a pulse front tilt in the ultrafast laser beam (Yang *et al.*, 2008b). Although any material should show a direction dependence due to a pulse front tilt, the effect was found to be almost negligible when processing borosilicate glass within the heat accumulation regime as evidenced by a directional coupler formed by arms written in opposite direction but showing a remarkably high peak coupling ratio of 99% (Eaton *et al.*, 2009).

As described in Section 11.3.2 wavelength is an important variable when processing high bandgap materials such as pure fused silica. Due to its large bandgap and low melting point, the increased fluence and enhanced multiphoton absorption provided by the second harmonic wavelength enabled stronger index contrast and lower loss waveguides in this material (Shah *et al.*, 2005). An infrared wavelength of 1.5 μm was applied to fused silica (Saliminia *et al.*, 2005), revealing a very wide energy processing window of 1–23 μJ in forming smooth waveguides compared to 0.5–2.0 μJ at 800-nm wavelength. Such a large processing window is desirable, but the added complexity of using an optical parametric amplifier has dissuaded researchers from adopting this approach for waveguide device fabrication.

11.4   Waveguide writing in polymers

Lab-on-a-chip (LoC) devices aim at miniaturizing and integrating standard laboratory processes on a single substrate for reduced cost and improved sensitivity (Whitesides, 2006). Traditionally, LoC device containing micro-fluidic channel networks have been fabricated with photolithographic techniques, which involve multi-step fabrication procedures requiring clean room environments. Femtosecond laser processing is a promising tool for fabricating surface microfluidic channels and buried optical waveguides, two important building blocks for LoCs. Although glasses such as fused silica have been proposed as substrates for LoC platforms fabricated by femtosecond lasers (Martinez-Vazquez *et al.*, 2007), thermoplastic polymers such as poly(methyl methacrylate) (PMMA) are potentially more attractive substrates since they offer similar optical transparency and biochemical compatibility but with significantly reduced costs.

11.4.1   Low repetition rate

Watanabe and coworkers demonstrated waveguides in PMMA using a low 1-kHz Ti:Sapphire laser amplifier (Sowa *et al.*, 2006; Watanabe *et al.*, 2006). They observed using microscopy the modification evolve from a negative refractive index change in the first 10 min after writing to a permanent and uniform positive refractive index with $\Delta n \sim 10^{-4}$ twenty minutes after the initial exposure (Fig. 11.19). To achieve the uniform and symmetric modification, a slit reshaping method was applied with a pulse energy of 185 nJ, NA of 0.55 and scan speed of 0.2 mm/s (Sowa *et al.*, 2006).

Scully's group also reported a positive refractive index contrast of $\Delta n \sim 10^{-3}$ using a similar 1-kHz femtosecond laser, but with a looser focusing condition provided by a 75-mm focal length singlet lens (Baum *et al.*, 2007). However, light was not launched into the structure to demonstrate waveguiding.

11.4.2   High repetition rate

Of the many exposure parameters required to optimize the laser–material interaction, the repetition rate plays the most important role in defining the regime of modification (Eaton *et al.*, 2008b). The benefits of processing within this heat accumulation regime has been demonstrated in fused silica and borosilicate glasses, as described in Section 11.3.2.

The first report of waveguide writing in PMMA within the cumulative heating regime of modification was with a 25-MHz stretched cavity oscillator providing 30-fs, 20-nJ pulses (Zoubir *et al.*, 2004). Buried structures were formed using the longitudinal writing configuration using a 0.25-NA

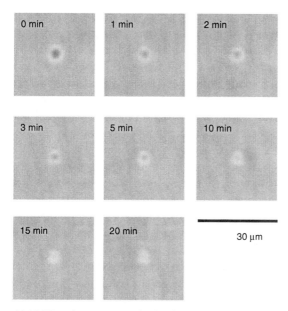

11.19 Time-lapse transmission images of the waveguide cross-section. The optical images of the exit surface of the waveguide cross-section when illuminated with a halogen lamp (Watanabe *et al.*, 2006).

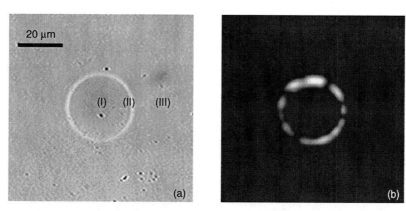

11.20 (a) Microsocope image of waveguide cross-section. (b) Guided modes at 633-nm wavelength (Zoubir *et al.*, 2004).

microscope objective, giving rise to a tubular refractive index morphology with annular waveguide modes (Fig. 11.20).

In our work, we exploit a variable repetition rate femtosecond laser to explore the vast intermediate range of conditions between the single pulse interaction regime at 1 kHz and the heat accumulation regime at high repetition rates (> 1 MHz). Using the 0.55-NA aspheric lens, optimal guiding structures at 100-kHz repetition rate were obtained with 20-mm/s scan speed using 1.2-μJ pulse energy, producing two vertically offset zones of depressed refractive index (Fig. 11.21), attributed to self-focusing and refocusing effects since the peak power was significantly above the critical power. These two zones were able to confine the 633-nm mode in between them, similar to the effect of using two separate scans to confine light between two zones of negative index change in crystals (Burghoff *et al.*, 2006). Propagation losses of 3 dB/cm were measured at the red wavelength, suitably low for sensing applications.

At 500 kHz, only a single annular zone was present (Fig. 11.22) which we attribute to diminished self-focusing from the 2-fold lower energy (600 nJ). Optimum guiding characteristics were obtained with 15-mm/s scan speed, although propagation losses were significantly higher and measured to be 6 dB/cm at 633-nm wavelength. As shown in Fig. 11.22, guiding only occurred around the outer ring of the modification, similar to the annular modes reported at 25-MHz oscillator (Zoubir *et al.*, 2004).

There are several mechanisms that may occur during femtosecond laser irradiation of PMMA which can influence the resulting refractive index morphology. These include degradation by unzipping or chain scission, crosslinking, and alteration of the absorption spectra through defects or impurities. In a previous study on phase gratings formed in PMMA by loosely focused 1-kHz repetition rate femtosecond laser pulses, the effects of laser-induced absorption changes, end groups and crosslinking were found to contribute a positive index change, although their contribution of $\Delta n \sim 10^{-4}$ was below the claimed refractive index increase of $\Delta n \sim 10^{-3}$ (Baum *et al.*, 2007). A time dependence of the refractive index change was observed and attributed to the outward diffusion of photoinduced monomer MMA, which is soluble in PMMA (Liu *et al.*, 2010).

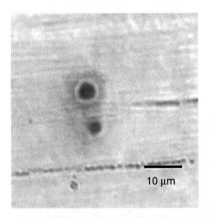 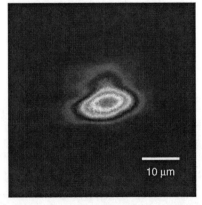

11.21 Cross-sectional microscope image (left) and mode profile (right) of optimum waveguide fabricated at 100 kHz repetition rate with 1.2 μJ pulse energy and 20 mm/s scan speed.

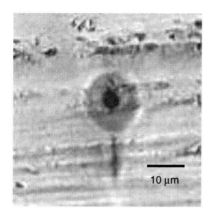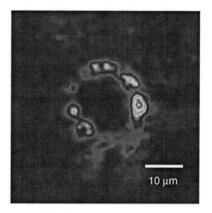

11.22 Cross-sectional microscope image (left) and mode profile (right) of optimum waveguide formed at 500 kHz repetition rate, 0.6 μJ pulse energy and 15 mm/s scan speed.

At higher repetition rates (100–500 kHz in this work), the interaction regime is observed to lead to a more violent laser–polymer interaction, resulting in an ablative-type mechanism such as thermal unzipping, leading to damaged structures lacking a positive refractive index increase at their cores. Unzipping is a thermal degradation mechanism, resulting in a breakdown of macromolecular chains into smaller monomer fragments, and therefore a decrease in density and refractive index. In this regime, therefore, we expect that the refractive index increase observed around the irradiated regions is due to material densification caused by the compressive stresses generated in the expansion of the irradiated volume. The waveguides fabricated in this regime are therefore more stable than those created at 1 kHz since they are related to material damages that are less prone to time decay.

In view of the results presented above on femtosecond laser waveguide writing in polymers, we can conclude that this class of materials is much less amenable to this microfabrication technique with respect to glasses. The disadvantages are low or negative refractive index change, high damping losses, and time instability. Although optical waveguides in PMMA may be suitable for certain sensing applications where propagation losses may not be a relevant issue, efficient light excitation and collection in LoCs would benefit from a more efficient device.

11.5   Conclusions

In glasses, femtosecond-laser writing now offers propagation losses as low as 0.2 dB/cm, but this is still substantially higher than the 1 dB/m loss typical with PLC technology relying on photolithographic techniques. The maximum refractive index contrast induced is $\Delta n = 0.01–0.02$ using high repetition rate femtosecond lasers, which is suitable for coupling to SMF, but still limits waveguide bends to a radius of 15–40 mm. Higher index contrast would lead to smaller mode sizes for tighter bends allowing for increased density of functions on a photonic chip. However, laser writing in out-of-plane architectures, exploiting the 3D volume of bulk glass, remains the most highly promising opportunity for dense optical integration enabled by femtosecond lasers.

In polymers, single mode waveguiding has been demonstrated in PMMA, an important substrate for biophotonic LoC devices. However, the propagation losses are relatively high at 4 dB/cm, with low positive index change demonstrated at 1 kHz repetition rate, and depressed refractive index changes at 100 kHz to 25 MHz repetition rates. Despite the non-ideal waveguide properties demonstrated by femtosecond laser microfabrication in polymers thus far, they are a promising alternative to glasses for commercial applications owing to their low cost, biochemical compatibility, mechanical flexibility and lack of brittleness.

11.6   Future trends

The rapid development of optimized recipes for femtosecond laser fabrication of waveguides enables high-quality passive and active devices in a variety of transparent materials in both 2D and 3D geometries. However, to further exploit the third spatial dimension, future work must address the issue of asymmetric refractive index profiles to enable out of plane evanescent coupled mode devices. The goal of writing a cylindrical waveguide with a symmetric, uniform and high refractive index contrast has still eluded researchers.

Higher resolution translation stages may result in more uniform structures for reduced propagation losses. However, the relatively high waveguide propagation loss of femtosecond laser-written waveguides is mainly attributed to the non-uniform refractive index modification resulting in scattering loss. By exploring a wider range of exposure parameters aided by the recent commercial development of higher power (> 20 W), shorter pulse duration, high repetition rate femtosecond lasers, more extreme temperatures giving increased heat diffusion and accumulation with faster cooling rates may be driven, yielding greater material densification over larger volumes, possibly with smoother refractive index profiles. Further development of real-time monitoring systems to study spectral emissions and infer glass temperatures may in the future be used to control laser exposure to avoid cracks, stress and other defects that lead to scattering loss. In the last few years, laser polarization, pulse duration, multiple scan lines and titled pulse fronts have been shown to play an important role in waveguide writing. The effect of other exposure parameters such as spatial beam profile, temporal pulse shape, bursts of pulses and external heat or stress treatment during writing, have yet to be studied and may offer improvement in waveguide losses and higher refractive index contrast.

In polymers, future work must be devoted to addressing the problem of high propagation losses and low or negative refractive index contrast shown to date. Post-thermal treatment has shown promise for improving waveguide properties and may be needed to produce high-quality waveguides in polymers for biophotonic applications.

The commercial success of the femtosecond laser waveguide writing in dielectrics in the context of displacing existing photolithographic technology is contingent upon addressing the remaining roadblocks of moderate refractive index contrast and waveguide losses described above. The femtosecond laser-writing technique is maskless, simple and flexible, making it ideal for rapid prototyping, refractive index trimming of existing optical circuits and sensor fabrication. With recent discoveries of easily tailored and predicted waveguide shapes enabled by cumulative heating and thermal diffusion, along with Bragg grating waveguide functionality in three-dimensional architectures, femtosecond laser fabrication now shows tremendous potential in fabricating tomorrow's integrated photonic circuits.

11.7   Sources of further information and advice

For an introductory overview of femtosecond laser waveguide writing, the reader is referred to the first review paper in the field by Itoh *et al.* (2006) entitled 'Ultrafast processes for bulk modification of transparent materials'. This paper provides an excellent overview of the fundamentals and gives an insightful classification of the regimes of femtosecond laser modification in fused silica glass based on the pulse energy (smooth isotropic refractive index change at low energy, nanogratings at moderate energy and voids at high energy).

In the high-impact *Nature Photonics* journal, Gattass and Mazur gave an excellent review of the field, where they emphasize the fundamentals in the femtosecond laser–material interaction (Gattass and Mazur, 2008). Della Valle *et al.* (2009) followed with an excellent review of photonic devices for telecommunications applications with the paper entitled 'Micromachining of photonic devices by femtosecond laser pulses'. This was followed by a review by Ams *et al.* (2009) entitled 'Ultrafast laser written active devices' discussing primarily active devices, including Bragg grating devices in bulk and fiber.

To date, there have been three books published relating to femtosecond laser fabrication of waveguides. The first book edited by Misawa and

Juodkazis (2006), *3D laser microfabrication: Principles and applications*, includes several chapters relevant to the field, including Chapter 2 by Gamaly *et al.* on the fundamentals of the femtosecond laser interaction in the bulk of a transparent material and Chapter 8 by Kazansky on the formation of nanogratings in glass. A second book edited by Sugioka *et al.* (2010), entitled *Laser precision microfabrication*, features a chapter by Sugioka and Nolte, providing an excellent introduction to the field. The recent book entitled *Femtosecond laser micromachining: Photonic and microfluidic devices in transparent materials* edited by Osellame et al. (2012) features 17 chapters, all relevant to the field.

Since the field's inception, the most relevant journals have traditionally been *Optics Letters*, *Optics Express*, *Journal of the Optical Society of America B* and *Applied Optics*, published by the Optical Society of America. Other important journals include *Applied Physics Letters* and *Journal of Applied Physics* by the American Institute of Physics, and *Photonics Technology Letters*, *Journal of Lightwave Technology* and *Journal of Selected Topics in Quantum Electronics* by the IEEE. With the recent trend towards real-world integrated devices with applications in interdisciplinary fields such as sensing, quantum optics and biophotonics, a growing percentage of recent publications have appeared in high-impact journals such as *Nature Photonics*, *Advanced Materials*, *Physical Review Letters* and *Lab on a Chip*. A simple way to find new advances in the field is by tracking papers that cite the original publication by Davis *et al.* (1996) and the review by Gattass *et al.* (Gattass and Mazur, 2008) using ISI Web of Science, Google Scholar or Scopus databases.

11.8   References

Agrawal, G. P. (1997), *Fiber-optic communication systems*. New York: John Wiley.

Allsop, T., Dubov, M., Mezentsev, V. and Bennion, I. (2010), 'Inscription and characterization of waveguides written into borosilicate glass by a high-repetition-rate femtosecond laser at 800 nm', *Applied Optics*, **49**, 1938–1950.

Ams, M., Marshall, G. D., Dekker, P., Piper, J. A. and Withford, M. J. (2009), 'Ultrafast laser written active devices', *Laser & Photonics Reviews*, **3**, 535–544.

Ams, M., Marshall, G. D., Spence, D. J. and Withford, M. J. (2005), 'Slit beam shaping method for femtosecond laser direct-write fabrication of symmetric waveguides in bulk glasses', *Optics Express*, **13**, 5676–5681.

Baum, A., Scully, P. J., Basanta, M., Paul Thomas, C. L., Fielden, P. R., Goddard, N. J., Perrie, W. and Chalker, P. R. (2007), 'Photochemistry of refractive index structures in poly(methyl methacrylate) by femtosecond laser irradiation', *Optics Letters*, **32**, 190–192.

Bhardwaj, V. R., Simova, E., Corkum, P. B., Rayner, D. M., Hnatovsky, C., Taylor, R. S., Schreder, B., Kluge, M. and Zimmer, J. (2005), 'Femtosecond laser-induced refractive index modification in multicomponent glasses', *Journal of Applied Physics*, **97**, 083102-1–083102-9.

Bruckner, R. (1970), 'Properties and structure of vitreous silica. I', *Journal of Non-Crystalline Solids*, **5**, 123–175.

Burghoff, J., Grebing, C., Nolte, S. and Tuennermann, A. (2006), 'Efficient frequency doubling in femtosecond laser-written waveguides in lithium niobate', *Applied Physics Letters*, **89**, 165–171.

Burghoff, J., Hartung, H., Nolte, S. and Tuennermann, A. (2007), 'Structural properties of femtosecond laser-induced modifications in lithium niobate', *Applied Physics A*, **86**, 165–170.

Chan, J. W., Huser, T., Risbud, S. and Krol, D. M. (2001), 'Structural changes in fused silica after exposure to focused femtosecond laser pulses', *Optics Letters*, **26**, 1726–1728.

Chan, J. W., Huser, T. R., Risbud, S. H., Hayden, J. S. and Krol, D. M. (2003a), 'Waveguide fabrication in phosphate glasses using femtosecond laser pulses', *Applied Physics Letters*, **82**, 2371–2373.

Chan, J. W., Huser, T. R., Risbud, S. H. and Krol, D. M. (2003b), 'Modification of the fused silica glass network associated with waveguide fabrication using femtosecond laser pulses', *Applied Physics A*, **A76**, 367–372.

Cheng, Y., Sugioka, K., Midorikawa, K., Masuda, M., Toyoda, K., Kawachi, M. and Shihoyama, K. (2003), 'Control of the cross-sectional shape of a hollow microchannel embedded in photostructurable glass by use of a femtosecond laser', *Optics Letters*, **28**, 55–57.

Crespi, A., Gu, Y., Ngamsom, B., Hoekstra, H. J. W. M., Dongre, C., Pollnau, M., Ramponi, R., Van Den Vlekkert, H. H., Watts, P., Cerullo, G. and Osellame, R. (2010), 'Three-dimensional Mach-Zehnder interferometer in a microfluidic chip for spatially-resolved label-free detection', *Lab on Chip*, **10**, 1167–1173.

Davis, K. M., Miura, K., Sugimoto, N. and Hirao, K. (1996), 'Writing waveguides in glass with a femtosecond laser', *Optics Letters*, **21**, 1729–1731.

Dekker, P., Ams, M., Marshall, G. D., Little, D. J. and Withford, M. J. (2010), 'Annealing dynamics of waveguide Bragg gratings: Evidence of femtosecond laser induced colour centres', *Optics Express*, **18**, 3274–3283.

Della Valle, G., Osellame, R. and Laporta, P. (2009), 'Micromachining of photonic devices by femtosecond laser pulses', *Journal of Optics A: Pure and Applied Optics*, **11**, 013001.

Dharmadhikari, J. A., Dharmadhikari, A. K., Bhatnagar, A., Mallik, A., Singh, P. C., Dhaman, R. K., Chalapathi, K. and Mathur, D. (2011), 'Writing low-loss waveguides in borosilicate (BK7) glass with a low-repetition-rate femtosecond laser', *Optics Communications*, **284**, 630–634.

Eaton, S. M., Chen, W., Zhang, H., Iyer, R., Ng, M. L., Ho, S., LI, J., Aitchison, J. S. and Herman, P. R. (2009), 'Spectral loss characterization of femtosecond laser written waveguides in glass with application to demultiplexing of 1300 and 1550 nm wavelengths', *IEEE Journal of Lightwave Technology*, **27**, 1079–1085.

Eaton, S. M., Ng, M. L., Bonse, J., Mermillod-Blondin, A., Zhang, H., Rosenfeld, A. and Herman, P. R. (2008a), 'Low-loss waveguides fabricated in BK7 glass by high repetition rate femtosecond fiber laser', *Applied Optics*, **47**, 2098–2102.

Eaton, S. M., Ng, M. L., Osellame, R. and Herman, P. R. (2011), 'High refractive index contrast in fused silica waveguides by tightly focused, high-repetition rate femtosecond laser', *Journal of Non-Crystalline Solids*, **357**, 2387–2391.

Eaton, S. M., Zhang, H., Herman, P. R., Yoshino, F., Shah, L., Bovatsek, J. and Arai, A. Y. (2005), 'Heat accumulation effects in femtosecond laser-written waveguides with variable repetition rate', *Optics Express*, **13**, 4708–4716.

Eaton, S. M., Zhang, H., Ng, M. L., Li, J., Chen, W.-J., Ho, S. and Herman, P. R. (2008b), 'Transition from thermal diffusion to heat accumulation in high repetition rate femtosecond laser writing of buried optical waveguides', *Optics Express*, **16**, 9443–9458.

Ehrt, D., Kittel, T., Will, M., Nolte, S. and Tunnermann, A. (2004), 'Femtosecond-laser-writing in various glasses', *Journal of Non-Crystalline Solids*, **345–346**, 332–337.

Ferrer, A., Diez-Blanco, V., Ruiz, A., Siegel, J. and Solis, J. (2007), 'Deep subsurface optical waveguides produced by direct writing with femtosecond laser pulses in fused silica and phosphate glass', *Applied Surface Science*, **254**, 1121–1125.

Fukuda, T., Ishikawa, S., Fujii, T., Sakuma, K. and Hosoya, H. (2004), 'Improvement on asymmetry of low-loss waveguides written in pure silica glass by femtosecond laser pulses', *Proceedings of SPIE*, **5279**, 21–28.

Gattass, R. R. and Mazur, E. (2008), 'Femtosecond laser micromachining in transparent materials', *Nature Photonics*, **2**, 219–225.

Glezer, E. N., Milosavljevic, M., Huang, L., Finlay, R. J., Her, T.-H., Callan, T. P. and Mazur, E. (1996), 'Three-dimensional optical storage inside transparent materials', *Optics Letters*, **21**, 2023–2025.

Hirao, K. and Miura, K. (1998), 'Writing waveguides and gratings in silica and related materials by a femtosecond laser', *Journal of Non-Crystalline Solids*, **239**, 91–95.

Hnatovsky, C., Taylor, R. S., Rajeev, P. P., Simova, E., Bhardwaj, V. R., Rayner, D. M. and Corkum, P. B. (2005a), 'Pulse duration dependence of femtosecond-laser-fabricated nanogratings in fused silica', *Applied Physics Letters*, **87**, 014104.

Hnatovsky, C., Taylor, R. S., Simova, E., Bhardwaj, V. R., Rayner, D. M. and Corkum, P. B. (2005b), 'High-resolution study of photoinduced modification in fused silica produced by a tightly focused femtosecond laser beam in the presence of aberrations', *Journal of Applied Physics*, **98**, 013517.

Hnatovsky, C., Taylor, R. S., Simova, E., Rajeev, P. P., Rayner, D. M., Bhardwaj, V. R. and Corkum, P. B. (2006), 'Fabrication of microchannels in glass using focused femtosecond laser radiation and selective chemical etching', *Applied Physics A*, **84**, 47–61.

Itoh, K., Watanabe, W., Nolte, S. and Schaffer, C. B. (2006), 'Ultrafast processes for bulk modification of transparent materials', *MRS Bulletin*, **31**, 620–625.

Johnston, T. F. (1998), 'Beam propagation (M2) measurement made as easy as it gets: The four-cuts method', *Applied Optics*, **37**, 4840–4850.

Juodkazis, S., Matsuo, S., Misawa, H., Mizeikis, V., Marcinkevicius, A., Sun, H. B., Tokuda, Y., Takahashi, M., Yoko, T. and Nishii, J. (2002), 'Application of femtosecond laser pulses for microfabrication of transparent media', *Applied Surface Science*, **197–198**, 705–709.

Juodkazis, S., Nishimura, K., Tanaka, S., Misawa, H., Gamaly, E. G., Luther-Davies, B., Hallo, L., Nicolai, P. and Tikhonchuk, V. T. (2006), 'Laser-induced microexplosion confined in the bulk of a sapphire crystal: Evidence of multimegabar pressures', *Physical Review Letters*, **96**, 166101.

Kanehira, S., Miura, K. and Hirao, K. (2008), 'Ion exchange in glass using femtosecond laser irradiation', *Applied Physics Letters*, **93**, 023112.

Little, D. J., Ams, M., Dekker, P., Marshall, G. D., Dawes, J. M. and Withford, M. J. (2008), 'Femtosecond laser modification of fused silica: The effect of writing polarization on Si-O ring structure', *Optics Express*, **16**, 20029–20037.

Liu, D., Kuang, Z., Perrie, W., Scully, P., Baum, A., Edwardson, S., Fearon, E., Dearden, G. and Watkins, K. (2010), 'High-speed uniform parallel 3D refractive index micro-structuring of poly(methyl methacrylate) for volume phase gratings', *Applied Physics B*, **101**, 817–823.

Low, D. K. Y., Xie, H., Xiong, Z. and Lim, G. C. (2005), 'Femtosecond laser direct writing of embedded optical waveguides in aluminosilicate glass', *Applied Physics A: Materials Science and Processing*, **81**, 1633–1638.

Marcinkevicius, A., Mizeikis, V., Juodkazis, S., Matsuo, S. and Misawa, H. (2003), 'Effect of refractive index-mismatch on laser microfabrication in silica glass', *Applied Physics A*, **76**, 257–260.

Martinez-Vazquez, R., Osellame, R., Cerullo, G., Ramponi, R. and Svelto, O. (2007), 'Fabrication of photonic devices in nanostructured glasses by femtosecond laser pulses', *Optics Express*, **15**, 12628–12635.

Mermillod-Blondin, A., Burakov, I. M., Stoian, R., Rosenfeld, A., Audouard, E., Bulgakova, N. and Hertel, I. V. (2006), 'Direct observation of femtosecond laser induced modifications in the bulk of fused silica by phase contrast microscopy', *Journal of Laser Micro/Nanoengineering*, **1**, 155–160.

Minoshima, K., Kowalevicz, A. M., Hartl, I., Ippen, E. P. and Fujimoto, J. G. (2001), 'Photonic device fabrication in glass by use of nonlinear materials processing with a femtosecond laser oscillator', *Optics Letters*, **26**, 1516–1518.

Misawa, H. and Juodkazis, S. (2006), *3D Laser microfabrication: Principles and applications*. Weinheim: Wiley-VCH.

Miura, K., Qiu, J. R., Inouye, H., Mitsuyu, T. and Hirao, K. (1997), 'Photowritten optical waveguides in various glasses with ultrashort pulse laser', *Applied Physics Letters*, **71**, 3329–3331.

Nasu, Y., Kohtoku, M. and Hibino, Y. (2005), 'Low-loss waveguides written with a femtosecond laser for flexible interconnection in a planar light-wave circuit', *Optics Letters*, **30**, 723–725.

Osellame, R., Chiodo, N., Della Valle, G., Cerullo, G., Ramponi, R., Laporta, P., Killi, A., Morgner, U. and Svelto, O. (2006), 'Waveguide lasers in the C-band fabricated by laser inscription with a compact femtosecond oscillator', *IEEE Journal of Selected Topics in Quantum Electronics*, **12**, 277–285.

Osellame, R., Chiodo, N., Maselli, V., Yin, A., Zavelani-Rossi, M., Cerullo, G., Laporta, P., Aiello, L., De Nicola, S., Ferraro, P., Finizio, A. and Pierattini, G. (2005), 'Optical properties of waveguides written by a 26 MHz stretched cavity Ti:sapphire femtosecond oscillator', *Optics Express*, **13**, 612–620.

Osellame, R., Hoekstra, H. J. W. M., Cerullo, G. and Pollnau, M. (2011), 'Femtosecond laser microstructuring: An enabling tool for optofluidic lab-on-chips', *Laser and Photonics Reviews*, **5**, 442–463.

Osellame, R., Maselli, V., Vazquez, R. M., Ramponi, R. and Cerullo, G. (2007), 'Integration of optical waveguides and microfluidic channels both fabricated by femtosecond laser irradiation', *Applied Physics Letters*, **90**, 231118–231123.

Osellame, R., Taccheo, S., Marangoni, M., Ramponi, R., Laporta, P., Polli, D., De Silvestri, S. and Cerullo, G. (2003), 'Femtosecond writing of active optical waveguides with astigmatically shaped beams', *Journal of the Optical Society of America B*, **20**, 1559–1567.

Osellame, R., Cerullo, G. and Ramponi, R. (2012) *Femtosecond laser micromachining: Photonic and microfluidic devices in transparent materials*. Berlin: Springer.

Pospiech, M., Emons, M., Steinmann, A., Palmer, G., Osellame, R., Bellini, N., Cerullo, G. and Morgner, U. (2009), 'Double waveguide couplers produced by simultaneous femtosecond writing', *Optics Express*, **17**, 3555–3563.

Ramirez, L. P. R., Heinrich, M., Richter, S., Dreisow, F., Keil, R., Korovin, A. V., Peschel, U., Nolte, S. and Tunnermann, A. (2010), Tuning the structural properties of femtosecond-laser-induced nanogratings, *Applied Physics A: Materials Science and Processing*, 100, 1–6.

Rayner, D. M., Naumov, A. and Corkum, P. B. (2005), 'Ultrashort pulse non-linear optical absorption in transparent media', *Optics Express*, 13, 3208–3217.

Sakakura, M., Shimizu, M., Shimotsuma, Y., Miura, K. and Hirao, K. (2008), 'Temperature distribution and modification mechanism inside glass with heat accumulation during 250 kHz irradiation of femtosecond laser pulses', *Applied Physics Letters*, 93, 231112–231114.

Saliminia, A., Vallee, R. and Chin, S. L. (2005), 'Waveguide writing in silica glass with femtosecond pulses from an optical parametric amplifier at 1.5 microns', *Optics Communications*, 256, 422–427.

Schaffer, C. B., Brodeur, A. and Mazur, E. (2001), 'Laser-induced breakdown and damage in bulk transparent materials induced by tightly focused femtosecond laser pulses', *Measurement Science and Technology*, 12, 1784–1794.

Schaffer, C. B., Garcia, J. F. and Mazur, E. (2003), 'Bulk heating of transparent materials using a high-repetition-rate femtosecond laser', *Applied Physics A*, A76, 351–354.

Shah, L., Arai, A., Eaton, S. M. and Herman, P. R. (2005), 'Waveguide writing in fused silica with a femtosecond fiber laser at 522 nm and 1-MHz repetition rate', *Optics Express*, 13, 1999–2006.

Shimotsuma, Y., Kazansky, P. G., Qiu, J. and Hirao, K. (2003), 'Self-organized nanogratings in glass irradiated by ultrashort light pulses', *Physics Review Letters*, 91, 247405-1–247405-4.

Siegel, J., Fernandez-Navarro, J. M., Garcia-Navarro, A., Diez-Blanco, V., Sanz, O., Solis, J., Vega, F. and Armengol, J. (2005), 'Waveguide structures in heavy metal oxide glass written with femtosecond laser pulses above the critical self-focusing threshold', *Applied Physics Letters*, 86, 121109.

Sowa, S., Watanabe, W., Tamaki, T., Nishii, J. and Itoh, K. (2006), 'Symmetric waveguides in poly(methyl methacrylate) fabricated by femtosecond laser pulses', *Optics Express*, 14, 291–297.

Streltsov, A. M. and Borrelli, N. F. (2002), 'Study of femtosecond-laser-written waveguides in glasses', *Journal of the Optical Society of America B*, 19, 2496–2504.

Sudrie, L., Couairon, A., Franco, M., Lamouroux, B., Prade, B., Tzortzakis, S. and Mysyrowicz, A. (2002), 'Femtosecond laser-induced damage and filamentary propagation in fused silica', *Physics Review Letters*, 89, 186601.

Sudrie, L., Franco, M., Prade, B. and Mysyrewicz, A. (1999), 'Writing of permanent birefringent microlayers in bulk fused silica with femtosecond laser pulses', *Optics Communications*, 171, 279–284.

Sugioka, K., Meunier, M. and Pique, A. (2010), *Laser precision microfabrication*. Berlin: Springer.

Ta'eed, V., Baker, N. J., Fu, L., Finsterbusch, K., Lamont, M. R. E., Moss, D. J., Nguyen, H. C., Eggleton, B. J., Choi, D.-Y., Madden, S. and Luther-Davies, B. (2007), 'Ultrafast all-optical chalcogenide glass photonic circuits', *Optics Express*, 15, 9205–9221.

Taylor, R., Hnatovsky, C. and Simova, E. (2008), 'Applications of femtosecond laser induced self-organized planar nanocracks inside fused silica glass', *Laser & Photonics Reviews*, 2, 26–46.

Thomson, R. R., Bockelt, A. S., Ramsay, E., Beecher, S., Greenaway, A. H., Kar, A. K. and Reid, D. T. (2008), 'Shaping ultrafast laser inscribed optical waveguides using a deformable mirror', *Optics Express*, **16**, 12786–12793.

Watanabe, W., Sowa, A. S., Tamaki, T., Itoh, K. and Nishii, J. (2006), 'Three-dimensional waveguides fabricated in poly(methyl methacrylate) by a femtosecond laser', *Japanese Journal of Applied Physics*, **45**, L765–L767.

Whitesides, G. M. (2006), 'The origins and the future of microfluidics', *Nature*, **442**, 368–373.

Will, M., Nolte, S., Chichkov, B. N. and Tunnermann, A. (2002), 'Optical properties of waveguides fabricated in fused silica by femtosecond laser pulses', *Applied Optics*, **41**, 4360–4364.

Yamada, K., Watanabe, W., Toma, T., Itoh, K. and Nishii, J. (2001), 'In situ observation of photoinduced refractive-index changes in filaments formed in glasses by femtosecond laser pulses', *Optics Letters*, **26**, 19–21.

Yang, W., Corbari, C., Kazansky, P. G., Sakaguchi, K. and Carvalho, I. C. (2008a), 'Low loss photonic components in high index bismuth borate glass by femtosecond laser direct writing', *Optics Express*, **16**, 16215–16226.

Yang, W., Kazansky, P. G., Shimotsuma, Y., Sakakura, M., Miura, K. and Hirao, K. (2008b), 'Ultrashort-pulse laser calligraphy', *Applied Physics Letters*, **93**, 171109.

Zhang, H. (2007), 'Bragg grating waveguides devices: Fabrication, optimization and application', PhD thesis, University of Toronto.

Zhang, H., Eaton, S. M. and Herman, P. R. (2006), 'Low-loss type II waveguide writing in fused silica with single picosecond laser pulses', *Optics Express*, **14**, 4826–4834.

Zhang, H., Eaton, S. M., Li, J., Nejadmalayeri, A. H. and Herman, P. R. (2007), 'Type II high-strength Bragg grating waveguides photowritten with ultrashort laser pulses', *Optics Express*, **15**, 4182–4191.

Zoubir, A., Lopez, C., Richardson, M. and Richardson, K. (2004), 'Femtosecond laser fabrication of tubular waveguides in poly(methyl methacrylate)', *Optics Letters*, **29**, 1840–1842.

12

Laser processing of optical fibers: new photosensitivity findings, refractive index engineering and surface structuring

S. PISSADAKIS, Foundation for Research and Technology – Hellas (FORTH), Institute of Electronic Structure and Laser (IESL), Greece

Abstract: The chapter presents a selective review on the laser processing of optical fibers, including new photosensitivity findings, refractive index engineering and surface structuring results, reported approximately during the last decade. Topics covered include the Bragg grating fabrication in standard germanosilicate and all-silica glass optical fibers using ultraviolet and infrared lasers with pulse durations from nanosecond to femtosecond, regenerated gratings and inscriptions into 'soft' glass fibers. The Bragg grating formation into photonic crystal and microstructured optical fibers is also discussed, while the last part of the chapter focuses onto the end-face, cladding and capillary surface structuring of optical fibers using lasers.

Key words: photosensitivity, refractive index engineering, fiber Bragg and long period gratings, photonic crystal fibers, ablation.

12.1 Introduction and historical overview

The manifold and complex material problem of photosensitivity referring to the underlying physical mechanisms and induced modifications of the physical properties of glasses/optical materials using electromagnetic or hard-particles radiation has remained an active research topic for more than five decades, exhibiting significant engineering impact (Primak and Kampwirth, 1968; Weeks, 1956, 1994). Focusing on glasses, numerous studies have been presented covering a different range of compositions, dopands and exposure conditions/approaches, illustrating the physical processes that take place during irradiation, as well as the corresponding optical, microscopic and mechanical properties modifications induced in the matrices. The favored material which has attracted intense and continuous academic interest is silicate glass, due to its high transparency over a broad band of wavelengths, its mechanical and chemical properties and radiation resistance. Moreover, silicate glasses are the backbone optical materials in numerous everyday

374

applications, in particular those in optical fiber communications and sensing. However, there have also been other matrixes such as those of phosphate, chalcogenide and fluoride glasses that have been studied with respect to their photosensitivity behavior and refractive index changes inscribed.

The invention of laser (Maiman, 1960) and the subsequent development of high power and photon energy laser sources revolutionized the field of optical glasses photosensitivity, efficiently substituting the hard radiation sources used in the decades of the 1950s and 1960s. High power and photon energy laser sources catalyzed light–matter interaction experiments, transferring photosensitivity experiments from the large installation radiation facilities into small-sized laboratories, while increasing engineering investigations and commercialization possibilities. Generally, a laser beam can be easily focused over submicron areas on the surface or in the volume of a glass, interfered, scanned, polarized and scattered, thus becoming an efficient, versatile and potentially low-cost tool for inducing and simultaneously studying photosensitive effects in optical materials. The above capabilities of laser radiation have prompted the evolution of photosensitivity processes in the engineering of refractive index of optical glassy materials over the last 25 years.

Early photosensitivity experiments using hard radiation such as X-rays were performed in silicate bulk glass samples, mostly studying macroscopic modifications induced related to compaction or optical absorption by means of coloration changes. However, there was a revision of the field with the development of high-power laser sources together with the use of silicate glasses as the backbone material for optical communications and sensing (Senior, 1992); and in microelectronics as lithographic mask material (Rothschild et al., 1997). The development of low-loss optical fibers and waveguides prompted the laser irradiation of such photonic components for either studying more efficiently photosensitivity mechanisms or for realizing new photonic devices. The principal photosensitivity product that attracted vast academic and industrial interest by modifying its magnitude, as well as by investigating the underlying physical background dominating its modifications, is that of refractive index change. Permanent refractive index changes formed inside optical fibers, planar waveguides and glass bulks for manipulating the guidance or propagation properties of light inside them, constituted the base for the realization of new photonic components. Therefore, the photo-inscription of diffractive and guiding structures such as Bragg and long-period gratings, waveguide channels and computer-generated holograms in fibers and bulks became reality.

Pristine fused silica is a high bandgap material (~ 9.3 eV) (Weinberg et al., 1979), exhibiting very low photosensitivity when exposed using standard ultraviolet laser sources, needing great radiation doses (Borrelli et al., 1997) for inducing significant modifications of its optical properties (Rothschild et al., 1989). The dopand of silica with Ge and other ion modifiers (such as boron) generates absorbing defect states while lowering the glass bandgap

(Cohen and Smith, 1958), hence, increasing interaction probability with resonant wavelength laser beams (Yuen, 1982). A milestone photosensitivity finding was presented by Hill *et al.* (1978), when during coupling of 488 nm C.W. Ar+ ion laser into a Ge-doped silicate fiber, 'self-recorded Bragg gratings' were formed into the Ge-doped fiber core; also named 'Hill gratings' (Hill, 2000). This 'holographic' recording led to gradually reduced transmission at the inscription wavelength, while the missing optical signal was easily detected in reflection. The Ge dopand was responsible for defect formation inside the silicate glass, while later experiments shown that these defects were excited by the 488 nm laser radiation through a higher order absorption process. The significance of the above experiment was twofold, firstly questioning the Ge-doped glass photosensitivity and refractive index engineering and secondly, demonstrating the first inscription of photosensitive photonic structures into optical fibers and waveguides. Whilst this first demonstration was quite impressive, its practical impact was rather limited since in-fiber Bragg back-scattering was achieved only for the inscription wavelength.

It took more than a decade for researchers to present externally written Bragg reflectors inside a Ge-doped optical fiber utilizing two-beam interferometry and 244 nm laser radiation (Meltz *et al.*, 1989) or alternatively point-by-point techniques (Malo *et al.*, 1993a). However, the Bragg grating inscription yield was boosted with the invention of the phase mask (Hill *et al.*, 1993; Malo *et al.*, 1993b), constituting the most reliable, versatile and cost-effective interference fringes generation approach. In the same period, the diffusion of hydrogen into silicate glass matrices under high pressure loading (Shelby, 1979) was a pre-conditioning technique that largely augmented refractive index changes formation (Lemaire *et al.*, 1993) using pulsed or CW lasers, rendering pristine or low co-dopand glasses highly photosensitive. Other significant findings of that early period refer to the demonstration of Type II (Archambault *et al.*, 1993) and Type IIA (Dong *et al.*, 1996; Xie *et al.*, 1993) gratings, exploring new kinds of photosensitivity of fiber materials, dependent upon glass codopands, exposure conditions and recording wavelengths. Shortly after, Bragg gratings were also recorded into pristine, all-silica fibers using 193 nm excimer laser radiation and standard phase mask technique (Albert *et al.*, 2002).

The use of femtosecond laser radiation of sub-bandgap photon energy, for ablating (Du, 1994; Stuart *et al.*, 1995) and then for modifying the structural and therefore the optical properties of transparent glasses (Davis *et al.*, 1996) revised the field of glass photosensitivity and refractive index engineering. Femtosecond laser sources provided extreme intensities concentration over sub-wavelength volumes, prompting multiphoton absorption and potentially low thermal dissipation effects, while succeeding in forming refractive index changes greater than $\times 10^{-3}$ in pure fused silica glass without hydrogenation or other pre-conditioning processes. Initially the research effort was focused on the formation of waveguide channels into glass blanks, but shortly after long-period gratings were inscribed into standard telecom optical fiber using 800 nm laser radiation (Kondo *et al.*, 1999). Then by using a custom phase mask design the first Bragg gratings were inscribed in pristine standard telecom fiber using femtosecond laser radiation by Mihailov *et al.* (2003).

Moreover, the invention of photonic crystal fiber (PCF) in 1996 by Russell *et al.* offered a new fiber platform for developing photonic devices (Knight *et al.*, 1996). Bragg (Eggleton *et al.*, 1999) and long-period gratings (Eggleton *et al.*, 2000) were also recorded into microstructured optical fibers (MOFs) containing Ge-doped silicate cores using standard ultraviolet laser sources. The inscription of periodic structures into PCFs/MOFs imposed new challenges related to the role of side beam scattering effects by the capillary structure (Marshall *et al.*, 2007; Pissadakis *et al.*, 2009a), the role of the photosensitivity mechanisms involved during recording, as well as the spectral effects related to the grating spectral diffraction behavior (Canning, 2008).

In the following sections a concise review will be given in the field of laser processing of optical fibers (Kashyap, 2010; Othonos and Kalli, 1999), illustrating new photosensitivity findings, as well as index engineering and structuring processes developed mainly during the last decade. Emphasis will be given to Bragg and long-period grating inscriptions using femtosecond and picosecond laser processing after exploiting multiphoton absorption effects, and inscriptions using deep ultraviolet lasers emitting close to the bandgap of silica and germanosilicate glasses. Fiber grating inscription based on thermo-plastic effects induced by CO_2 lasers will be referenced where necessary but they will not be emphasized; also gratings in polymer fibers will not be examined. Grating inscriptions will refer to both standard and MOFs drawn from silicate glasses, either containing sensitization dopands or being pristine. Photosensitivity processes and grating inscriptions will also be presented for fibers drawn from other non-silicate 'soft' glasses, such as phosphate and ZBLAN glasses. For consistency purposes a short introduction to glass photosensitivity and to different types of optical fiber photosensitivity mechanisms will be given. This review will then focus on the surface structuring of optical fibers using laser radiation by means of ablation, including ablative structuring of MOFs. Finally, the prospects of optical fiber photosensitivity and structuring will be discussed in the last section. Since the context of this update focuses on photosensitivity and laser inscription techniques, the theory related to Bragg and long-period gratings and their scattering behavior will not be covered; that has been extensively covered by several other authors (Erdogan, 1997; Kashyap, 1999; Othonos and Kalli, 1999).

12.2   Glass photosensitivity using laser beams

Photosensitivity constitutes a material modification/damage process occurring at intensities well below or close to the ablation threshold of the material exposed, excluding any direct material removal from the exposed target. However, volume damage processes can occur as filamentation (Sudrie *et al.*, 2002), ion precipitation (Takeshima *et al.*, 2004) and phase changes (Chan *et al.*, 2001) during the irradiation processes. While in ablation the surface topology is correlated with exposure conditions (Bäuerle, 2000), in photosensitivity the material modified is studied *in situ* or in post-exposure mode, utilizing a variety of optical (optical density and reflectivity measurements, monitoring of diffractive effects, pump-probe, photoluminescence) and structural (Raman spectroscopy, hardness and volume modification measurements, electron spin resonance, annealing demarcation) probing methods. *In situ*, online methods monitor both permanent and temporal photorefractive processes and changes, but more importantly post-exposure methods monitor the yield of the permanent microscopic and macroscopic products with respect to their magnitude and stability after the radiation stimulus has been ceased.

It is difficult to elaborate a generalized photosensitivity theory covering and accurately describing the majority of physical phenomena and products obtained during the interaction between a laser beam of specific photon energy and intensity and an optical material of given bandgap and microstructural properties. The material and exposure parameters affecting such manifold and complex interaction, but also the products emerging, are several and cross-dependent, rendering the elaboration of a common interpretation route impractical. The great challenge refers to the correlation between the refractive index and absorption changes induced with possible physical mechanisms that are activated during laser exposure. For specific optical glasses and exposure conditions the puzzle of such correlation has progressed adequately, allowing better understanding of the underlying physical effects involved and a straightforward exploitation of the photosensitivity and refractive index modification process (David, 2011; Hill and Meltz, 1997; Skuja, 1998). Nonetheless, interpretations become more laborious in the case of optical fiber structures where there can be substantial material differences compared to the bulk glass samples with respect to the photosensitivity effects triggered due to drawing induced defects (Friebele, 1976; Ky *et al.*, 1998), limited heat dissipation volume and implications arising from the cylindrical fiber geometry (Lemaire, 1995).

In general, the matching between the photon energy and intensity, and the bandgap of the material but also with the defect states that may exist within that energy gap, define both the order of interaction by means of single- or multiphoton absorption; but also its photo-thermal (Schaffer *et al.*, 2003) or photochemical (Fokine, 2002b) nature. Single-photon processes rely on the direct ionization of pre-existing defects or defect transformation processes, being linearly dependent upon the absorption cross-section of the above states at the exposure wavelength, while they can occur even at low intensities. Single-photon photosensitivity processes reach a saturation plateau after exhausting or fully transforming the defect states that can be excited, often leading to bleaching of the exposed absorption band. They exhibit linear or sublinear dependency upon the intensity of the exposure and their evolution is dominated by the population of initial defect states in resonance with the exposure wavelength. Single-photon photosensitivity effects occur within the linear absorption length of the material at the irradiation wavelength.

Higher-order absorption photosensitivity processes are accordingly triggered at substantially greater intensities than those of single-photon absorption events. They occur in highly transparent and low defect concentration optical materials, for irradiations using photons of energy lower than the material bandgap, which are not resonant with intra-band defect states. Their yield in terms of occurrence and, thus, products are proportional to the power of the intensities used, rendering them highly sensitive to spatial (i.e. interference, scattering, aberrations) and temporal (i.e. pulse broadening) photon densities, and to incubation damage (i.e., colour centre generation). Upon intensity figures, and existence/or excitation of seed electrons in the conduction band, high-order nonlinear absorption can lead to multiphoton and avalanche ionization (Schaffer et al., 2001; Stuart et al., 1995, 1996), promoting rapid structural modification of the irradiated material through plasma formation. Thus, the evolution of higher order absorption processes versus dissipated energy is not 'directly' dependent upon a specific number and type of pre-existing color centers/defects. Multiphoton ionizations triggered are greater in energy than the bandgap of the material, photo-dissociating the majority of the bonds of the exposed glass. The last leads to significant modifications induced inside the glass usually translated to high refractive index changes (Streltsov and Borrelli, 2002) and phase changes demarcated at temperatures often close to the T_g of the glass matrix (Chan et al., 2001).

12.3   Correlation of underlying photosensitivity mechanisms with refractive index changes

There are still standing queries related to the correlation of the photosensitive refractive index changes inscribed into optical glass or fiber, with the underlying physical mechanisms triggered by the laser stimulus. The ionization and defect transformation effects that take place during irradiation lead to transient and permanent electronic and microscopic structural modifications inside the glass matrix, so that their overall superposition constitutes macroscopically the refractive index engineering process. Several models have been proposed trying to correlate the photosensitive refractive index changes induced in an optical material or fiber to microscopic material modification mechanisms that might be activated. Four mechanisms/models have proven to be applicable to the majority of the experimental results obtained; however, none of them can independently justify in whole the refractive index engineering and evolution observed during grating inscription in an optical fiber. These photosensitivity hypotheses presented and supported by a significant amount of experimental data are: (a) the color-center model, (b) the volume modification model, (c) the stress-relief/generation model and (d) the phase-changes model.

12.3.1   Color-center model

The color-center model was firstly proposed by Hand and Russell (1990), and supported by several others reporting similar results for the case of exposed germanosilicate (Atkins *et al.*, 1993; Dong *et al.*, 1995; Williams *et al.*, 1992) and other types of glasses (Roman and Winick, 1993). In short, Hand, Russell and co-workers proposed that optical absorption changes related to the Ge oxygen deficiency center defects, peaking at the 240 nm spectral band (Yuen, 1982), and translated to GeE' centers by the laser irradiation (Nishii *et al.*, 1995), are translated into refractive index changes by employing the Kramers–Kronig parity transformation (see Eq. [12.1]).

$$\Delta n(\lambda) = \frac{1}{2\pi} P \int_0^\infty \frac{\Delta\alpha(\lambda')}{1-(\lambda'/\lambda)^2} d\lambda' \qquad [12.1]$$

In this hypothesis ion displacements were disregarded, letting only electronic rearrangements be accounted (Othonos and Kalli, 1999). Results related to the color-center model had been also reported for fused silica exposed to 193 nm excimer laser radiation (Rothschild *et al.*, 1989), where a part of the refractive index changes was attributed to SiE' center generation peaking at 215 nm (Hosono *et al.*, 1996).

Other most recent examples related to the photosensitivity of silver doped and undoped phosphate glasses (Pissadakis and Michelakaki, 2008; Pissadakis *et al.*, 2004) showed that the color-center model can be used for obtaining a reliable estimation of the minimum refractive index changes inscribed inside an optical matrix using laser radiation of the same wavelength but of different pulse duration. In short, the color-center model succeeded in describing partially the photosensitivity of Ge-doped silicate glasses (including hydrogenated species) where strong absorption precursors prompt efficient color-center transformations, predicting index changes of few parts of 10^{-4} (Dong *et al.*, 1995; Tsai *et al.*, 1997). However, the photosensitivity of low-defect concentration optical matrices such as fused silica (Albert *et al.*, 1999), or refractive index changes induced into glasses for prolonged exposures after the color-center annihilation or transformation has been saturated, cannot be described accurately by this model.

12.3.2   The volume modification model

The volume modification model is one of the most fundamental hypotheses in glass photosensitivity, formed since the early years of investigations (Primak, 1972; Primak and Kampwirth, 1968). Correlated refractive index (n) and volume (V) changes are well described by the Lorentz–Lorentz equation (see Eq. [12.2]), accounting the molar refractivity $R = 4/3\pi N\alpha$ (N number of molecules, α polarizability) of the exposed glass matrix (Born and Wolf, 1999).

$$\Delta n \Big/ n = \frac{(n^2 + 2)(n^2 - 1)}{6n} \left\{ \frac{V}{\Delta V} \frac{\Delta R}{R} - 1 \right\} \frac{\Delta V}{V} \qquad [12.2]$$

Both for the cases of exposed fused silica (Borrelli *et al.*, 1997; Fiori and Devine, 1986; Rothschild *et al.*, 1989) and germanosilicate glasses (Borrelli *et al.*, 1999; Cordier, 1997) volume modification effects have been observed mostly in the form of compaction/densification (Douay *et al.*, 1997). In that model, the silicon or germanium-oxygen deficiency bonds forming higher order defect rings and voids are dissociated by the laser radiation to lower order structures, but also to modifications of these bonds' intermediate oxygen angle, leading to microscopic void annihilation and macroscopic volume compaction (Piao *et al.*, 2000). Refractive index changes Δn induced in silicate glasses are well described by a universal power law rule of the form:

$$\Delta n \infty \left(\frac{NF^2}{\tau} \right)^b \qquad [12.3]$$

where F is the energy density, N the number of pulses, τ the pulse duration and b the power index laying between 0.5 and 0.7 (Borrelli *et al.*, 1999).

Such densification effects in silica glass have also been investigated using 157 nm excimer (Smith and Borrelli, 2006), 213 nm, quintupled Nd:YAG (Schenker, 1994) laser and 800 nm femtosecond (Bellouard *et al.*, 2006) laser radiation. Opposite sign, thus volume dilation effects have been measured in the case of hydrogenated silica glass (Smith *et al.*, 2001), as well as for the case of phosphate (Michelakaki and Pissadakis, 2009; Yliniemi *et al.*, 2006b) and fluoride (Sramek *et al.*, 2000) glasses under 193 nm irradiation. Especially, for the case of phosphate glass the role of the P–O bond and other ion modifiers existing in the glass dominate the progression of negative index changes, in direct dependence on the accumulated energy density doses delivered into the glass. The volume modification model has been successfully used for describing refractive index changes induced in silicate glasses under irradiation using deep ultraviolet lasers (Albert *et al.*, 1999; Borrelli *et al.*, 1999; Pissadakis and Konstantaki, 2005b); and partially for the refractive index changes induced by infrared femtosecond irradiations (Streltsov and Borrelli, 2002).

12.3.3   The stress-relief model

The third model, that of stress-relief, was proposed by Limberger and Fonjallaz (Fonjallaz *et al.*, 1995) for describing the refractive index changes induced in high Ge-doped silicate glass fibers exposed using pulsed 240 nm laser radiation. In the stress-relief model, the compaction induced by the irradiation in Type I gratings generates axial and radial stresses in the bright fringes of interference; which contribute negative refractive index changes in the overall photosensitivity through the photoelastic effect (Frocht). Stress birefringence, or photoelasticity, refractive index changes Δn in isotropic materials are described by the equation:

$$\Delta n_\sigma = \frac{1}{2} n_x^3 (q_{11} - q_{12})(P_{xx} - P_{yy}) \qquad [12.4]$$

where n_x is the refractive index of the unstressed material, q_{11} and q_{12} are stress optical coefficients and P_{xx}, P_{yy} arestress components (Born and Wolf, 1999). Under this scheme the refractive index changes associated with compaction are considered as inelastic, while those associated with secondary stress-relief are considered elastic.

Moreover, for B/Ge-codoped fibers where compressive or tensile, radial and axial stresses are formed due to the large difference between the core–cladding compositions and thus thermal expansion coefficients (Kim *et al.*, 2000; Raine *et al.*, 1999), exposure to ultraviolet radiation can relieve these stresses, contributing with according sign to the refractive index changes. Such hypothesis as the above, employing stress relaxation and high compaction, has already been used to better understand the Type IIA (see Section 12.4.3) Bragg grating photosensitivity, in B/Ge-codoped fibers (Ky *et al.*, 2003); but also stress reduction in inscriptions in hydrogenated fibers (Ky *et al.*, 1999; Limberger *et al.*, 2007). The stress-relief model is volume modification driven and can be considered as a secondary mechanism following extensive exposed glass volume changes of the exposed glass.

12.3.4   The phase-changes model

Exposures performed under extreme intensities and energy densities can induce phase changes inside the exposed glass, and the refractive index changes formed cannot be described accurately by any of the models described above or their combination. Such phase or damage changes may be of the form of filamentation (Watanabe *et al.*, 2003), void formation (Kazansky, 2007; Taylor *et al.*, 2008), extreme compaction (Mihailov *et al.*, 2003), crystallization/amorphization (Fisette, 2006) and ion-migration (Pissadakis *et al.*, 2004), generated by the combination of rapid transformation effects such as shock waves, melting and resolidification and crack-propagation. Photoluminescence (Dianov *et al.*, 1996; Nishikawa *et al.*, 1992) and micro-Raman spectroscopy (Chan *et al.*, 2001; Dianov *et al.*, 1997b; Fletcher *et al.*, 2009) are the most favorable tools for investigating such abrupt material changes, together with topological investigation techniques such as scanning electron microscopy, atomic force microscopy, or micro-/nano-indentation (Aashia *et al.*, 2009; Bellouard *et al.*, 2006).

12.4   Types of photosensitivity in optical fibers

It is much easier to categorize different photosensitivity mechanisms based on their final products by means of refractive index changes magnitude and their growth trend versus exposure conditions, sensitization steps or the thermal stability of the inscribed optical and structural changes (Othonos and Kalli, 1999). The need for categorization became obvious after the boom of photosensitivity findings and grating recording approaches presented during the 1990s for the case of the germanosilicate optical fibers. Commonly, there are four main types of optical fiber photosensitivity initially applied for describing the behavior observed in germanosilicate fibers; however, these types are now used for categorizing the photosensitivity behavior observed in other glasses and dopands (see Table 12.1). The specific characteristics accounted for defining the types of optical fiber photosensitivity are related to the evolution of refractive index changes observed during grating exposure, the sign of refractive index changes, and then the thermal stability/dynamics of the reflectors inscribed. However, there are also other subcategories of optical fiber photosensitivity categorized in terms of the specific nature of laser–matter interaction (Fokine, 2002a; Lemaire *et al.*, 1993), pre-exposure sensitization steps (Canagasabey and Canning, 2003; Chen *et al.*, 2002a) or the post-exposure thermal characteristics (Canning *et al.*, 2009; Lindner *et al.*, 2009) that are observed in the grating structures inscribed. There have been examples where the same optical fiber, under exposure with a fixed laser wavelength but utilizing different strain (Kukushkin, 2007) or intensity/energy density conditions (Sozzi *et al.*, 2011)

Table 12.1 Types of optical fiber photosensitivity

Type	Refractive index changes characteristics	Other characteristics	Fiber material characteristics	Radiation sources	References
Type I	Average and modulated refractive index changes follow a monotonic versus accumulated energy density dose. Refractive index changes reach a saturation plateau for prolonged exposures. Inscriptions under low tension.	For Ge-doped fibers exposed using standard C.W. and nanosecond pulse duration laser sources the thermal stability of Type I refractive index changes is low to medium, laying well below the T_g of the exposed glass. Progressive decay versus temperature.	NA 0.1–0.2 6.6% Ge-concentration	488 nm Ar+ ion laser	(Hill et al., 1978)
				244 nm Ar+ ion pumped excimer laser	(Meltz et al., 1989)
			Δn=0.0013 between core and cladding	248 nm excimer laser	(Hill et al., 1993)
				262 nm quadrupled Nd:YLF	(Armitage, 1993)
			Hydrogenated Ge-doped fibers		(Lemaire et al., 1993)
			Low Ge-concentration fibers (3–7%)	Nanosecond pulse duration 248 nm and 193 nm excimer lasers	(Albert et al., 1995) (Albert et al., 1999)
			B/Ge and hydrogenated fibers	255 nm doubled copper vapor laser	(Zhang et al., 1999)
			Hydrogenated SMF-28 fiber	157 nm excimer laser	(Chen et al., 2001)
			Silica glass fibers	193 nm excimer laser	(Albert et al., 2002)
			Silica glass and SMF-28 fibers	800 nm femtosecond laser	(Mihailov et al., 2003)
			Hydrogenated silica glass and SMF-28 fibers GF1B Nufern 10% Ge	264 nm femtosecond radiation	(Zagorulko et al., 2004)
				213 nm, 150ps Nd:YAG laser	Pissadakis and Konstantaki, 2005,
			SMF-28 and all-silica Sumitomo Z-Fiber	248 nm, 500fs laser	(Livitziis and Pissadakis, 2008)
			SMF-28 fiber	157 nm, excimer laser	(Dyer et al., 2008)
			SMF-28e fiber	244 nm, Ar+ ion laser	(Limberger and Violakis, 2010)

Type	Properties	Fiber	Laser	References
Type II	High reflectivities reaching 100% over short grating lengths. Single pulse grating recordings at high energy densities using excimer lasers. Multi-pulse recordings using 800 nm femtosecond radiation. Filamentation gratings. Extreme thermal stability up to temperatures close to the T_g of the fiber glass	High Ge-concentration doped fibers	248 nm excimer laser	(Archambault et al., 1993)
		SMF-28 fibers	193 nm excimer laser	(Dyer et al., 1994)
			800 nm femtosecond Ti:sapphire laser	(Smelser et al., 2005)
Type IIA	Average and modulated refractive index changes follow a non-monotonic versus accumulated energy density dose. Possibility of negative refractive index changes. Turning points can occur for modulated refractive index changes. Refractive index changes reach a saturation plateau for prolonged exposures. Type IIA behavior can be enhanced for inscriptions in strained fibers. Improved thermal stability compared to Type I refractive index changes. Flat-top thermal stability up to an annealing temperature, then fast decay.	High Ge-concentration fibers	243 nm dye laser	(Xie et al., 1993, Niay et al., 1994)
		B/Ge-codoped fiber	193 nm excimer laser	(Dong et al., 1996, Groothoff and Canning, 2004)
				(Riant and Haller, 1997)
			193 nm excimer laser	(Canning et al., 1998, Bazylenko et al., 1998)
		B/Ge and Sn/Ge codoped fibers	248 nm excimer laser	(Ky et al., 1998, Ky et al., 2003)
		28% Ge-doped fibers	248 nm excimer laser	(Tsai et al., 1999)
		B/Ge-codoped fiber	213 nm, 150ps quintupled Nd:YAG laser	(Pissadakis and Konstantaki, 2005)
		B/Ge-codoped and high-Ge fibers	248 nm, 500fs laser	(Violakis et al., 2006)
		Highly strained SMF-28 fiber	248 nm excimer laser	(Kukushkin, 2007)
		B/Ge-codoped fiber	213 nm, quintupled Nd:YAG laser at KHz repetition rate	(Gagné and Kashyap, 2010)

Continued

Table 12.1 Continued

Type	Refractive index changes characteristics	Other characteristics	Fiber material characteristics	Radiation sources	References
Type IA	Abnormal growth effects in hydrogenated Ge-doped fibres for prolonged exposures. Nonmonotonic growth of modulated refractive index changes. Average index changes increase positively with accumulated energy dose, exhibiting extreme spectral shifts (10 nm<) of the Bragg wavelength.	Thermal stability similar to the grating exhibiting Type I photosensitivity. Reduced thermal sensitivity compared to Type I counterparts.	Hydrogenated B/Ge and SMF-28 fibers	244 nm doubled Ar⁺ ion laser	(Liu *et al.*, 2002) (Kalli *et al.*, 2006)

may exhibit different types of photosensitivity behavior, as well as refractive index changes and thermal stability measures thereof.

12.4.1   Type I photosensitivity

Type I photosensitive gratings are the most common category, being fabricated in low and medium concentration Ge concentration fibers under modest exposure intensities, using either CW or pulsed (Albert *et al.*, 1995) laser sources. Their refractive index changes evolution during recording and follows a monotonic power law trend, while usually reaching saturation after the defect states have been exhausted for the single-photon absorption process; multiphoton absorption inscriptions in Ge-doped fiber can also be of Type I under controlled intensities for avoiding volume damage effects. Hydrogenated germanosilicate (Lemaire *et al.*, 1993) fibers under low accumulated energy dose exposures also follow a Type I photosensitivity evolution. Pristine and hydrogen loaded, all-silica fibers, under 193 nm excimer laser exposure or ultraviolet and infrared femtosecond laser irradiation can also exhibit Type I photosensitivity changes (Albert *et al.*, 2002; Smelser *et al.*, 2005; Zagorulko *et al.*, 2004). Type I refractive index changes are usually associated with a single underlying photosensitivity mechanism (Othonos and Kalli, 1999), or with co-occurring photosensitivity processes that contribute to the overall refractive index changes under the same sign.

12.4.2   Type II photosensitivity

Volume damage gratings induced under extreme exposure conditions by means of high energy densities (i.e. ~ 1 J/cm^2 at 248 nm) (Archambault *et al.*, 1993) or intensities (several TW/cm^2 at 800 nm) (Smelser *et al.*, 2005) are classified as Type II. In such Type II photosensitivity gratings refractive index engineering is associated with phase changes induced (Archambault, 1994) as a result of the rapid heating and resolidificaton of the exposed material, which in turn induce thermal fictive effects in the exposed glass. Type II gratings were initially demonstrated as 'single-pulse' gratings (Archambault *et al.*, 1993); however, similar damage modifications can be obtained under multi pulse femtosecond laser exposures (Mihailov *et al.*, 2003). Since such phase changes are formed above the T_g or even the melting point of the exposed glass, their thermal stability is massively increased compared to the standard Type I photosensitive gratings. Type II photosensitivity has also been observed in exposures of silica blanks (Zhang *et al.*, 2006) using sub-MHz repetition laser sources, due to heat accumulation effects.

12.4.3   Type IIA photosensitivity

The irradiation through a phase mask of non-hydrogenated, high Ge-concentration and B/Ge-co-doped optical fibers at energy densities below the phase changes threshold revealed the complex nature Type IIA photosensitivity. First indications of the behavior of such photosensitivity were presented by Xie *et al.* (1993); however, a clearer demonstration of Type IIA behavior was reported by Dong *et al.* (1996) when a B/Ge co-doped fiber was exposed to 193 nm excimer laser radiation. In Type IIA photosensitivity both the average and modulated refractive index changes probed by the Bragg reflector inscribed follow a non-monotonic trend. Initially, the modulated refractive index changes increase similar to Type I photosensitivity, while after reaching a short plateau point, follow a declining trend leading to a turning point when the Bragg grating strength minimizes or even vanishes. Upon energy dose provided to the system the grating strength grows again, reaching a final plateau of saturation. The average refractive index changes induced into the fiber red-shift in wavelength up to the turning point of the Bragg grating strength, and then become either stable or negative/blue-shifted (Dong *et al.*, 1996; Riant and Haller, 1997). In the broader family of Type IIA photosensitivity are included most of the complex refractive index changes induced in fiber and planar waveguide samples, where underlying photosensitivity processes are manifold and negative component refractive index changes emerge (Canning *et al.*, 1998; Wiesmann *et al.*, 1999). Type IIA photosensitivity inscriptions in Ge-doped fibers withstand much higher temperatures than standard Type I counterparts, exhibiting minor decay up to 800°C in specific cases (Groothoff and Canning, 2004).

12.4.4   Type IA photosensitivity

The most recent kind of distinct photosensitivity behavior is that of Type IA, observed in hydrogenated germanosilicate fibers under prolonged exposures using CW or pulsed laser sources, however, applying intensities which can form Type IIA gratings in the same pristine fibers (Liu *et al.*, 2002; Simpson *et al.*, 2004). These Type IA gratings undergo significant red-shifts of the Bragg wavelength, mostly greater than 10 nm for the whole duration of the exposure. Such massive wavelength shifts correspond to average refractive index changes of the order of few parts of $\times 10^{-2}$, substantially modifying the guiding properties of the exposed fiber, by means of higher mode cut-off and fundamental mode confinement. The modulated refractive index behavior of Type IA gratings resembles that of Type IIA photosensitivity, where after a rapid increase the grating strength decreases to a minimum turning point, followed by a slower increase toward stabilization (Kalli *et al.*, 2006).

12.5   Grating fabrication in standard, germanosilicate optical fibers

After the adoption of the phase mask approach (Hill, 1993; Malo *et al.*, 1993b) as the main and most reliable Bragg grating recording method and the emergence of long-period gratings (Vengsarkar *et al.*, 1996), the importance of understanding and simultaneously optimizing the photosensitivity processes available became quite obvious. In addition, the possibilities opened relating to commercialization of Bragg and long-period grating devices intensified the efforts in increasing germanosilicate glasses' photosensitivity yield by means of refractive index changes and simplification of the inscription process.

There were three main axes toward the increase of the photosensitivity of germanosilicate glasses: the doping of the fiber core with ion modifiers such as boron or tin and the fiber drawing under special conditions for increasing defects and softening the glass; the hydrogen loading process; and finally the use of standard deep ultraviolet sources of photons close to the bandgap of the glass. These three approaches in combination or independently to each other, led to increase of the figure-of-merit of germanosilicate glass photosensitivity by almost three orders of magnitude from the $\times 10^{-5}$ levels demonstrated during the 1980s, to levels up to $\times 10^{-2}$. However, all these approaches included either complex processing steps, issues of repeatability and reliability or increased operational cost, rendering them less attractive for immediate commercialization. The above needs, conditions and prospects turned the research effort into alternative approaches, where the intensity and photon energy of the laser radiation used could trigger nonlinear absorption effects, bypassing the traditional single-photon photosensitivity paths. In the following subsections a review of the grating recording methods and related results will be presented focused on germanosilicate glass fibers, categorized upon the inscription wavelengths and pulse durations of the lasers used. In the last subsection the regenerated gratings will be reviewed, irrespective of recording wavelengths and pulse durations; such a type of gratings tend to constitute a new category themselves.

12.5.1   Ultraviolet nanosecond and picosecond laser inscriptions

A significant demonstration of the straightforward inscription of Bragg reflectors into low-photosensitivity fibers was presented by Albert *et al.* (1995), after exploiting two-photon absorption effects and laser radiation of high photon energy (Fig. 12.1a). In that work, 193 nm excimer laser radiation (6.42 eV photon) was used to inscribe refractive index changes of ~ 10^{-3} in a hydrogen unsensitized SMF-28 standard telecom fiber, while the same

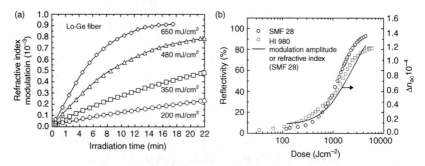

12.1 (a) Growth of refractive index modulation amplitude in Low-Ge fiber (SMF-28 Corning) resulting from irradiation through a phase mask with ArF laser at 50 pulses/s with pulses of different energy density (after Albert *et al.*, 1995). (b) Comparison of spectral reflectivity of FBG in HI-980 and SMF-28 fiber versus 157 nm laser dose. Fluence per pulse ~ 58 mJ/cm². The solid line represents the modulation amplitude of refractive index for SMF 28 fiber. (After Dyer *et al.*, 2008. Used with permission from Optical Society of America [OSA].)

irradiation produced less than half the index changes in a high Ge-doped fiber. Such refractive index changes were enough to produce Type I short and strong Bragg reflectors, in standard telecom fibers, while simultaneously avoiding the obstacles of hydrogenation. In a similar manner, Herman *et al.* exploited the much shorter photon wavelength of 157 nm (7.9 eV) for inscribing long-period gratings in pristine SMF-28 fibers (Chen *et al.*, 2001), providing energy to the system above the germanosilicate core glass bandgap, while allowing single-photon ionization processing (Dyer *et al.*, 2001) and refractive index changes of the order of ~ 4×10^{-4}. Compaction was the primary underlying physical mechanism in the 157 nm exposures, while cladding absorption defined a rather narrow envelope for optimum inscription conditions. The above two deep ultraviolet wavelengths were also used for locking photosensitivity in hydrogen loaded fibers (Chen *et al.*, 2002b), as well as for Bragg grating amplification (Dyer *et al.*, 1994). Later, Dyer *et al.* (2008) used 157 nm nanosecond laser radiation and a custom-made CaF_2 phase mask for recording Bragg reflectors in low-defect SMF-28 and Hi-980 Corning fibers (Fig. 12.1b), leading to refractive index changes of ~ 2.8×10^{-4}, extending the work of Herman and Chen (Chen *et al.*, 2001).

In 2005 there was presented the first inscription of Bragg reflectors in a germanosilicate fiber using 213 nm, 150 ps frequency quintupled Nd:YAG laser radiation, targeting the low absorption spectral valley formed between the Ge oxygen deficiency centers peaking at 5 eV, and the GeE' centers absorption band tail initiated at this wavelength regime (Pissadakis and Konstantaki, 2005b). Spectro-photometric measurements of 9% mole Ge-doped fiber performs presented by Archambault (1994) reveal that the absorption at the valley of the 213 nm wavelength is more than 10× lower

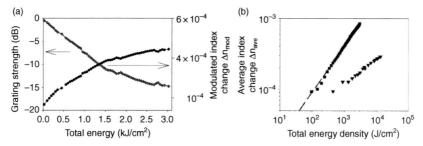

12.2 (a) Bragg grating strength (black cycles) and modulated refractive index changes Δn_{mod} (gray diamonds) versus total energy density for Nufern GF1B fiber using 213 nm Nd:YAG laser radiation, using 42 mJ/cm² energy density per pulse. (b) Comparative graph on the average refractive index changes Δn_{ave} versus total energy density for Bragg grating recording using 213 nm, 150 ps 213 nm Nd:YAG (black squares) and 248 nm, 34 ns excimer (inverse triangles) laser radiation. Energy density for 248 nm exposure: 360 mJ/cm². Dashed line: power law regression using Eq. [12.3] for average refractive index changes induced by 213 nm laser radiation.

that that measured at the 242 nm germanium-oxygen deficiency band, heavily exploited in grating recording using 248 nm excimer lasers.

Average index changes $\Delta n \approx 8.5 \times 10^{-4}$ were obtained after 2 hours' exposure of the Nufern GF1B fiber with a cumulative energy density of 2.9 kJ/cm² (Fig. 12.2a and 12.2b), while employing an elliptical Talbot interferometer (Pissadakis and Reekie, 2005). The Talbot interferometer configuration imposed adequate separation between the same mask and the fiber, allowing high energy density delivery in the fiber complex, without significant absorption in the phase mask due to the SiE' centers formation located at the 5.8 eV band (Skuja, 1998). The refractive index growth data followed the universal power law trend applied in the case of compaction in silicate glasses, exhibiting a b factor of ~ 0.66 (Borrelli et al., 1999; Schenker, 1994), described by Eq. [12.3]. The GW/cm² magnitude intensities used in the inscription (Pissadakis and Konstantaki, 2005b) in conjunction with the high two-photon absorption coefficient of the germanosilicate core led to a nonlinear absorption coefficient similar or even higher than the linear absorption at this wavelength (Kalachev et al., 2005), promoting two-photon effects. Exposures performed using different energy densities in the Nufern GF1B fiber support further the above assertion, by leading to refractive index changes such that their ratio is proportional to the second power of the corresponding ratio of the energy densities per pulse (see Fig. 12.3).

Using 213 nm wavelength and 150 ps pulse duration the recording of Type IIA Bragg reflectors in the commercial PS1250/1500 Fibercore B/Ge co-doped fiber was also presented, leading to refractive index changes of 9.0×10^{-4} (Pissadakis and Konstantaki, 2005a). Again, the refractive index

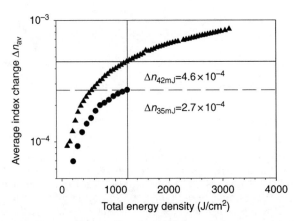

12.3 Average refractive index data Δn_{ave} for Bragg gratings inscribed in Nufern GF1B optical fiber using 213 nm, 150 ps Nd:YAG laser radiation, for different energy densities per pulse. Triangles: 42 mJ/cm^2. Circles: 35 mJ/cm^2. The vertical line defines the iso-energy point of ~ 1.2 kJ/cm^2, where the two gratings are measured.

changes data referring to the Type IIA evolution regime were well fitted to a power law growth (see Eq. [12.3]), with b factor 0.65 (Fig. 12.4).

Both the results obtained for the Nufern GF1B and the PS1250/1500 Fibercore optical fibers during Bragg grating inscriptions using 213 nm, 150 ps Nd:YAG radiation, provide evidence that the primary underlying physical mechanism was that of volume compaction. The first surprising finding that supports the compaction model emerges from the power law fitting of the refractive index data for both fibers used and exposed using the 213 nm, 150 ps. Both fibers exhibit almost identical b-factor value after fitting their photosensitivity data using Eq. [12.3]; similar b-values have been also reported by Borrelli (Borrelli *et al.*, 1999) and Schenker (1994). Another experimental finding in support of the compaction model for the 213 nm, 150 ps irradiation is obtained by producing Bragg reflectors using 248 nm, nanosecond excimer laser radiation, utilizing extreme energy densities (~ 1 J/cm^2 per pulse). By comparing the data for the 213 and 248 nm exposures under the dose conditions of Fig. 12.5, one can see that initially the 248 nm exposure exhausts 242 nm peaking germanium-oxygen deficiency centers and their corresponding refractive index contribution; then follows an identical growth trend with the 213 nm exposure. Since the 213 nm, 150 ps radiation does not primarily rely on the exhausting of color centers for inducing refractive index changes, both laser wavelengths produce the same type of changes in the glass (Pissadakis and Konstantaki, 2005b), after specific dose threshold has been surpassed.

Shortly after the above reports on 213 nm photosensitivity, 211 nm 250 fs radiation was used for long-period grating recording in hydrogenated

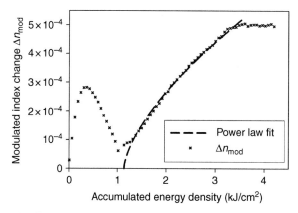

12.4 Index modulation Δn_{mod} versus accumulated energy, for a Type IIA Bragg grating exposure in the B/Ge-codoped PS1250/1500 Fibercore optical fiber, of 72 000 pulses and a pulse energy density of 60 mJ/cm².

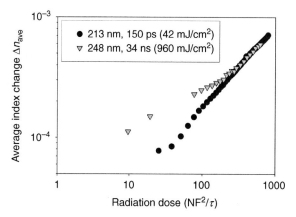

12.5 Average refractive index data Δn_{ave} versus radiation dose for comparative Bragg grating exposures using 213 nm, 150 ps (circles) and 248 nm, 34 ns (triangles) laser radiation.

SMF-28 and B/Ge co-doped fibers, following a clear Type I photosensitivity behavior for both fibers exposed (Kalachev *et al.*, 2005). Recently 30 KHz, 213 nm Nd:VO$_4$, 7 ns laser radiation was used for recording Bragg ratings in hydrogen-free, SMF-28 fiber, achieving refractive index changes of the order of 10^{-3} (Gagné and Kashyap, 2010). Gagné and Kashyap supported that the primary physical mechanism dominating the photosensitivity of SMF-28 fiber using 213 nm nanosecond radiation was of a single-photon absorption nature, prompting color-center generation. The single-photon color-center generation was associated with the longer pulse duration employed.

12.5.2   Infrared femtosecond laser inscriptions

There have been early investigations on the use of femtosecond laser radiation in the refractive index engineering of optical fibers, starting from the work of Saifi *et al.* (1989), who measured permanent refractive index changes of few parts of $\times 10^{-5}$ induced in a twin core, germanosilicate optical fiber, directional coupler by exposure to 620 nm, 100 fs laser radiation. Saifi *et al.* quantified these changes by measuring changes in the beat-length of the directional coupler, by power coupling between the two cores, while attributing for the first time the photosensitivity obtained in multiphoton generated structural changes (Griscom, 2011). A decade later, Cho *et al.* (1999) observed increase of the SiE' centers after the formation of plasma channelling in a multimode silica fiber using 790 nm, 110 fs Ti:Sapphire laser radiation. The formation of plasma channelling induced a permanent double cladding structure into the multimode silica fiber of 2×10^{-2} refractive index contrast. Then, Fertein *et al.* (2001) measured 6×10^{-3} refractive index changes using in the cavity length a Bragg gratings Fabry-Perot in Corning fiber SMF-28, exposed using 800 nm femtosecond laser, at a focal spot of 0.4 mm.

The landscape of fiber photosensitivity and grating recording was redrawn after the first Bragg reflectors inscribed in SMF-28 and all-silica depressed cladding optical fibers by Mihailov *et al.* (2003, 2004). In this work, both SMF-28 and all-silica fibers were exposed using a modified design phase mask for forming short length and highly scattering Bragg reflectors (Fig. 12.6), which maintained their reflectivities in temperatures greater than 1000°C.

The refractive index changes were a product of a four/five-photons (depending upon the material bandgap) nonlinear absorption at intensities of 1.2×10^{13} W/cm^2. That group exposed hydrogenated SMF-28 fibers using the same setup and conditions revealing reduction of compaction effects in the silica glass cladding (Limberger *et al.*, 2007), while the gratings exhibited annealing behavior similar to those inscribed using ultraviolet radiation (Smelser *et al.*, 2004). By varying exposure energy densities pure Type I-IR or Type II-IR gratings could be formed in SMF-28 fibers, where Type I-IR are possibly associated with color-center accumulation, and Type II-IR are products of extreme ionization and plasma formation processes. Type I, ultrabroad bandwidth, chirped Bragg reflectors were fabricated using Ti:Sapphire 800 nm, 1 KHz femtosecond laser radiation, in hydrogenated and pristine standard telecom fibers, using highly chirped phase masks, achieving FWHM bandwidths greater than 200 nm (Bernier *et al.*, 2009).

Alternatively to the use of custom design phase masks for controlling spatial dispersion and beam de-condensation effects that can detrimentally affect the high intensities required during femtosecond laser refractive index engineering (Mihailov *et al.*, 2004), point-by-point Bragg grating

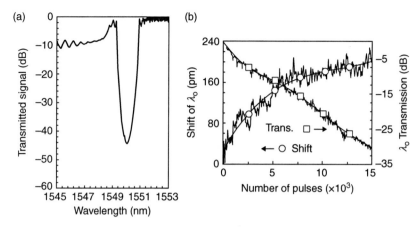

12.6 (a) Reflectivity of a high-order grating written in a SMF-28 fiber with the 2.142 µm pitch mask and 800 nm and 120 fs Ti:Sapphire laser radiation. (After Mihailov *et al.*, 2003. Used with permission from Optical Society of America [OSA].) (b) Variation of the reflectivity and resonant wavelength of a grating recorded using 800 nm and 120 fs Ti:Sapphire laser radiation with the number of incident IR pulses (300 mJ pulse, 10 Hz). The squares denote reflectivity; circles, wavelength shift. (After Mihailov *et al.*, 2004. Used with permission from IEEE.)

recordings were performed in standard telecom fibers (Martinez *et al.*, 2004). The nonlinear refractive index engineering at sub-wavelength volumes using 800 nm femtosecond lasers (Glezer *et al.*, 1996) offers significant advantages in the point-by-point Bragg grating processing, allowing ease in tailoring the topology of the periodic structure as well as its polarization characteristics. Martinez *et al.* inscribed such reflectors operating in the first diffraction and also in higher orders, using high magnification objectives and an 800 nm, 1 kHz, 150 fs Ti:Sapphire laser (Fig. 12.7). The Bragg reflectors inscribed were almost 25 dB in strength, exhibiting typical birefringence of ~ 3×10^{-5} due to the highly asymmetrical nature of the inscription process. Using the same method Martinez *et al.* (2006) also presented Bragg grating inscription in a jacket unstripped standard telecom fiber, succeeding in grating strengths greater than 25 dB. The above was a significant improvement compared to the previous art, where special coatings were used for inscribing Bragg reflectors through the fiber jackets using 244 nm laser radiation (Chao *et al.*, 1999). Infrared femtosecond laser point-by-point Bragg grating inscription technique was also used for recording apodized geometry structures in SMF-28e fibers by adopting a slanted scanning technique with respect to the fiber core (Williams *et al.*, 2011b) or complex sampled and phase-shifted Bragg reflectors (Marshall *et al.*, 2010).

Different kinds of Bragg reflectors were inscribed in standard telecom fibers exploiting filamentation effects, induced under Ti:Sapphire 800 nm

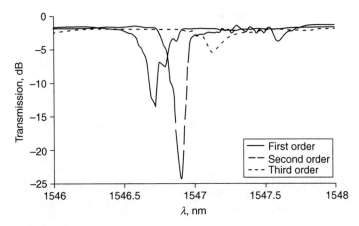

12.7 Transmission spectra measured in gratings of first, second and third orders fabricated using 800 nm femtosecond radiation and point-by-point technique. (After Martinez *et al.*, 2004. IET copyright permission is acknowledged.)

femtosecond laser radiation (Bernier *et al.*, 2011). High energy density inscriptions can lead to filamentation generation and propagation inside the fiber complex, inducing refractive index changes as high as $\times 10^{-2}$ (Sudrie *et al.*, 2002; Yamada *et al.*, 2001). Due to the nature of filament generation and propagation, the modifications induced in the hosting glass matrix are spatially localized in typical dimensions below a few microns, allowing high refractive index contrasts, thus increased diffraction efficiencies. Bernier *et al.* (2011) reported grating inscription by cross-sectioning the fiber core with the periodic filament generated by a phase mask, inducing refractive index changes of $\sim 2.5 \times 10^{-3}$.

12.5.3   Ultraviolet femtosecond laser inscriptions

While in IR femtosecond exposures several photons were required for covering the large bandgap of germanosilicate glasses and fused silica, at the expense of tight focusing and narrow envelope inscription conditions, the use of ultraviolet picoseconds and femtosecond sources could alleviate these tight inscription conditions exploiting lower-order, nonlinear effects. The scaling down rule of the damage fluence threshold of an optical material versus pulse duration (Boyd, 2003; Stuart *et al.*, 1995) applies also in photosensitivity, questioning the efficiency of traditional laser wavelengths such as 248 and 266 nm into the inscription of higher index changes at lower energy doses. The first inscriptions of Bragg gratings in hydrogenated SMF-28 fibers were presented using a low repetition rate, frequency quadrupled 264 nm Nd:glass laser (Dragomir *et al.*, 2003), at maximum intensities of 77 GW/cm^2.

Comparative results on the Bragg grating inscription process in SMF-28, phosphorus doped and all-silica fibers using 1 KHz repetition rate, tripled 267 nm, Ti:Sapphire femtosecond laser with results obtained using 157 and 193 nm ultraviolet excimer lasers (Zagorulko *et al.*, 2004). Zagorulko *et al.*'s study revealed that the photosensitivity of hydrogenated, low-Ge content silicate glass fibers using femtosecond, ultraviolet laser resembles that obtained using 157 nm excimer laser, leading to similar refractive index changes (few parts of 10^{-3}) and growth rates, while suffering less from saturation effects (Fig. 12.8).

The fabrication of long-period gratings in hydrogenated SMF-28 fiber variants using 352 nm, emitted from a frequency tripled Nd:glass laser and point-by-point exposure, was reported by Dubov *et al.* (2005). In this report a three-photon absorption process was triggered at intensities of the order up to 2000 GW/cm^2, while cladding damage was visible denoting asymmetrical absorption across the fiber cross-section.

Strong Bragg reflectors were also fabricated in SMF-28 fibers for modest accumulated energy densities (Livitziis and Pissadakis, 2008), using a double phase mask interferometer and 248 nm, 500 fs laser radiation (see Fig. 12.9a). These Type I gratings saturated at accumulated energy densities as low as 3.5 kJ/cm^2, reaching refractive index changes up to ~ 7×10^{-4}, while exhibiting thermal durability up to 900°C (Fig. 12.9b). The enhanced thermal stability of these SMF-28 fiber gratings, compared to those recorded using 193 nm excimer laser radiation (Albert *et al.*, 1995), was associated

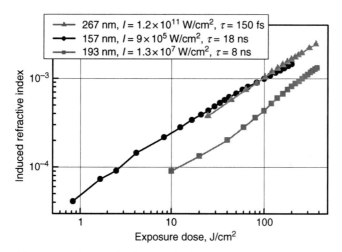

12.8 Dependence of refractive-index changes on exposure dose of an H$_2$-loaded SMF-28 fiber for inscriptions using 267, 157 and 193 nm laser wavelengths and corresponding pulse durations. (After Zagorulko *et al.*, 2004. Used with permission from Optical Society of America [OSA].)

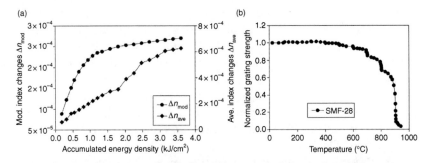

12.9 (a) Refractive index modulation Δn_{mod} (circles) and average index Δn_{ave} (diamonds) changes versus accumulated energy density for grating exposure in SMF-28 optical fiber using 248 nm, 500 fs laser radiation. (b) Isochronal thermal annealing results for a Bragg grating recorded in SMF-28 optical fiber, using 248 nm, 500 fs laser radiation.

with the higher two-photon absorption coefficient of the Ge-doped core (Dragomir, 2002), which in turn may lead to a substantial increase of the local temperature levels in the bright fringes.

Bragg and long-period grating inscriptions using femtosecond lasers produced mostly Type I and Type II photosensitive refractive index changes, while no Type IIA had been produced either by 800 nm Ti:Sapphire or quadrupled Nd:glass lasers, for prolonged irradiations and energy density doses into germanosilicate glasses fibers irrespective of Ge concentration and codopands. The first Type IIA Bragg grating recording utilizing femtosecond laser radiation was presented by Violakis *et al.* (2006) using 248 nm, 500 fs hybrid excimer/dye laser and phase mask configuration in contact mode. The fibers exposed were the B/Ge co-doped, PS1250/1500 from Fibrecore and a Hi-Ge from FiberLogix (Fig. 12.10a,b).

Violakis *et al.* presented comparative data with similar inscriptions using 248 nm, 34 ns excimer laser radiation, where the femtosecond grating recording was saturated for doses 10× times smaller than those required in the nanosecond recording; while both Type I and Type IIA refractive index changes slopes were accelerated for the femtosecond exposure case (Fig. 12.11). In addition, annealing studies revealed that the Type IIA gratings recorded using 248 nm, 500 fs radiation demarcated at 700°C, 100°C higher than the nanosecond counterparts, due to structural changes formed by the irradiation (Violakis *et al.*, 2006).

In later studies performed by the same group (Pissadakis *et al.*, 2006), comparative Type IIA Bragg grating recordings were demonstrated using 5 ps, 500 and 120 fs, 248 nm radiation together with micro-Raman studies, investigating the effect of exposure conditions in the photosensitivity growth characteristics, by performing exposures for fixed intensity and energy density and different pulse durations (Fig. 12.12a). Exposures performed under fixed intensities

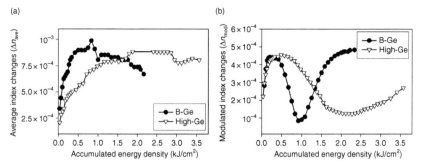

12.10 (a) Index modulation Δn_{mod} and (b) average index Δn_{ave} changes versus accumulated energy density, for Type IIA Bragg grating exposure of B-Ge and High-Ge optical fibers using 248 nm, 500 fs excimer laser radiation.

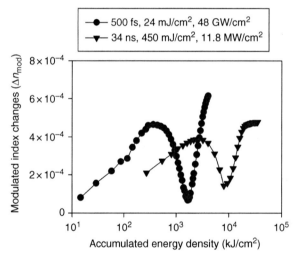

12.11 Comparative graph for index modulation Δn_{mod} changes recorded for grating inscription in PS1250/1500 B-Ge fiber using 248 nm, 500 fs and 34 ns excimer laser radiation.

for different pulse durations illustrated that the effect of the energy density plays a predominant role in the triggering of the Type IIA photosensitivity mechanism, while intensity rather affects its progression speed (Fig. 12.12b).

Additionally, the 248 nm femtosecond laser irradiation modifies the characteristics of the Boson peak (Hehlen *et al.*, 2002) in Raman spectra obtained (Dianov *et al.*, 1997b) from the boron-germanosilicate glass core, shifting this to longer wave numbers (Fig. 12.13), while exhibiting similarities to the response of silicate glasses subjected to hydrostatic pressure above the plasticity level (Inamura *et al.*, 1997).

The modification of the Boson peak constitutes a strong indication that the primary underlying mechanism behind Type IIA photosensitivity is that

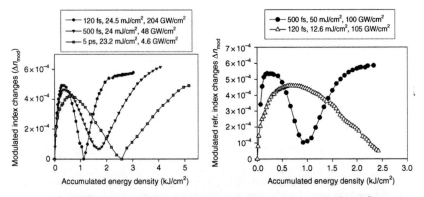

12.12 (a) Index modulation Δn_{mod} changes versus accumulated fluence, for Bragg grating exposure in B-Ge optical fiber using 248 nm, 120 fs, 500 fs and 5 ps laser radiation, of fixed fluence. (b) Index modulation Δn_{mod} changes versus accumulated fluence, for grating exposures of fixed intensity in the same optical fiber using 120 and 500 fs laser pulse duration.

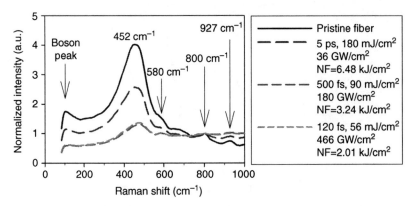

12.13 Micro-Raman scattering spectra for unexposed and exposed B-Ge doped fiber cores, using different pulse durations, for isochronal exposures using $N = 36\,000$ pulses. NF: total accumulated energy fluence in J/cm². The micro-Raman scattering spectra were obtained using 473 nm line of an Ar+-ion laser, while they have been normalized to the strength of the Si-O 800cm⁻¹ peak.

of compaction in the bright fringes of interference, and secondary that of stress relief; that was also described by Ky *et al.* (2003b).

12.5.4   Regenerated gratings

Bragg grating regeneration and post-exposure amplification were first observed in Type IIA gratings fabricated using 193 nm, excimer laser radiation in a B/Ge co-doped optical fiber (Dong and Liu, 1997). In such grating type, the exposure was ceased before reaching Type IIA saturation, and then the fiber reflector was subjected to annealing processes, where amplification strength was observed. Similar abnormal thermal annealing behavior had also been observed in Bragg gratings fabricated in OH-flooded, F-depressed cladding fiber at temperatures within the range of 1000°C (Fokine, 2002a), attributed to molecular water formation and diffusion into the silica glass matrix, and subsequent oxygen reduction. Then, thermal regeneration was also observed in a hydrogenated, F, P and B/Ge co-doped fiber with high concentration of both B and Ge ions that had been exposed to 193 nm, for forming a Type I Bragg reflector (Bandyopadhyay et al., 2008).

Canning et al. (Bandyopadhyay et al., 2008) speculate that the grating writing process leaves a structural signature in the exposed glass matrix, which is not demarked by the thermal treatment, even though the refractive index changes may erase. Instead such structural signature imprinted in the glass may constitute a catalytic base for assisting chemical reactions with hydrogen for forming possible species of hydrite or hydroxyl groups that can induce local stresses in the exposed fringe level (Canning et al., 2008b). Further, the above regenerated gratings exhibited extreme durability to thermal treatment, withstanding temperatures up to 1000°C (Fig. 12.14). Similar regeneration results were obtained for gratings fabricated in high Ge content fibers under strained annealing (Lindner et al., 2011), while surviving under similar temperature conditions as those presented in (Bandyopadhyay et al., 2008).

Thermal regeneration behavior has been observed for Type IIA Bragg gratings fabricated using femtosecond and picosecond 248 nm laser radiation (Pissadakis et al., 2006), for exposures ceased before reaching the Type IIA saturation level (Fig. 12.15).

Figure 12.15a shows index modulation Δn_{mod} changes versus accumulated energy density for Bragg grating exposure in B-Ge optical fiber using 248 nm, 120 fs, 500 fs and 5 ps laser radiation, of similar fluence. Figure 12.15b shows normalized grating strength, during regeneration annealing at isothermal temperature steps of 100°C, of the previous Bragg gratings until erasure point is reached.

Three different Bragg gratings exposures performed in the PS1250/1500 Fibercore B/Ge codoped fiber using 5 ps, 500 fs and 120 fs while keeping similar energy densities, ceased when reaching the same refractive index change level at the Type IIA photosensitivity regime. Accordingly all gratings were annealed from room temperature up to 600°C, reaching full demarcation. These three gratings exhibited almost identical regeneration characteristics, amplifying in strength with increasing temperature, reaching the maximum strength point at 550°C approximately; then erasing until the

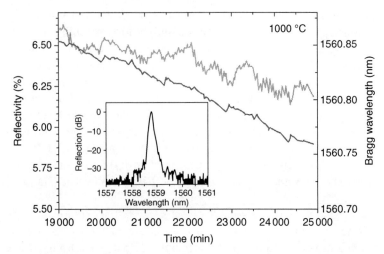

12.14 Annealing behavior of refractive index modulation and Bragg wavelength of regenerated grating after 2 weeks at 1000°C. Inset: grating spectrum (Lindner *et al.*, 2011).

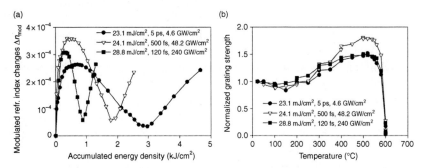

12.15 (a) Index modulation Δn_{mod} changes versus accumulated energy density for Bragg grating exposure in B-Ge optical fiber using 248 nm, 120 fs, 500 fs and 5 ps laser radiation, of similar fluence. (b) Normalized grating strength, during regeneration annealing at isothermal temperature steps of 100°C, of the previous Bragg gratings until erasure point is reached.

temperature of 600°C. Such behavior reveals that the glass composition and thermal history dominates the specific regeneration process, while the exposure conditions play rather a secondary role. Therefore, the regeneration characteristics may be tuned by changing the pre-exposure thermal history of the glass (i.e. rapid heating and cooling process), for modifying the population and type of fictive defects in the matrix, that can be later seeded by the irradiation process (Bandyopadhyay *et al.*, 2008).

12.6 Grating fabrication in standard, all-silica optical fibers

While in germanosilicate glass fibers there was significant progress related to grating inscription and refractive index engineering, the formation of photosensitive refractive index changes in all-silica fibers remained a rather hard task, thus frustrating the optimum exploitation of those fibers in photonic devices development due to lack of inscription processes. The photosensitivity of the high quality silicate glass at the spectral band from 5 to 8 eV is dependent upon low concentration of defects such as oxygen deficiencies, Si-Si wrong bonds and SiE' centers peaking at 5.8 eV, with significant tail up to 6.5 eV (Skuja, 1998). The state-of-the-art was limited to irradiations using 193 nm excimer laser and several hundred thousand pulses for inducing refractive index changes of the order of 10^{-5} (Rothschild et al., 1989); being insufficient for forming strong Bragg reflectors in all-silica fibers.

The formation of the first strong Bragg reflectors in depressed cladding all-silica fibers using 193 nm laser radiation (Albert et al., 2002) that have been subjected to hydrogen flooding and OH defect generation (Lancry et al., 2007), was the first demonstration of photo-engineering of refractive index ($\Delta n \sim 0.5 \times 10^{-4}$) in such fibers (Fig. 12.16). The formation mechanism of those reflectors was classified as of chemical nature where Si-OH species initially formed by the high temperature flooding are photo-activated for forming water species in the bright fringes of interference, while simultaneously inducing photorefractive effects and stress generation into the matrix (Fokine, 2002a; Smith et al., 2001). In the work of Albert et al. pristine F-depressed cladding fiber was exposed, producing Bragg grating reflectors of smaller diffraction efficiency than the OH-flooded one, with corresponding refractive index changes of few parts of $\times 10^{-5}$. Shortly after the demonstration of Albert and Fokine (Albert et al., 2002), strong Bragg reflectors were recorded in non-sensitized silica glass, optical fibers by Mihailov et al. (2004) employing a 800 nm femtosecond Ti:Sapphire laser, expanding the tools available for refractive index engineering of all-silica glass fibers.

Since the Bragg grating fabrication in all-silica fibers was successful using 800 nm, a fundamental question arose relating to the combination of ultraviolet radiation with photon energy comparable to the bandgap of the silica glass with femtosecond pulse duration for improving further the recording yield. In such a combination the order of multiphoton absorption would lower to 2–3 photons, alleviating focusing requirements, while accessing centers closer to the Urbach tail of the glass (Kühnlenz et al., 2000). First Bragg gratings inscription attempts in pristine, all-silica fibers were presented in (Zagorulko et al., 2004) using frequency tripled 800 nm Ti:Sapphire laser emitting at 267 nm, while using a phase mask in contact mode. However,

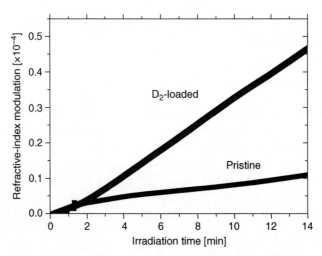

12.16 Growth in refractive-index modulation for deuterium-loaded and pristine all-silica, fluorine depressed cladding fibers. (After Albert *et al.*, 2002. Used with permission from Optical Society of America [OSA].)

in this work the yield of Bragg grating inscription in the non-hydrogenated silica glass fiber was low, forming reflectors less than 0.1 dB in strength after accumulated energy density doses of ~ 50 kJ/cm². This may be associated with the photon energy of 267 nm which is ~ 4.6 eV marginally above the nominal bandgap of silica glass if considered in a two-photon absorption scheme (Dragonmir *et al.*, 2002).

By adopting slightly shorter wavelength at 248 nm and a substantially different inscription setup, strong Bragg gratings were inscribed in hydrogen unloaded, low-OH concentration, all-silica Sumitomo Z-fiber by Livitziis *et al.* (Livitziis and Pissadakis, 2008), using 500 fs laser pulses. Livitziis *et al.* used a dual phase mask interferometer detaching the phase mask element from the fiber (Fig. 12.17a) and allowing the delivery of sub-TW/cm² intensities in the fiber complex, without suffering from two-photon absorption and color-center accumulation in the phase mask (Taylor *et al.*, 1988).

The combination of the 248 nm wavelength (of ~5 eV photon energy) and the extreme intensities delivered in the fiber, boosted two-photon absorption effects, and led to the formation of modulated refractive index changes of the order of ~ 2.2 × 10⁻⁴ (and of 5 × 10⁻⁴ average refractive index changes) for intensities of (≈ 440 GW/cm2) at the bright fringes of the interference pattern (Fig. 12.17b). Inscriptions using different energy densities verify the two-photon absorption nature of the inscription process, where the refractive index changes obtained are dependent upon the second power of the inscription intensity. Also, the appearance of low cladding mode notches in the transmission spectra of the gratings recorded using 248 nm, 500 fs laser

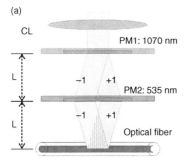
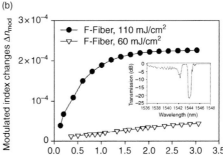

12.17 (a) Schematic diagram of the double phase mask interferometer. CL: cylindrical lens. PM1, 2: phase mask elements with periodicity annotated. (b) Refractive index modulation Δn_{mod} changes versus accumulated energy density for grating exposures in all-silica Sumitomo Z-fiber, using 248 nm, 500 fs laser radiation, for two different energy densities. Inset: transmission spectrum of the Bragg reflector inscribed using 110 mJ/cm² energy density.

radiation, indicates that refractive index changes have also been inscribed in the radiation resistant fluorine doped cladding (Hosono, 1999).

A useful comparison of the trend and the yield of the 248 nm, 500 fs inscription, with one performed using 193 nm, 10 ns laser radiation in the same all-silica fiber, presented in Fig. 12.18, illustrates that the photosensitivity processes induced by these two wavelengths are different (Fig. 12.18a). The two-photon mediated photosensitivity process of the 248 nm, 500 fs radiation progresses faster with respect to energy density dissipated in the fiber; the intermediate germanium-oxygen deficiencies state at 5 eV can also assist such accelerated behavior. The corresponding exposure using 193 nm radiation exhibits different slope shape; the defects accessed are possibly of a broader nature close to the band-gap of the material. However, subsequent annealing of the Bragg reflectors produced using those lasers and pulse durations denotes that the type of modifications produced are of the same type (Fig. 12.18b).

A standard Talbot interferometer was used for recording Bragg reflectors inside a Al/Yb co-doped all-silica standard fiber using 262 nm femtosecond laser pulses, emitted by a frequency tripled Ti:Sapphire laser (Becker *et al.*, 2008). This report re confirmed that by increasing the spacing between the phase mask and the fiber energy dissipation constraints are alleviated, leading to efficient inscription inside all-silica glass fibers using ultraviolet femtosecond lasers. Long exposures of energy doses greater than 2 MJ/cm², using CW frequency doubled Ar⁺ ion laser radiation were used for forming Bragg reflectors in unsensitized Ge-free, Bi-Al, silica glass fibers (Violakis *et al.*, 2011), reaching refractive index changes up to 3.5×10^{-4}. The possible mechanism of grating formation using such continuous wave (CW) exposures can be two-photon absorption-related compaction, progressing

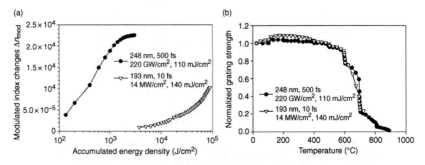

12.18 (a) Refractive index modulation Δn_{mod} changes versus accumulated energy density for grating exposures in all-silica Sumitomo Z-fiber, using 248 nm, 500 fs and 193 nm, 10 ns laser radiation. (b) Annealing study of the reflectors produced using the above wavelengths and pulse durations.

at slow rates due to the lack of sufficient intensities; pre-existing color centers exhausting are saturated during the early phases of inscription.

Other interesting results included the Bragg grating inscription in N-doped core silica-glass fibers using 193 nm laser radiation and standard phase mask setup (Butov *et al.*, 2006; Dianov *et al.*, 1997a), where a clear Type IIA photosensitivity behavior was monitored during inscription (Fig. 12.19a). The Type IIA gratings recorded in the N-doped fiber exhibited thermal regeneration characteristics, withstanding temperatures up to 1200°C (Fig. 12.19b). The authors of this work (Butov *et al.*, 2006) justify the extreme thermal stability and regeneration effects upon a nitrogen-species mobility model, where the photo-excited nitrogen can diffuse and be re-trapped from the core to the cladding area forming structural corrugations. In later work on N-doped silica fibers gratings recorded using 193 nm excimer laser, Lanin *et al.* (2007) observed similar thermal and non-thermal regeneration effects, of non-reversible nature, upon hydrogen diffusion into the pre-exposed fiber, due to thermochemical reaction of hydrogen toward the formation of Si-NH and Si-OH species.

12.7   Grating fabrication in phosphate and fluoride glass fibers

12.7.1   Phosphate glass fiber gratings

The photosensitivity of phosphorus codoped silicate glass fibers has been investigated since the 1990s, when 193 and 240 nm pulsed laser radiation was used for inducing transient and permanent refractive index in those fibers (Canning *et al.*, 1995). The permanent residue of refractive index changes obtained for hydrogen loaded, phosphorus doped silica glasses was of the

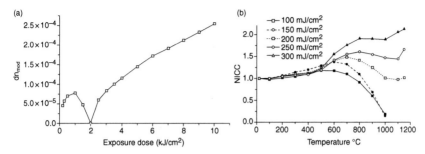

12.19 (a) Effective refractive index modulation in a Bragg grating
(λ_B = 1540 nm) as a function of exposure dose of a 193 nm wavelength
laser radiation, the power density per pulse being 100 mJ/cm², pulse
duration 8ns and repetition rate 100 Hz. (b) Isochronal annealing of Type
IIA Bragg gratings written in an N-doped fiber under different regimes.
(After Butov *et al.*, 2006.)

order of 10^{-3}, under 193 nm irradiation; unloaded fibers were quite less photosensitive. It was also known that phosphorus dopand suppresses the 242
nm germanium-oxygen deficiencies band when inserted in germanosilicate
glasses, acting as a passivator for such types of wrong bonds, thus reducing
photosensitivity (Dong *et al.*, 1994).

The valence state of phosphorus allowing the formation of single and double bonds with oxygen, defines the micro-coordination of phosphate glasses
while leading to a linear-like polymerization chain built in combination
with oxygen and other ion modifiers; a micro-coordination structure which
is substantially different from that of silicate glasses. Due to this versatile
phosphorus–oxygen bond state, phosphate glasses exhibit interesting optical, chemical and mechanical properties, dependent upon the ratio between
phosphorus and oxygen ion concentrations, and the incorporation of other
matrix modifiers. Phosphate glasses are highly transparent and durable to
ultraviolet radiation, have a high solubility of rare-earth ions and exhibit low
T_g points (Ehrt *et al.*, 1994). Rare-earth doped phosphate glasses have been
drawn into standard and MOFs and planar waveguides for realizing high
gain amplifiers (Hwang *et al.*, 1999) and lasers (Li *et al.*, 2005) of short physical lengths. Therefore, the straightforward inscription of Bragg reflectors in
those guiding structures is required for realizing lasing devices and sensors
(Strasser, 1996). The first studies of the photosensitivity of pure phosphate
glasses were presented in the mid-1990s, mostly focused on fluorophosphates
slab samples including Ce⁺ ion modifiers while being irradiated using 248
and 193 nm excimer laser radiation (Ebendorff-Heidepriem and Ehrt, 1996;
Ebendorff-Heidepriem *et al.*, 2002). In these studies spectroscopic modifications were solely examined, rather than refractive index engineering.

Generally, the exposure of phosphate glasses to ultraviolet laser and
X-ray radiation results in color-center generation extended from the visible

to the ultraviolet and bandgap spectral regime, associated with the transformation of the phosphorus–oxygen bond and the related defects, with most prominent the PO hole center in the visible band and the PO_4 electron center peaking at the 242 nm wavelength (Ebeling *et al.*, 2002; Ehrt *et al.*, 2000). Further to these first photosensitivity studies carried out using short wavelength sources and single-photon absorption excitation, there were presented refractive index engineering demonstrations using infrared (Chan *et al.*, 2003) and near ultraviolet femtosecond lasers in pristine and silver doped phosphate glass substrates (Watanabe, 2001), respectively.

A study focusing on the refractive index engineering of the commercially available rare-earth doped IOG-1 Schott glass including the irradiation of pristine and silver ion-exchanged samples using 248 nm nanosecond excimer laser was presented in 2004 (Pissadakis *et al.*, 2004). This study revealed that the refractive index changes obtained in a pristine glass slab for accumulated energy density doses of 12 kJ/cm^2 was of the order of a few parts of $\times 10^{-5}$, while the addition of silver boosted the photosensitivity of the glass by almost three orders of magnitude. The same glass exposed using 213 nm, 150 ps Nd:YAG laser and an elliptical Talbot interferometer exhibited similar refractive index changes (few parts $\times 10^{-5}$); however, non-monotonic growth effects were monitored during grating inscription (Pappas and Pissadakis, 2006).

Other investigations in the same IOG-1 glass using 193 nm laser radiation led to greater photosensitivity effects that were directly exploited for waveguide laser development (Yliniemi *et al.*, 2006a, 2006b). Most importantly the work of Yliniemi *et al.* revealed that phosphate glass undergoes structural changes under 193 nm laser irradiation including volume modification, confirmed by utilizing micro-Raman spectroscopy and atomic force microscopy. Similar findings related to volume modifications induced by 193 nm (Michelakaki and Pissadakis, 2009) and 248 nm, 500 fs (Pissadakis and Michelakaki, 2008) laser radiation were presented by Michelekaki *et al.* where Knoop micro-hardness measurements were used for monitoring non-monotonic volume dilation effects versus energy density dose, under a single-photon absorption mechanism (Fig. 12.20a).

From the variation of the Knoop hardness data obtained for 193 nm excimer laser exposures in bulk phosphate glass samples, the corresponding changes in the elastic modulus were evaluated, confirming the volume dilation induced by the irradiation process (Fig. 12.20b). Such atypical radiation-induced volume modification effects were attributed to PO bond transformation from single to double and subsequent cleaving upon the conditions of the irradiation process.

The aforementioned investigations on the photosensitivity of bulk phosphate glass samples constituted a solid background for attempting Bragg grating inscription in all-phosphate glass fibers. The first Bragg gratings in a rare-earth doped all-phosphate glass fiber were demonstrated by Albert *et al.*

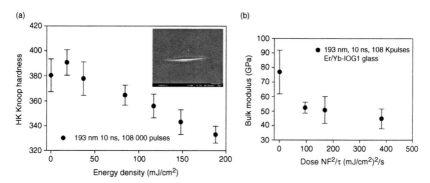

12.20 Knoop hardness versus energy density of IOG1 phosphate glass exposed using 193 nm, 10 ns excimer laser radiation. Inset: Knoop hardness indentation imprint on a phosphate glass sample. Elastic modulus versus radiation dose of IOG1 glass exposed using 193 nm, 10 ns excimer laser radiation. The phosphate glass for the exposure conditions above undergoes volume dilation.

(2006) using 193 nm excimer laser and standard phase mask inscription technique. Albert *et al.* inscribed strong, Type I Bragg reflectors in a phosphate glass fiber, obtaining average refractive index changes of 5×10^{-4}. These gratings under low temperature annealing, amplified more than 65% in strength; this behavior was attributed to slow stress relaxation effects accelerated by the thermal treatment (Albert *et al.*, 2006; Rodica Matei *et al.*, 2007). Grobnic *et al.* (2007) followed with positive refractive index, Type I Bragg grating inscriptions in a similar phosphate glass fiber using 800 nm, femtosecond laser radiation, reaching refractive index changes greater than 1.5×10^{-3}. These reflectors exhibited rather standard thermal stability, without post-exposure amplification effects, while maintaining their strengths up to temperatures of 400°C. Notably, the results of Grobnic *et al.* (2007) oppose earlier observations of Chan *et al.* (2003) where negative refractive index changes formed waveguides in IOG-1 glass under 800 nm, femtosecond irradiation; these negative refractive index changes demarcated at substantially lower temperatures.

Recently, Sozzi *et al.* exposed the same fiber as Grobnic *et al.* (2007) using 248 nm, 500 fs laser radiation and a double phase mask interferometer (Sozzi *et al.*, 2011). Sozzi *et al.* found that at high energy densities for such pulse duration, a photosensitivity mechanism similar to Type IIA was activated in the phosphate glass fiber, following non-monotonic growth of both the average and modulated refractive index changes (Fig. 12.21a). The last was the first demonstration of Type IIA-like photosensitivity growth in a non-silicate, soft glass matrix optical fiber. The average refractive index changes obtained were greater than $\times 10^{-3}$, for accumulated energy doses of 6.5 kJ/cm^2. These anomalous growth Bragg gratings, maintained the greater part of their strength up to 377°C (Fig. 12.21b). Inscriptions performed at lower energy

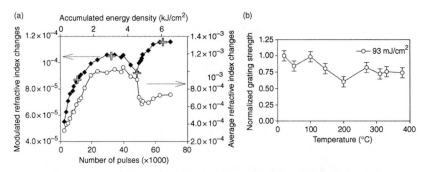

12.21 (a) Index modulation Δn_{mod} (diamonds) and average refractive index Δn_{ave} changes (circles) versus accumulated energy density, for grating exposure of the phosphate glass fiber using 248-nm 500-fs excimer laser radiation. The cross-points denote the exposure instances (number of pulses for fixed energy density) where Knoop micro-indentation measurements were performed. (b) Isochronal annealing results for Bragg gratings recorded in the phosphate glass fiber, using 248-nm, 500 fs laser radiation. Circles: 93 mJ/cm² energy density, 6.5 kJ/cm² accumulated energy dose. (After Sozzi *et al.*, 2011.)

densities followed a Type I photosensitivity trend, indicating that the specific non-monotonic photosensitivity mechanism is possibly not of linear-dependence. Sozzi *et al.* further investigated the origin of the specific photosensitivity mechanism using micro-indentation Knoop hardness measurements, performed in exposed and side polished phosphate glass fibers.

These micro-indentation measurements provided further insight into the volume modifications that the glass undergoes under 248 nm, 500 fs irradiation, showing that for prolonged exposures volume dilation takes place, driving negative refractive index changes. These results are in agreement with those presented by Michelakaki *et al.* where the Knoop hardness of another phosphate glass initially hardens and then dilates under 193 nm, 10 ns and 248 nm, 120 fs laser exposures (Michelakaki and Pissadakis, 2009). The results of Knoop hardness variation, exhibit a universal character (Fig. 12.22), irrespective of glass composition and irradiation conditions, similarly to the compaction model studied extensively for silica glass (Borrelli *et al.*, 1997; Primak, 1972). The non-monotonic photosensitivity behavior observed for this phosphate fiber for Bragg grating inscription using 248 nm, 500 fs radiation can be described as a mirror counterpart of the Type IIA mechanism of heavily doped germanosilicate fibers (Ky *et al.*, 2003). The fundamental difference between the mechanisms occurring in the phosphate and the germanosilicate glass fibers is the sign of refractive index changes induced in the bright fringe of interference: in the case of the phosphate glass these changes are negative due to volume dilation, generating compressive stresses.

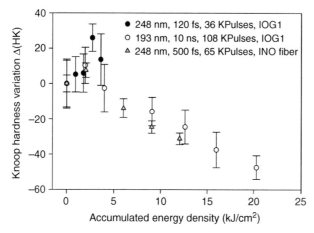

12.22 Knoop hardness variation results versus accumulated energy density for different phosphate glasses, geometries (IOG1 slabs and INO fiber), laser wavelengths and pulse durations.

12.7.2   Fluoride glass fibers

While fluoride glasses were first realized in 1975 (Poulain *et al.*, 1975), and shortly after their spectroscopic and active ion-host characteristics established them as promising materials for lasing devices (Zhu and Peyghambarian, 2010), their laser photosensitivity has been investigated in detail for almost two decades after the initial invention. The structure of fluoride glasses where the oxygen ion has been replaced by that of fluorine allows low phonon energies and thus great transparencies both at the Urbach tail band and the far-infrared regime, sketching an efficient bandwidth from 200 nm to 8 µm approximately. Poignant *et al.* recorded Bragg gratings in a ZBLAN fiber doped with Ce^{+3} ions as photosensitivity enhancer, reaching refractive index changes of few parts of $\times 10^{-4}$ after exposure to 246 nm pulsed radiation (Poignant *et al.*, 1994; Taunay *et al.*, 1994).

Since the photosensitivity observed was underlined by a specific sensitizer, potentially the mechanism of refractive index changes could be that of color-center formation; however, that was not concluded from the experiment (Taunay *et al.*, 1994). Sramek *et al.* (2000) exposed fluorozirconate glass slabs of different compositions to 193 nm excimer laser radiation, and found that surface expansion occurs at heights of tens of nanometers for accumulated energy doses lower than 1 kJ/cm^2 (Fig. 12.23). The last authors justified this laser-induced expansion after considering similar crystal structures such as b-BaZr$_2$F$_{10}$ and BaZrF$_6$, which lead to density ratios close to unity after comparison with the fluorozirconate glasses exposed; thus exhibiting

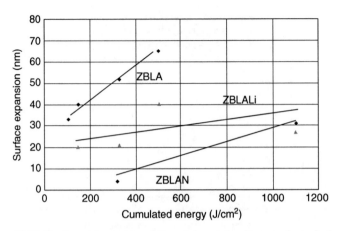

12.23 Surface expansion of fluorozirconate glasses photo-induced at 193 nm as a function of the total energy impinging the glass. (After Sramek *et al.*, 2000.)

low structural defect population. The underlying photosensitivity mechanism speculated for 193 nm excitation, refers to Coulombian bond breaking between ion modifiers such as Ba with unpaired fluorine, changing the glass micro-coordination with respect to the Zr_2F_{13} unit blocks.

Moreover, Zeller *et al.* (2005) exposed undoped fluoride glasses to 193 nm excimer laser radiation and investigated the radiation-induced glass disorder by examining changes in the Urbach tail of the irradiated specimens (John *et al.*, 1986; Urbach, 1953). Zeller and co-authors further quantified the depth of radiation damage into the exposed glasses by gradually polishing them and re-evaluating refractive index changes by means of Kramers–Kronig transformation of the color centers formed. Cerium doped ZBLAN fluoride bulks were also exposed by Williams *et al.* (1997) using 248 nm, excimer lasers, revealing that the Ce^{3+} excitation was of a two-step absorption nature, exploiting the $4f^1$–$5d^1$ transition leading to quadratic dependence upon intensity. In the same study, it was also found that the Cerium ion color centers created were simultaneously bleached by the 248 nm excitation.

Similar to other Bragg grating inscription in low defect glass optical fibers Grobnic *et al.* (2006) used 800 nm femtosecond laser for recording ~ 2.0×10^{-4} refractive index changes in an undoped ZBLAN fiber; nonetheless, without referring to the sign of the refractive index changes. These gratings resisted to temperatures up to 240°C, quite close to the T_g of the glass at 265°C. An illustrative study on the photosensitivity of ZBLAN fibers was presented by Bernier *et al.* (2007), where negative refractive index changes were identified locally in the fiber core and cladding areas, using fiber profilometry (Fig. 12.24). Also, positive refractive index changes were traced at the core–cladding interface, and at other areas of the cladding, providing

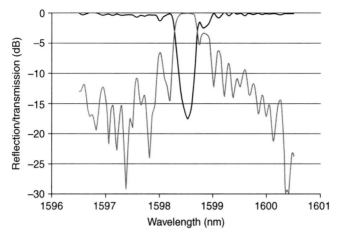

12.24 Measured transmission and reflection spectra of a Bragg grating written in an undoped-ZBLAN LVF fiber at 0.9 mJ during a 2 s exposure using 800 nm femtosecond radiation. (After Bernier *et al.*, 2007. Used with permission from Optical Society of America [OSA].)

an initial evidence that the sign of the refractive index changes is directly dependent upon the dissipated intensity at microscopic level. Moreover, these refractive index engineering results appear to be in agreement with the volume dilation measurements in bulk samples presented by Sramek *et al.* (2000) and the Urbach tail investigations of Zeller *et al.* (2005) where a radiation-induced disordering of the glass is described.

12.8   Microstructured optical fiber (MOF) gratings

12.8.1   Bragg and long-period grating inscriptions

Photonic crystal fibers (PCFs) fuse unique light guiding properties together with useful microfluidic structures, offering great flexibility in developing photonic devices implemented with novel and high performance functionalities. Even though there was a standardized photosensitivity and refractive index engineering background related to the inscription of Bragg and long-period gratings in standard optical fibers and most of the glass materials used for drawing them, the emergence of PCFs and MOFs imposed new inscription issues as well as challenges (Knight *et al.*, 1996; Russell, 2006). The capillary structure surrounding the guiding area affects the light penetration for side-illumination as this is principally required for periodic structure inscription using laser radiation. Specifically, the geometrically complex core and cladding shape of PCFs and MOFs induces significant scattering and refraction effects that in general reduce the average laser intensities delivered into the fiber core. Such reduced power dissipation into the fiber core during the side-illuminated grating inscription process deteriorates the photosensitivity yield by means of the magnitude of the refractive index changes formed into the guiding area and the accumulated energy densities needed for reaching saturation of the inscription. These scattering and refraction effects can dominate over the type of the photosensitivity process occurring, as well as the order of the underlying photon absorption. Many PCF and MOF designs are drawn from a single type of optical glass, often of low defect concentration such as fused silica, exhibiting low photosensitivity using conventional exposure methods. Therefore, additional issues arise during the grating recording in those high quality glass MOFs, rendering the inscription process a difficult and multiparameter materials science and light propagation problem.

Most of the methods used for inscribing gratings in standard optical fibers have been used for similar inscriptions in MOFs counterparts, revealing interesting grating recording behaviors and spectral responses. Initially, the safe path of Ge-based photosensitivity and hydrogenation was followed for forming Bragg and long-period gratings inside MOFs, by doping a small socket of the guiding area with Ge-ions.

Three years after the demonstration of PCF, Eggleton et al. (1999) presented the first Bragg and long-period grating inscription in such a solid core fiber (Fig. 12.25), which contained a Ge-doped photosensitive socket,

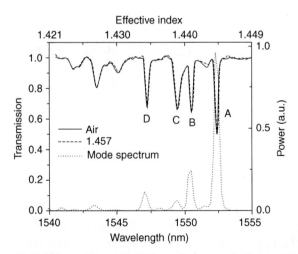

12.25 Measured transmission spectrum of a Bragg grating written in PCF (solid curve) before and (dashed curve) after application of the external index; the dotted curve shows the computer mode spectrum when beam-propagation modeling was used. (After Eggleton *et al.*, 1999. Used with permission from Optical Society of America [OSA].)

using a 242 nm frequency-doubled excimer-pumped dye laser. For alleviating side-illumination scattering issues, Eggleton *et al.* hydrogen loaded the Ge-doped solid core PCF, reaching grating strengths greater than 50%. The same group presented a more detailed analysis of Bragg grating inscription and spectral resonance effects in several MOFs of different geometry and dopands, including excitation of cladding modes (Eggleton *et al.*, 2000; Kerbage *et al.*, 2000).

Kakarantzas *et al.* (2002) presented the first long-period grating recording in a solid core all-silica PCF by using 10.6 μm CO_2 laser radiation, and a scanning mirror beam focusing setup, following the technique inaugurated by Davis *et al.* (1998) in standard optical fibers. In this demonstration, the glass PCF was heated by the CO_2 laser above the fictive temperature of silica, resulting in the formation of both geometrical and micro-coordination changes in the rapidly cooling glass and succeeding in inscribing short length, high strength LPGs and rocking filters (Kakarantzas *et al.*, 2003).

Accordingly, the first Bragg grating inscription in an all-silica, solid core PCF was reported by Groothoff *et al.* (2003) using a 193 nm, excimer laser of standard spatial coherence and a phase mask in contact mode (Fig. 12.26). The Type I refractive index changes reported in that demonstration were of the order of 2.4×10^{-4}, after an exposure of cumulative fluence of 200 kJ/cm^2, leading to significant grating strengths. The refractive index modification mechanism was assumed to be that of two-photon resulting primarily in the dissociation of the Si-O bond, inducing matrix compaction. Such experimental observation is similar to the results presented by Rothschild (Rothschild

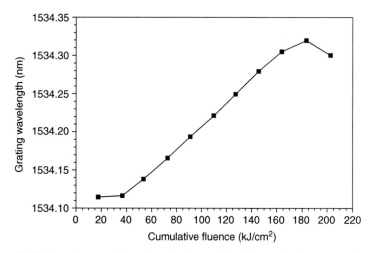

12.26 Grating wavelength evolution as functions of 193-nm cumulative fluence. (After Groothoff *et al.*, 2003. Used with permission from Optical Society of America [OSA].)

et al., 1989), Devine (Fiori and Devine, 1986) and Borrelli (Borrelli *et al.*, 1997). Also, these authors found that there were non-thermal, post-exposure matrix relaxation effects, resulting in spectral modifications of the inscribed reflectors. Canning *et al.* further exploited the above writing approach for developing DFB laser in an Er-doped silicate glass PCF, reaching slope efficiencies of 12.5% (Canning *et al.*, 2003; Groothoff *et al.*, 2005); but also in the recording of Bragg reflectors in a Fresnel type MOF (Martelli *et al.*, 2005).

Other groups used 193 nm excimer laser radiation for recording Bragg gratings in hydrogenated phosphosilicate and germanosilicate PCFs, reaching refractive index changes greater than 3×10^{-3} for the phosphosilicate matrix, and one order of magnitude higher changes for the germanosilicate fibers (Beugin *et al.*, 2006). In the work of Beugin *et al.* (2006) the PCF section was spliced both sides to a standard optical fiber prior to hydrogenation, for encapsulating hydrogen under the loading pressure into the capillaries (Sørensen *et al.*, 2005), maintaining high concentrations of hydrogen species into the PCF core during the grating recording process, while reducing the out-diffusion rate (Liou *et al.*, 1997).

Violakis *et al.* (Violakis and Pissadakis, 2007b) used a high spatial coherence 193 nm excimer laser for recording strong Bragg reflectors in a commercially available (Fig. 12.27a), solid core PCF, reaching high refractive index changes ($\Delta n = 3 \times 10^{-4}$). The use of an increased spatial coherence laser allowed the formation of high contrast interference fringes along the whole cross-section of the exposed fiber (Othonos and Lee, 1995), a condition that is significant for maintaining high intensities necessary for nonlinear absorption photosensitivity processes. In the demonstration of Violakis *et al.* (Violakis and Pissadakis, 2007b) the Bragg reflectors recorded using an energy density of

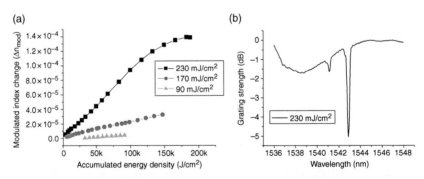

12.27 (a) Modulated refractive index versus accumulated energy density curves of Bragg grating inscriptions realized in the ESM-12–01, PCF, using energy density of 230, 170 and 90 mJ/cm² using 193 nm, 10 ns laser radiation. (b) Transmission spectrum of a grating inscribed in the above fiber using 230 mJ/cm² energy density per pulse and 187 kJ/cm² accumulated energy density.

230 mJ/cm² exhibited a broad wide spectral valley observed at the left side of the fundamental grating notch, related with broadband coupling to radiation modes. This kind of radiation mode coupling was also confirmed when visible laser was injected into the PCF fiber and out-coupled by the inscribed Bragg grating. Such a spectral feature that is routinely observed in the case of surface relief gratings imprinted on planar waveguides (see Fig. 12.27b) may be related to a shallow, compaction relief periodic feature induced in the microstructured fiber capillaries (Schenker, 1994).

Fu *et al.* (2005) exposed a silica glass solid core PCF using a frequency tripled Ti:Sapphire 267 nm femtosecond laser for recording Bragg reflectors; however, the fiber was hydrogenated for increasing photosensitivity, similar to the example of Zagorulko *et al.* (2004). Using this approach the refractive index changes inscribed in the PCF exposed were ~ 3 × 10⁻⁴ under accumulated energy density reaching 100 kJ/cm², while in a standard all-silica fiber three times greater index changes were recorded (1.1 × 10⁻³) using 54.5 kJ/cm² (Zagorulko *et al.*, 2004).

Similarly, Brambilla *et al.* (2006) used 264 nm, 220 fs laser radiation for inscribing 1 cm long and 20 dB strong LPGs into an all-silica solid core PCF. Different types of hydrogenated PCF fibers were also exposed to femtosecond and picosecond, 248 nm laser radiation using standard phase mask and double phase mask interferometer (Fig. 12.28), leading to refractive index changes of 1.2 × 10⁻⁴, for energy doses of ~ 18 kJ/cm² (Pissadakis *et al.*, 2008).

Other demonstrations included the use of Ti:Sapphire 800 nm, femtosecond laser radiation for Bragg grating inscription in pristine and tapered PCFs (Mihailov *et al.*, 2006); and frequency tripped Ti:Sapphire laser at 262 nm for comparative recordings in unloaded and hydrogen-loaded PCFs (Wang *et al.*, 2009b). All of the above demonstrations led to the formation

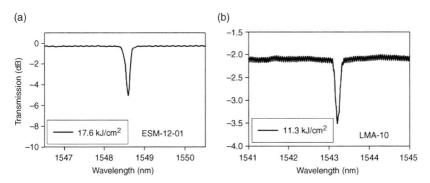

12.28 Transmission spectra of Bragg gratings fabricated in (a) ESM-12–01 and (b) LMA-10 hydrogenated optical fibers, having 1 cm length, by using 248 nm, 5 ps laser radiation. Accumulated energy densities appear in the spectra insets.

of Type I photosensitivity refractive index changes and inscription behavior; however, there were examples where Type IIA Bragg reflectors were inscribed in MOF and PCF using 193 nm excimer lasers. Cook *et al.* (2008) exposed a 6.7%wt Ge-doped core, 12-ring PCF for completing Type IIA saturation after typical accumulated doses of 20 kJ/cm². Others (Pissadakis *et al.*, 2009b) exposed a 36%wt Ge-doped, small-core, highly nonlinear PCF, for observing complex Type IIA photosensitivity effects for the two guiding modes supported by the fiber (Fig. 12.29a). Both scattering modes exhibited post-fabrication, amplification effects after storing at room temperature due to stress relaxation of the small dimension 1.38 μm, highly doped core; while surviving up to 800°C following a mixed Type I and IIA decay behavior. (see Fig. 12.29b)

In parallel, there is interest in the inscription of Bragg reflectors in small and collapsed core microstructured optical fibers. Due to the specific core geometry and dimensions (< 5 μm), collapsed core fibers exhibit a guiding mode profile largely extending in the inter-capillary areas, allowing high overlap with the refractive index medium included in those. The last renders them ideal platforms for the development of high sensitivity in-fiber refractometers and biological sensors, prompting the inscription of Bragg reflectors in their minimal size cores.

Initially, there were reports on the Bragg grating inscription in hydrogenated, Ge-socket collapsed core microstructured optical fibers using CW, 244 nm, frequency doubled Ar⁺ ion laser radiation (Huy *et al.*, 2006; Phan Huy *et al.*, 2007) (see Fig. 12.31b, c). Becker *et al.* (2009) used 267 nm, femtosecond laser radiation for recording arrays of Bragg reflectors in

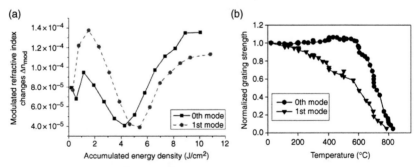

12.29 (a) Index modulation Δn_{mod} changes versus accumulated energy density, monitored for the 0th and 1st order guiding modes of the MOF, for Type IIA Bragg grating exposure using 193 nm, 10 ns excimer laser radiation. (b) Isochronal annealing results for Type IIA Bragg gratings recorded in high-Ge microstructured fiber, using 193 nm excimer laser radiation. Exposure conditions for the ESM-12–01 fiber: 193 nm, 230 mJ/cm² energy density, 200 kJ/cm² accumulated energy dose. The ESM-12–01 fiber has been annealed at shorter isothermal intervals of 10 min.

all-silica collapsed core fibers (Fig. 12.30b,c), reaching refractive index changes of 6×10^{-4}.

Recently, Ge-doped socket, collapsed core fibers were exposed using high coherence, 193 nm, excimer laser radiation (Fig. 12.31), leading to visible compaction effects in the exposed core, resulting in significant light out-scattering for shorter wavelengths (Konstantaki *et al.*, 2011). The structural changes induced in the collapsed core and surrounding cladding are under further investigation using micro-Raman analysis.

Novel Bragg grating inscription approaches, while exploiting the absorption or transparency of the infiltrated/encapsulated material into the PCF

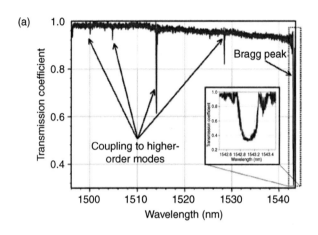

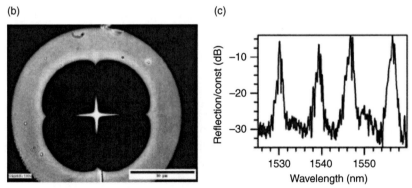

12.30 (a) Transmission spectrum of FBG photo-written in the three-hole fiber over a spectral window of 50 nm, showing spectral resonances toward high-order modes. The inset shows a zoom on the Bragg wavelength itself. (After Phan Huy *et al.*, 2007. Used with permission from Optical Society of America [OSA].) (b) Target fiber IPHT-256b1 with grating arrays. Microscope picture. The black bar is the scale for 50 μm. (c) Reflection spectrum of a fiber Bragg grating array with four gratings. (After Becker *et al.*, 2009. Used with permission from IEEE.)

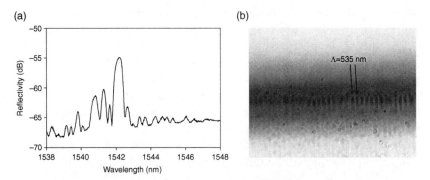

12.31 (a) Reflection spectrum of a Bragg reflector inscribed into a Hi-Ge socket, collapsed core fiber using phase mask and 193 nm, 10 ns excimer laser. Exposure conditions: 150 mJ/cm², 24 K pulses. (b) Optical microscope photo of the collapsed fiber core after the above grating inscription.

capillaries, have been presented for all-solid PCFs. Jin *et al.* (2007) attempted Bragg grating inscriptions using 248 nm excimer laser, in a hydrogenated, all-solid silica PCF containing into the core surrounding capillaries 1%wt Ge-doped silica glass, reaching ~ 1 dB reflectivities, and cladding mode inversion at the red-side of the spectrum. In another fashion, Bigot *et al.* (2009) designed an all-solid PCF, where the capillaries contained a low-photosensitivity, phosphosilicate glass, while the grating host material was a F/Ge co-doped core. In such an arrangement, the low photosensitivity glass infused into the capillaries did not impose absorption issues for side irradiation at the recording wavelength, allowing significant intensities reaching the fiber core and thus inducing refractive index modulation.

12.8.2   Studies on the issue of the side-illumination scattering during grating inscription

In all of the above Bragg and long-period grating demonstrations utilizing ultraviolet laser sources, there was not any special preparation related to the relative rotational position of the fibers with respect to the side-illuminating laser beam, therefore, the photosensitivity and refractive index engineering results presented there were not absolutely accurate or repeatable. The first study on the impact of the capillary structure and its rotational or transversal position along the fiber axis with respect to the laser beam on the Bragg grating recording processing was presented by Marshall *et al.* (2007).

Experimentally, Marshall *et al.* utilized 267 nm, femtosecond radiation coupled under different conditions into the PCF under exposure for different relative orientations and positions, and accordingly the 1.9 eV photoluminescence (Stathis and Kastner, 1984) induced was measured through the PCF fiber core. The same group used a finite element method for simulating the propagation of a small size beam for the aforementioned side-illuminating conditions evaluating an average field intensity (Fig. 12.32), and correlating such measurements with those obtained from photoluminescence experiments for different angles of rotation (Marshall *et al.*, 2007). Similar investigations were presented by Canning *et al.* (2008a) using finite difference time domain (FDTD) simulations for estimating the intensities reaching the PCF core for a variety of rotational angles during Bragg grating side-illumination, for 193 nm excimer laser radiation. Geernaert *et al.* (2008) inscribed Bragg gratings in a highly asymmetric germanium-doped, birefringent microstructured optical fiber using a 248 nm excimer laser that had been subjected to hydrogenation; thus, strong single-photon absorption effects dominated its photosensitivity. Geernaert *et al.* found that the rotational state of the fiber around its axis during this single-photon absorption inscription did not significantly affect the reflection strength of the Bragg grating, while having a greater impact in its wavelength.

In another approach the FDTD method was used again for illustrating in greater detail the average, maximum and minimum intensities reaching the core of all-silica standard and microstructured optical fibers (Fig. 12.33), leading to better insights with respect to the impact to the material photosensitivity (Pissadakis *et al.*, 2009a). In that study, the wavelength investigated was that of 248 nm, for 5 ps pulse duration and for propagation into hydrogenated silica glass fibers. The simulation results obtained depicted that while the mean intensities occurring in a side-illuminated PCF were

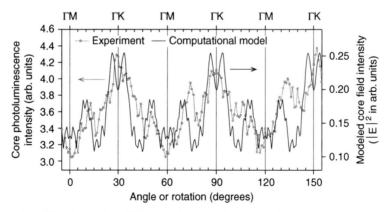

12.32 TE polarization PCF core photoluminescence intensity and computational model wholecore field intensity as a function of PCF rotation angle for the wavelength of 267 nm. (After Marshall *et al.*, 2007. Used with permission from Optical Society of America [OSA].)

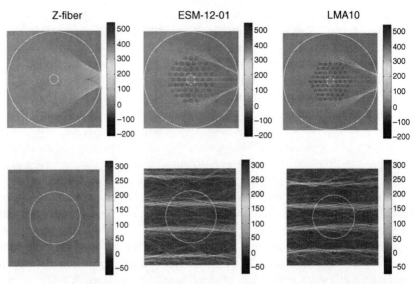

12.33 Simulation results of the spatial distribution of the Pointing vector S using the FDTD method for the cases of Z-fiber, ESM-12–01 and LMA10 for 0° rotation angle. The outer and inner white circles define the fiber cladding and core, respectively. The grey-scale bar corresponds to the scale of the values of the Poynting vector S measured in GW/cm². The dark circles define the spatial boundaries of the individual capillaries. (Upper row) Simulation of the side-illumination corresponding to views of the whole fiber areas with dimensions 126 μm × 126 μm. (Lower row) Simulation of the side-illumination corresponding to the vicinity of the fiber core of the above optical fibers, for core diameters 9 μm (Z-fiber), 12 μm (ESM-12–01) and 10 μm (LMA10).

30%–40% lower than those found in a standard all-silica fiber, the maximum intensities that can be found in the core of a side-illuminated PCF can even be 200% greater (Fig. 12.34).

Scattering can greatly affect the triggering of specific photosensitivity processes, especially those of two-photon absorption which are quadratically dependent upon the local intensities, and dominate the refractive index engineering. In this study, the standard and PCF all-silica fibers exposed reconfirmed that the Bragg gratings recorded in the PCF were of lower strength than those formed in the standard all-silica fiber (Fig. 12.35).

The issue of laser beam scattering for side-illumination during Bragg grating recording has been also addressed by infiltrating inside the fiber capillaries liquids of high transparency at the irradiation wavelength and of refractive index value close to that of silica; or by suitably designing the fiber capillaries' geometry for scattering less. This approach can lead to improved recording results, in the case where the fiber core is highly photosensitive, with linear absorption coefficient substantially higher than that of

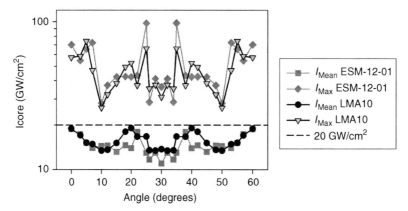

12.34 Intensity figure in the core area I_{core} versus rotational angle with respect to the ΓK axis for the ESM-12–01 and LMA10 microstructured optical fibers. I_{Mean}: Averaged intensity. I_{Max}: Maximum intensity. The I_{Mean} values have been evaluated for the circular areas presented in Fig. 12.33. The I_{Max} figure has been sampled for a square patch of 0.2 μm side, around the maximum points of intensity. The dash line refers to input pulse intensity as that measured before reaching the fiber complex.

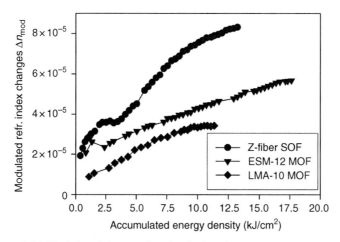

12.35 Modulated Δn_{mod} refractive index changes versus accumulated energy density for grating recordings in the standard optical fiber (Z-fiber), and in the PCFs, ESM-12–01 and LMA10. The refractive index data have been normalized to the corresponding mode confinement factor of each fiber.

the infiltrated liquid. Sørensen *et al.* (2007) infiltrated organic solvents into a hydrogenated, Ge-doped core PCF for reducing the deteriorating effect of extensive side-illuminating scattering during Bragg grating recording using 248 nm pulsed laser radiation. Methanol and n-heptane reduced scattering effects which allowed significant amount of photons to reach the Ge-doped

PCF core and induce significant refractive index changes, improving the grating strengths obtained by several decibels.

Following a different route, Baghdasaryan *et al.* (2011) used the FDTD method for simulating different candidate geometries with respect to the d/λ factor and hole diameter for reducing side-illumination scattering for 800 nm and a specific pulse duration. The results obtained in Baghdasaryan *et al.* (2011) provide a first indication that such optimization is possible without sacrificing the guiding properties of the custom-made PCF. Petrovic, Allsop, *et al.* (Allsop *et al.*, 2011; Petrovic and Allsop, 2010) examined the effect of the numerical aperture of the focusing optics used during grating inscription in PCF using femtosecond lasers, further studying the effect of the specific recording conditions in the annealing and spectral characteristics of the inscribed gratings, thus providing a correlation with the photosensitivity mechanisms triggered.

12.9 Laser machining of optical fibers

The modification and structuring of optical materials surfaces have been constantly progressed since the early years after the development of the laser sources; greatly benefiting from the emergence of high power lasers (Bäuerle, 2000). While progress in the laser surface structuring and ablation field has continued since the early 1980s for modifying the surface of planar samples (Endert *et al.*, 1995), the structuring of optical fibers progressed at a later time. The last is mainly attributed to three parameters: (i) optical fibers are species of small physical dimensions, needing precise translation and beam focusing systems for achieving machining over their limited end-face or curved cladding; (ii) optical fibers are mostly drawn from the highly transparent silicate glasses exhibiting poor surface etching quality using most of the nanosecond laser sources available at that period (Ihlemann *et al.*, 1992); (iii) the potential and advantages of optical fiber surface micro-structuring were not illustrated at that time. Therefore, the examples of such microstructuring were quite few, mainly focused on the melting of micro lenses onto fiber end-faces (Paek and Weaver, 1975; Presby *et al.*, 1990) using CO_2 lasers and the drilling of via holes into optical fibers or fiber-like structures (Gower, 2000, 2001).

The state-of-the-art in the optical fiber machining field was enriched after high spatial accuracy, ultraviolet nanosecond (mainly 157 nm excimer laser) and infrared femtosecond laser micromachining systems became available (Gattass and Mazur, 2008; Vainos *et al.*, 1996), in conjunction with the demonstration of promising results in the high quality etching of highly transparent optical materials (Dyer *et al.*, 2003; Jackson *et al.*, 1995; Jürgen *et al.*, 2007). Together, the emergence of PCFs (Knight *et al.*, 1996) and the lab-on-chip biosensing and microfluidic architectures (Haeberle and Zengerle, 2007; Oosterbroek and Berg, 2003) further underlined the advantages and possible functionalities that a laser-machined optical fiber can offer in developing advanced photonic (Monat *et al.*, 2007) and optofluidic (Psaltis *et al.*, 2006) systems. The cylindrical geometry of optical fibers and the small area of the fiber end-face frustrate the use of standard photolithographic techniques to be applied in structuring them. In addition, the use of laser beams for such surface modification offers distinct advantages compared to other approaches: via hole opening and drilling can be driven along the whole diameter of the fiber at reasonable etching times, while rotational machining such as lathing can be easily performed; real-time in-fiber spectral monitoring of the etching is also possible.

The favorable lasers for silicate glass optical fiber machining are these of 157 nm, excimer nanosecond and 800 nm, Ti:Sapphire femtosecond lasers. The 157 nm excimer laser radiation offers single-photon absorption interaction for the silicate glasses irrespective of dopands and impurities concentration, while etching rates achieved are of the order of a few tens of nanometers per pulse, solely dominated by volume absorption within shallow depths (Dyer *et al.*, 2003; Jürgen *et al.*, 2007; Li *et al.*, 2007). Etching quality is of highest quality for wide area pits even after several pulse exposures, where incubation effects are in favor of the etching quality. On the other hand, ablation of silicate glasses using 800 nm, or other near-infrared femtosecond lasers are dominated by multiphoton absorption triggering avalanche ionization and plasma formation effects over minimal volumes defined by the intensity pattern of the focused beam (Liu *et al.*, 1997; Sugioka *et al.*, 2005). Etching rates are generally greater than those obtained using ultraviolet sources, allowing machining in shorter time intervals of deeper trenches. The etching quality of the structures is highly dependent upon the specific parameters of the exposure, where heat accumulation effects (Eaton *et al.*, 2005; Schaffer *et al.*, 2003) can substantially affect the surface quality obtained. In general, surface or volume structuring using near-infrared femtosecond sources is accompanied by post-fabrication chemical conditioning or etching process for achieving improved roughness of the exposed areas or for etching deep channels into the materials volume (He *et al.*, 2010). Moreover, there are two main approaches in the laser machining of optical fibers: that of end-face and that of cladding surface/volume structuring. Examples of the two above machining fashions will be reviewed below, describing the etching conditions and structure design and the photonic functionality imparted into the etched optical fiber; the new approach of 'in-fiber' structuring will be also presented.

12.9.1   End-face machining

The most frequent structures formed onto the cleaved end-face of standard optical fibers were these of diffraction gratings, micro lenses, interferometers and Fabry-Perot cavities. These optical structures combine relative straightforward etching processing protocol, while providing significant photonic operations such as tailored guiding mode out-/in-coupling, spectral analysis and optical phase reading. The simplest micromachining approach followed by several groups was that of local heating of cleaved fiber ends using 10.6 μm CO_2 laser beams, for forming end-firing, convex micro lenses (Abdelrafik *et al.*, 2001; Presby *et al.*, 1990); however, this achieved rather low control over the shape of the lenses formed (Fig. 12.36a). Li *et al.* (2007) used a 157 nm excimer laser for etching diffraction gratings of 2 μm period, micro lenses and Mach–Zehnder step-shifted interferometer onto the end-face of multimode and standard telecom fibers. In this work energy densities greater than 4 J/cm² were employed for etching smooth pits and features onto the cleaved silicate glass fiber facet, reaching 5 μm flat depth pits, after single-photon absorption interaction.

Parabolic end-face lenses were formed using lathing machining schemes (Fig. 12.36b) while utilizing Schwarzschild reflecting objective and 157 nm radiation (Dou *et al.*, 2008). Excimer laser 157 nm radiation was also employed by Machavaram *et al.* (Machavaram, 2006) for end-face pit and sharp tip machining in several fiber types (including sapphire fibers), in combination with selective chemical etching post-processing for improving surface smoothness and achieve specific shaping. The same wavelength was used by Ran *et al.* (2009) for initially forming flat-top, high smoothness, ~ 30 μm deep pits on SMF-28 optical fiber, and then enclosing these etched cavities into the fiber line by fusion splicing, forming high finesse Fabry-Perot interferometers. This in-fiber Fabry-Perot etalon exhibited spectral fringe contrast of 30 dB, allowing strain measurements at high temperatures and modulation rates. Using a more complex splicing, cleaving and re-ablating process an end-face refractometer was presented by the above group providing resolution of 1130.887 nm/RI units (Zengling *et al.*, 2009).

Using different laser wavelength (785 nm) and pulse duration (125 fs) than nanosecond 157 nm excimer laser radiation, 33° angle (see Fig. 12.37), conical shape pits were formed into optical fibers for side-firing out-coupling

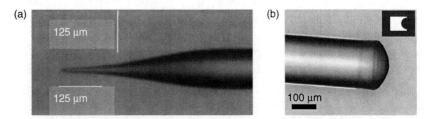

12.36 (a) Optical micro lens fabricated in an optical fiber using CO_2, 10.6 mm laser radiation, following a two-step process. (After Abdelrafik *et al.*, 2001.) (b) An optical microscopy image of a micro lens-tipped optical fiber using 157 nm, excimer laser radiation. (After Li *et al.*, 2007.)

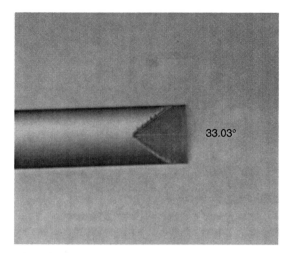

33.03°

12.37 Optical microscope image of the optical fiber tip microstructured by using a femtosecond laser. (After Sohn *et al.*, 2010. Used with permission from Optical Society of America [OSA].)

(Sohn *et al.*, 2010). The relatively rough ablated surface was polished to better quality using post-exposure arc discharging. Alternatively, by adapting a suitable length splicing buffer fiber to a PCF, and subsequent machining holes through the end-face of this buffer layer, Wang *et al.* (2010) presented a selective infiltration technique of the surrounding micro-capillary structure.

12.9.2   Structuring through the cladding

The structuring of the fiber end-face using laser beams led to the development of optical fiber structures exhibiting purely optical functionalities. The structuring through the cladding area of optical fibers using laser processes allowed the implementation of microfluidic functionalities in addition to the photonic structures developed until then. For a long period, the related state-of-the-art was rather primitive where simple etched holes (Gower, 2000) and jacket removal (Barnier *et al.*, 2000) only had been presented. Then the use of 157 nm and 800 nm-femtosecond laser micromachining systems prompted the versatile structuring of small size photonic devices such as optical fibers.

Deeper through-out trenches, of several tens of micrometers' length, were ablated in SMF-28 optical fibers using 157 nm excimer laser radiation by several authors for forming rectangular shape air cavities (Fig. 12.38a) and observing Fabry-Perot interferometric spectral modulations (Fig. 12.38b) in transmission and reflection mode (Machavaram *et al.*, 2007; Ran *et al.*, 2007,

(a)

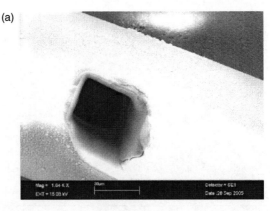

(b)

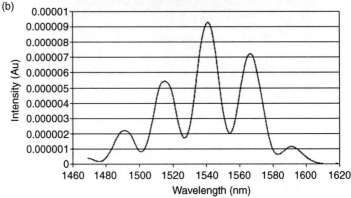

12.38 SEM picture of an ablated cavity through the diameter of the optical fiber at 25× 10⁴ J/m². Fabry–Perot interference fringes produced from cavities micromachined at an energy density of 25× 10⁴ J/m² in SMF-28 with 47.5 μm cavity length. (After Machavaram *et al.*, 2007.)

2008); others used 800 nm, 1 KHz femtosecond radiation for ablating similar long length cavities (Li *et al.*, 2008). Such demonstrations were also reported for in-fiber etalon etching in endlessly single mode PCFs allowing fringe probing for wider wavelength bandwidths (Ran *et al.*, 2007; Rao *et al.*, 2007). In most of these Fabry-Perot in-fiber interferometers a primary concern was related to the flatness and parallelism of the opposite walls constituting the oscillating cavities along the longitudinal axis of the fiber.

Zhou *et al.* (2007) used 800 nm femtosecond laser volume damage micromachnining, in combination with HF acid selective chemical etching for forming narrow slot openings and boreholes in the cladding of optical fibers, suitably positioned within the length of pre-inscribed photosensitive Bragg gratings. These inscribed slots allowed the infiltration of these microfluidic fiber Bragg gratings using fluids with different refractive indexes, developing refractometers of 945 nm/RIU sensitivity. A similar micromachined hole design in a standard optical fiber was used for refractometric measurements where the light was refracted and out-coupled from the cylindrical shape ring upon its refractive index contrast with the fiber core (Wang *et al.*, 2009a). Selective chemical etching of femtosecond laser-exposed Bragg gratings using a phase mask was used for forming deep periodic structures into a standard optical fiber (Fig. 12.39) and realizing infiltrated refractometer by monitoring the shift and power change of the Bragg peak scattered (Yang *et al.*, 2011).

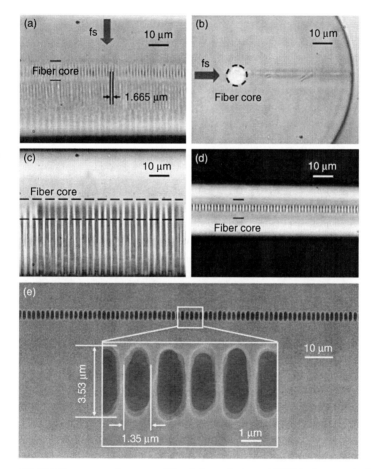

12.39 Microscope images of the fs laser-induced FBG before and after chemical etching. (a) Side view normal to the beam axis and (b) cross-section before etching. (c) Side view normal to the beam axis and (d) view from output side of laser beam after etching. (e) SEM image on the fiber grating surface after etching. (After Yang *et al.*, 2011. Used with permission from Optical Society of America [OSA].)

The laser etching of the cladding of microstructured and photonic crystal fibers provided a novel approach in the optimum exploitation of the micro-/optofluidic properties of these fibers. A first example was presented by van Braken *et al.* (2007), using an 800 nm femtosecond laser for drilling micro-channels reaching the photonic bandgap periodic structure of a hollow core PCF; also in the same work micro-channels were drilled in collapsed core fiber claddings. Such micro-channels were exploited for side pressurized C_2H_2 gas infiltration and optical absorption measurements at the 1.5 μm band. Trench etching of the fiber cladding using femtosecond infrared laser radiation was used as an alternative method for selectively filling of the capillaries of a solid core PCF, by removing an arc slice of the out-cladding and reaching capillary channels lying underneath, exposing them for liquid immersion (Yiping *et al.*, 2010). Controlled hole drilling with respect to the depth toward the fiber core was achieved using 800 nm femtosecond radiation, for developing long-period rejection filters in solid core all-silica PCF (Liu *et al.*, 2010).

CO_2 lasers emitting at 10.6 μm do not provide structuring of high spatial resolution and most applications were presented on thermal lensing formation onto fiber end-faces. Nonetheless, after the demonstrations of Davis *et al.* (1998) and Kakarantzas *et al.* (2002), there several other examples were reported that laser radiation could be used for fine cladding structuring of standard and microstructured fibers. Wang *et al.* (2006) used CO_2 laser radiation for etching grooves onto standard optical fiber claddings for forming strongly scattering LPGs, while later presenting the thermal cladding deformation in hollow core PCFs demonstrate the first LPG in such a fiber design (Wang *et al.*, 2008). Another group used a CO_2 laser beam in a lathing setup for inducing periodic, helical deformations outside of a standard optical fiber for realizing torque sensitive band rejection filters (Oh *et al.*, 2004).

12.9.3   'In-fiber' structuring using lasers

The use of laser radiation for modifying/etching the inner walls/capillaries of PCFs and MOFs has not been broadly investigated yet. In such an approach the laser beam passes through the cladding area and interacts selectively with the channels enclosed; the fiber core can be also modified. The main problems related to the in-fiber etching approach are those of extensive damage of the structure due to the explosive nature of ablation, deteriorating transmission signal and increasing losses; as well as the difficulty in controlling the exposure conditions and gas infiltration pressure for maintaining etching near the fiber core. Shujing *et al.* (2010) exploit the debris products of in-fiber ablation generated by 800 nm, femtosecond laser beam for selectively filling the capillaries of a solid core PCF and form a kind of damaged long-period gratings in those fibers; while achieving diffraction efficiencies of more than 20 dB over few (< 15) grating periods.

Violakis *et al.* (Violakis and Pissadakis, 2007a) attempted for the first time the etching of Bragg gratings into the capillaries of solid core PCFs infiltrated with fluorinated gas, using 193 nm excimer laser radiation. For simple gas infiltration inside the fiber capillaries under atmospheric pressure, some exposure results indicated that the existence of SF_6 inside the fiber capillaries can promote relief structure generation. While SF_6 exhibits high selectivity over silica glass etching, attacking preferentially silicon (Chuang, 1981), its photo-dissociation inside the PCF channels could offer controlled etching of the capillary walls (Batool *et al.*, 2004). However, the method did not lead to easily reproducible results, due to extensive redeposition of ablated glass and reaction products inside the capillaries (Batool *et al.*, 2004). Also, the uncontrolled SF_6 gas pressure conditions, where no buffer gas was used, while maintaining high pressures, increased recombination rates of excited species of fluorine.

Gratings fabricated in SF_6 infiltrated PCFs, exhibited increased strengths and faster saturation under specific exposure conditions by means of energy density and number of pulses (Fig. 12.40). Nonetheless, SEM pictures of cleaved fibers at the grating exposed section, revealed extensive debris redeposition and hole blocking. The last can be responsible for the saturation of the Bragg grating etching process earlier than inscriptions performed in empty PCF using identical conditions.

Recently, Sozzi *et al.* (forthcoming) have employed fluorinated organic liquids infiltrated inside MOFs and PCFs for etching Bragg and long-period gratings into their capillary structure using 248 nm, 5 ps laser radiation. By suitably tailoring the absorption coefficient of the infiltrated medium with

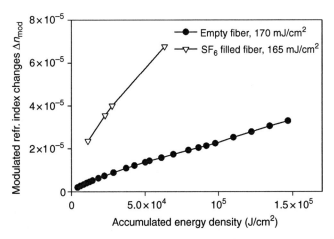

12.40 Bragg grating recording in empty and SF_6 gas infiltrated ESM-12–01 optical fiber using 193 nm, 10 ns excimer laser radiation and a phase mask.

(a)

(b)

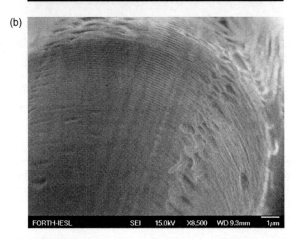

12.41 SEM pictures from relief Bragg gratings etched inside the capillaries of a grapefruit optical fiber, using 248 nm, 5 ps excimer laser radiation and fluorinated gas infiltration. (a) Far view of the etched fiber cross-section. (b) Zoomed view in a single capillary area. The MOF fiber used was provided by ACREO. The fiber has not been washed after etching process for detaching from the machined area debris deposition.

respect to the capillary structure and geometry, poor etching and non-repeatability issues reported by Violakis *et al.* can be minimized, allowing efficient in-fiber periodic structuring (Fig. 12.41).

12.10   Future trends and prospects

The methods, results and developments presented above, primarily related to optical fiber photosensitivity and the mirror-field of laser structuring, reveal immense progress with respect to the founding experiments of Weeks, Primak and Hill, reported more than three decades ago. This great but unpredictable progress was fertilized by three basic parameters: the continuous development of new and efficient laser sources and processing methods, the deeper understanding of the underlying laser–matter interactions, and the evolution of the optical fiber from a simple optical cable to a versatile and efficient photonic device platform. While the two former aforementioned parameters catalyzed the material science of laser photosensitivity and structuring, the last prompted the innovation effort for integrating the results emerging from fundamental and applied studies into the development of real photonic devices.

Reviewing the results over the last three decades, one may conclude that the optical fiber photosensitivity field has reached maturity mainly by solving two cross-related problems: the inscription of large and controllable refractive index changes into almost any fiber available, irrespective of material, dopand concentration and pre-conditioning, by using fast and reasonably priced techniques, suitable for mass production applications. The last statement is quite accurate and describes well the case of standard optical fibers, where the laser methods and yields for the inscription of Bragg and long-period gratings have fulfilled even the most demanding industrial or academic expectation. The use of high photon energy (ultraviolet) and/or ultrafast pulse duration (femtosecond and picosecond) laser sources can bridge the bandgap of most optical materials by exploiting specific defect annihilation or multiphoton absorption and correlated effects, allowing inscription of exploitable refractive index changes. Notably, the fact that the laser sources needed for performing such material modifications are commercially available providing high intensity and energy density outputs, at relatively modest costs, prompts the broader exploitation and improvement of the photosensitivity approaches investigated.

Alternative methods leading to the inscription of refractive index changes into high bandgap optical materials, exhibiting specific optical, annealing, radiation resistance or mechanical properties can be issues for future investigation. These methods may include the use of more than two laser sources irradiating the fiber sample simultaneously or with specific pulse delays, for triggering or suppressing photosensitivity mechanisms and thus controlling the type and properties of refractive index changes inscribed (Obata *et al.*, 2002). Another approach of further interest is that of filamentation, for inscribing Bragg or long-period gratings in undoped silicate or other soft glass fibers, boosting diffraction efficiency and thermal stability of the reflectors (Bernier *et al.*, 2011). The topic of thermally regenerated and photochemically induced gratings can also be further investigated/improved with respect to the amplification characteristics of the gratings and the underlying physiochemical mechanisms involved (Fokine, 2002a; Lindner *et al.*, 2011). Such regenerated gratings may serve as durable radiation sensors (Gusarov *et al.*, 2007) operating in harsh environments. Another route toward the improvement of the thermal stability and radiation behavior of fiber grating structures can be that of the adoption of alternatives to Ge ion dopands for modifying the material properties of the glasses used for fiber drawing, and accordingly the type of defects and glass transformation properties of the fibers (Butov *et al.*, 2006; Shen *et al.*, 2004).

Even though the grating inscription processing in standard optical fibers is based on well-established techniques and the photosensitivity mechanisms are understood to a great extent, there is a lot of room for new investigations and further improvements related to the grating inscription in microstructured and PCFs using laser radiation. The main issue arising from the capillary structure surrounding the solid or hollow guiding core is that of beam scattering for side-illuminations during grating recording. This important issue has been addressed by several groups while employing a number of simulation methods and experimental techniques (Canning *et al.*, 2008a; Geernaert *et al.*, 2008; Marshall *et al.*, 2007; Pissadakis *et al.*, 2009a). All these studies have agreed as to the deleterious role of the capillary structure constituting the PCFs and MOFs; however, they have not yet addressed the exact growth of the induced refractive index changes for different photosensitivity models (single- or multiphoton), and the overlap of these index changes with the guiding modes in the fiber core. Moreover, recently there has been effort made in the optimization of the Bragg grating recording conditions, either by liquid filling the fiber capillaries, or by designing a fiber capillary structure that results in reduced scattering during side-illumination, without significantly affecting the guiding properties (Baghdasaryan *et al.*, 2011). The optimization of the combination between the PCF/MOF geometry and the recording wavelength and pulse duration for maintaining high photon fluencies reaching the fiber core is a hot topic that will address a challenging technical problem with direct practical impact. Such studies may also include the design of PCF/MOF with tailored spatial characteristics, dopand concentration within the fiber core or cladding for maximizing the overlap with the side-illuminating beam, for alleviating inscription complexity and repeatability obstacles. The infiltration of highly photosensitive polymeric matrices inside all-silica glass PCFs, and the recording of Bragg gratings in those infiltrated capillaries, and not into the highly transparent glass, constitutes a promising approach which can be directly exploited in the future for rendering the grating inscription in such fibers a rather easier task (Kakarantzas *et al.*, 2011).

The structuring of optical fibers using deep, ultraviolet and infrared femtosecond lasers exhibits greater prospects with respect to novel future developments. Among others, there are two future laser machining directions that may lead to the development of novel optical fiber devices for sensing and actuating applications. First, the laser micro-/nanostructuring of standard and microstructured optical fibers, for generating complex and functional fluidic and diffracting elements in the cladding and core areas of those. Second, the cladding of standard but mostly of microstructured optical fibers can constitute a 'free-space playground' where the laser etching techniques described above can be used for inscribing relief or perforated structures and channels which can either allow controlled liquid/gas infiltration, fluid-trapping or localized mode perturbation. Such photostructuring engineering, together with photosensitivity changes inscribed and other selective chemical etching methods, will allow the realization of complex and versatile optofluidic circuits, which can constitute the 'lab-in-fiber' approach, following the example of the 'lab-on-chip'. Such 'lab-in-fiber' photonic devices can find sensing applications in biology, instrumentation or medicine, implementing more than one functionality into a single photonic element.

Other advanced laser processing approaches can include new etching techniques such as laser-induced backside wet etching (Böhme *et al.*, 2002) for structuring the inner walls of MOFs and PCFs, without affecting the out-cladding area, by infiltrating the fiber capillaries using highly absorbing liquids at the wavelength of the exposure. Other similar material sputtering techniques that can be applied in the case of MOFs and PCFs are those of laser chemical vapour deposition (LCVD) and laser-induced forward transfer (LIFT) for creating relief hetero-material structures inside their capillaries (Arnold and Pique, 2007; Klini *et al.*, 2008).

Non-ablative methods are described in the use of femtosecond laser for inscribing relief (Do Lim *et al.*, 2009) or photo-polymerizing organic–inorganic relief optical elements onto the end-face of optical fibers may also attract academic and industrial interest. Such 'on-fiber' hybrid structures can be photo-polymerized by self-guiding processes (Soppera *et al.*, 2009) or by utilizing external femtosecond laser beams for creating a variety of guiding, refracting and diffractive elements by multiphoton absorption processing, at sub-wavelength resolutions (Malinauskas *et al.*, 2011; Williams *et al.*, 2011a). By applying specific chemical affinity processes onto these organic–inorganic structures, new kinds of chemical and biological sensing probes can be realized, exhibiting the advantages and functionalities of planar geometries into the cylindrical optical fiber topology.

12.11 Conclusions

The use of laser radiation for photosensitization, refractive index engineering and structuring of optical fibers has been one of the fastest growing topics in photonic science and technology during the last three decades. The majority of the scientific background and technological innovation accumulated during this period has found several commercialization paths in the service of everyday applications ranging from telecommunications and sensing to health and metrology. The refractive index engineering of glass optical fiber that dominated the field has reached maturity, leading to an increase of the photo-induced refractive index yields by more than four orders of magnitude, reaching levels of $\times 10^{-2}$ or greater. Simultaneously, new technological challenges are opening in the mirror-field of surface micro-/nanostructuring and ablation of standard and microstructured optical fibers, using high intensity and photon energy laser beams, realizing and embedding microfluidic operations into the fiber geometry. The critical idea is to judiciously combine and complement the fields of laser processing, photosensitivity and surface structuring in order to bring a powerful fabrication tool into the realization of future fiber optic and photonic devices.

12.12   Acknowledgments

The author would like to acknowledge the contribution of his colleagues Maria Konstantaki, Georgios Tsibidis, Paul Childs and Aashia Rahman for their scientific work in the fields of fiber photosensitivity and optical fiber structuring. Moreover, the author would like to thank the vast work of the talented PhD and MSc students Christos Pappas, Georgios Violakis, Michalis Livitziis, Irene Michelakaki and Michele Sozzi for their valuable findings during their dissertation studies. Several parts of the work presented above have been funded by research projects of the European Commission, while fiber and glass samples have been kindly provided by Sumitomo Europe, Schott USA, IPHT-Jena, Fibercore Ltd and ACREO SA. Finally, the author would like to thank the Optical Society of America (OSA) and the Institution of Engineering and Technology (IET) for kindly providing copyrighted material included in this chapter.

12.13   References

Aashia, R., Madhav, K. V., Ramamurty, U. and Asokan, S. (2009), 'Nanoindentation study on germania-doped silica glass preforms: Evidence for the compaction-densification model of photosensitivity', *Optics Letters*, **34**, 2414–2416.

Abdelrafik, M., Renaud, B. and Frederic Van, L. (2001), 'Two-step process for micro-lens-fibre fabrication using a continuous CO_2 laser source', *Journal of Optics A: Pure and Applied Optics*, **3**, 291.

Albert, J., Fokine, M. and Margulis, W. (2002), 'Grating formation in pure silica-core fibers', *Optics. Letters.*, **27**, 809–811.

Albert, J., Hill, K. O., Johnson, D. C., Bilodeau, F., Mihailov, S. J., Borrelli, N. F. and Amin, J. (1999), 'Bragg gratings in defect-free germanium-doped optical fibers', *Optics Letters*, **24**, 1266–1268.

Albert, J., Malo, B., Hill, K. O., Bilodeau, F., Johnson, D. C. and Theriault, S. (1995), 'Comparison of one-photon and 2-photon effects in the photosensitivity of germanium-doped silica optical fibers exposed to intense arf excimer-laser pulses', *Applied Physics Letters*, **67**, 3529–3531.

Albert, J., Schulzgen, A., Temyanko, V. L., Honkanen, S. and Peyghambarian, N. (2006), 'Strong Bragg gratings in phosphate glass single mode fiber', *Applied Physics Letters*, **89**, 101127.

Allsop, T., Kalli, K., Zhou, K., Smith, G. N., Komodromos, M., Petrovic, J., Webb, D. J. and Bennion, I. (2011), 'Spectral characteristics and thermal evolution of long-period gratings in photonic crystal fibers fabricated with a near-IR radiation femtosecond laser using point-by-point inscription', *Journal of the Optical Society of America B*, **28**, 2105–2114.

Archambault, J.-L. (1994), 'Photorefractive gratings in optical fibres' [Online], University of Southampton.

Archambault, J.-L., Reekie, L. and Russell, P. S. J. (1993), '100% reflectivity Bragg reflectors produced in optical fibres by single excimer laser pulses', *Electronics Letters*, **29**, 453–455.

Armitage, J. R. (1993). Fiber bragg reflectors written at 262nm using a frequency quadrupled diode-pumped Nd3+-YLF laser. *Electronics Letters*, **29**, 1181–1183.

Arnold, C. B. and Pique, A. (2007), 'Laser direct-write processing', *MRS Bulletin*, **32**, 9–12.

Atkins, R. M., Mizrahi, V. and Erdogan, T. (1993), '248 nm induced vacuum UV spectral changes in optical-fiber preform cores: Support for a color center model of photosensitivity', *Electronics Letters*, **29**, 385–387.

Baghdasaryan, T., Geernaert, T., Berghmans, F. and Thienpont, H. (2011), 'Geometrical study of a hexagonal lattice photonic crystal fiber for efficient femtosecond laser grating inscription', *Optics Express*, **19**, 7705–7716.

Bandyopadhyay, S., Canning, J., Stevenson, M. and Cook, K. (2008), 'Ultrahigh-temperature regenerated gratings in boron-codoped germanosilicate optical fiber using 193 nm', *Optics Letters*, **33**, 1917–1919.

Barnier, F., Dyer, P. E., Monk, P., Snelling, H. V. and Rourke, H. (2000), 'Fibre optic jacket removal by pulsed laser ablation', *Journal of Physics D: Applied Physics*, **33**, 757–759.

Batool, S., Parviz, P. and Mohamad Amin, B. (2004), 'SF 6 decomposition and layer formation due to excimer laser photoablation of SiO_2 surface at gas–solid system', *Journal of Physics D: Applied Physics*, **37**, 3402–3408.

Bäuerle, D. (2000), *Laser processing and chemistry*. Berlin; London: Springer.

Bazylenko, M. V., Moss, D. and Canning, J. (1998). Complex photosensitivity observed in germanosilica planar waveguides. *Optics Letters*, **23**, 697–699.

Becker, M., Bergmann, J., Brückner, S., Franke, M., Lindner, E., Rothhardt, M. W. and Bartelt, H. (2008), 'Fiber Bragg grating inscription combining DUV sub-picosecond laser pulses and two-beam interferometry', *Optics Express*, **16**, 19169–19178.

Becker, M., Fernandes, L., Rothhardt, M., Bruckner, S., Schuster, K., Kobelke, J., Frazao, O., Bartelt, H. and Marques, P. V. S. (2009), 'Inscription of fiber Bragg grating arrays in pure silica suspended core fibers', *IEEE Photonics Technology Letters*, **21**, 1453–1455.

Bellouard, Y., Colomb, T., Depeursinge, C., Dugan, M., Said, A. A. and Bado, P. (2006), 'Nanoindentation and birefringence measurements on fused silica specimen exposed to low-energy femtosecond pulses', *Optics Express*, **14**, 8360–8366.

Bernier, M., Faucher, D., Vallée, R., Saliminia, A., Androz, G., Sheng, Y. and Chin, S. L. (2007), 'Bragg gratings photoinduced in ZBLAN fibers by femtosecond pulses at 800 nm', *Optics Letters*, **32**, 454–456.

Bernier, M., Gagnon, S. and Vallée, R. (2011), 'Role of the 1D optical filamentation process in the writing of first order fiber Bragg gratings with femtosecond pulses at 800nm', *Optical Materials Express*, **1**, 832–844.

Bernier, M., Sheng, Y. and Vallée, R. (2009), 'Ultrabroadband fiber Bragg gratings written with a highly chirped phase mask and infrared femtosecond pulses', *Optics Express*, **17**, 3285–3290.

Beugin, V., Bigot, L., May, P., Lancry, M., Quiquempois, Y., Douay, M., Melin, G., Fleureau, A., Lempereur, S. and Gasca, L. (2006), 'Efficient Bragg gratings in phosphosilicate and germanosilicate photonic crystal fiber', *Applied Optics*, **45**, 8186–8193.

Bigot, L., Bouwmans, G., Quiquempois, Y., Le Rouge, A., Pureur, V., Vanvincq, O. and Douay, M. (2009), 'Efficient fiber Bragg gratings in 2D all-solid photonic bandgap fiber', *Optics Express*, **17**, 10105–10112.

Böhme, R., Braun, A. and Zimmer, K. (2002), 'Backside etching of UV-transparent materials at the interface to liquids', *Applied Surface Science*, **186**, 276–281.

Born, M. and Wolf, E. (1999), *Principles of optics: Electromagnetic theory of propagation, interference and diffraction of light.* Cambridge: Cambridge University Press.

Borrelli, N. F., Allan, D. C. and Modavis, R. A. (1999), 'Direct measurement of 248- and 193-nm excimer-induced densification in silica–germania waveguide blanks', *Journal of the Optical Society of America B*, **16**, 1672–1679.

Borrelli, N. F., Smith, C., Allan, D. C. and Seward, T. P. (1997), 'Densification of fused silica under 193-nm excitation', *Journal of the Optical Society of America B*, **14**, 1606–1615.

Boyd, R. W. (2003), *Nonlinear optics.* Amsterdam and London: Academic Press.

Brambilla, G., Fotiadi, A. A., Slattery, S. A. and Nikogosyan, D. N. (2006), 'Two-photon photochemical long-period grating fabrication in pure-fused-silica photonic crystal fiber', *Optics Letters*, **31**, 2675–2677.

Butov, O. V., Dianov, E. M. and Golant, K. M. (2006), 'Nitrogen-doped silica-core fibres for Bragg grating sensors operating at elevated temperatures', *Measurement Science and Technology*, **17**, 975–979.

Canagasabey, A. and Canning, J. (2003), 'UV lamp hypersensitisation of hydrogen-loaded optical fibres', *Optics Express*, **11**, 1585–1589.

Canning, J. (2008), 'Fibre gratings and devices for sensors and lasers', *Laser & Photonics Reviews*, **2**, 275–289.

Canning, J., Bandyopadhyay, S., Stevenson, M., Biswas, P., Fenton, J. and Aslund, M. (2009), 'Regenerated gratings', *Journal of the European Optical Society – Rapid Publications*, **4**, 09052.

Canning, J., Groothoff, N., Buckley, E., Ryan, T., Lyytikainen, K. and Digweed, J. (2003), 'All-fibre photonic crystal distributed Bragg reflector (PC-DBR) fibre laser', *Optics Express*, **11**, 1995–2000.

Canning, J., Groothoff, N., Cook, K., Martelli, C., Pohl, A., Holdsworth, J., Bandyopadhyay, S. and Stevenson, M. (2008a), 'Gratings in structured optical fibres', *Laser Chemistry*. **2008**, 239417.

Canning, J., Moss, D., Aslund, M. and Bazylenko, M. (1998), 'Negative index gratings in germanosilicate planar waveguides', *Electronics Letters*, **34**, 366–367.

Canning, J., Sceats, M. G., Inglis, H. G. and Hill, P. (1995), 'Transient and permanent gratings in phosphosilicate optical fibers produced by the flash condensation technique', *Optics Letters*, **20**, 2189–2191.

Canning, J., Stevenson, M., Bandyopadhyay, S. and Cook, K. (2008b), 'Extreme silica optical fibre gratings', *Sensors*, **8**, 6448–6452.

Chan, J. W., Huser, T. R., Risbud, S. H., Hayden, J. S. and Krol, D. M. (2003), 'Waveguide fabrication in phosphate glasses using femtosecond laser pulses',. *Applied Physics Letters*, **82**, 2371–2373.

Chan, J. W., Huser, T., Risbud, S. and Krol, D. M. (2001), 'Structural changes in fused silica after exposure to focused femtosecond laser pulses', *Optics Letters*, **26**, 1726–1728.

Chao, L., Reekie, L. and Ibsen, M. (1999), 'Grating writing through fibre coating at 244 and 248 nm', *Electronics Letters*, **35**, 924–926.

Chen, K. P., Herman, P. R. and Tam, R. (2002a), 'Strong fiber Bragg grating fabrication by hybrid 157- and 248-nm laser exposure', *IEEE Photonics Technology Letters*, **14**, 170–172.

Chen, K. P., Herman, P. R., Zhang, J. and Tam, R. (2001), 'Fabrication of strong long-period gratings in hydrogen-free fibers with 157-nm F-2-laser radiation', *Optics Letters*, **26**, 771–773.

Chen, K. P., Wei, X. and Herman, P. R. (2002b), 'Strong 157 nm F2-laser photosensitivity-locking of hydrogen-loaded telecommunication fibre for 248 nm fabrication of long-period gratings', *Electronics Letters*, **38**, 17–19.

Cho, S.-H., Kumagai, H., Midorikawa, K. and Obara, M. (1999), 'Fabrication of double cladding structure in optical multimode fibers using plasma channeling excited by a high-intensity femtosecond laser', *Optics Communications*, **168**, 287–295.

Chuang, T. J. (1981), 'Multiple photon excited SF[sub 6] interaction with silicon surfaces', *Journal of Chemical Physics*, **74**, 1453–1460.

Cohen, A. J. and Smith, H. L. (1958), 'Ultraviolet and infrared absorption of fused Germania', *Journal of Physics and Chemistry of Solids*, **7**, 301–306.

Cook, K., Pohl, A. A. P. and Canning, J. (2008), 'High-temperature type IIa gratings in 12-ring photonic crystal fibre with germanosilicate core', *Journal of the European Optical Society – Rapid Publications*, **3**, 08031.

Cordier, P. (1997), 'Evidence by transmission electron microscopy of densification associated to Bragg grating photoimprinting in germanosilicate optical fibers', *Applied Physics Letters*, **70**, 1204.

David, L. G. (2011), 'Trapped-electron centers in pure and doped glassy silica: A review and synthesis', *Journal of Non-Crystalline Solids*, **357**, 1945–1962.

Davis, D. D., Gaylord, T. K., Glytsis, E. N., Kosinski, S. G., Mettler, S. C. and Vengsarkar, A. M. (1998), 'Long-period fibre grating fabrication with focused CO_2 laser pulses', *Electronics Letters*, **34**, 302–303.

Davis, K. M., Miura, K., Sugimoto, N. and Hirao, K. (1996), 'Writing waveguides in glass with a femtosecond laser', *Optics Letters*, **21**, 1729–1731.

Dianov, E. M., Golant, K. M., Khrapko, R. R., Kurkov, A. S., Leconte, B., Douay, M., Bernage, P. and Niay, P. (1997a), 'Grating formation in a germanium free silicon oxynitride fibre', *Electronics Letters*, **33**, 236–238.

Dianov, E. M., Plotnichenko, V. G., Koltashev, V. V., Pyrkov, Y. N., Ky, N. H., Limberger, H. G. and Salathe, R. P. (1997b), 'UV-irradiation-induced structural transformation of germanosilicate glass fiber', *Optics Letters*, **22**, 1754–1756.

Dianov, E. M., Starodubov, D. S. and Frolov, A. A. (1996), 'UV argon laser induced luminescence changes in germanosilicate fibre preforms', *Electronics Letters*, **32**, 246–247.

Do Lim, S., Kim, J. G., Lee, K., Lee, S. B. and Kim, B. Y. (2009), 'Fabrication of a highly efficient core-mode blocker using a femtosecond laser ablation technique', *Optics Express*, **17**, 18449–18454.

Dong, L. and Liu, W. F. (1997), 'Thermal decay of fiber Bragg gratings of positive and negative index changes formed at 193 nm in a boron-codoped germanosilicate fiber'. *Applied. Optics.*, **36**, 8222–8226.

Dong, L., Archambault, J. L., Reekie, L., Russell, P. S. J. and Payne, D. N. (1995), 'Photoinduced absorption change in germanosilicate preforms: Evidence for the color-center model of photosensitivity', *Applied Optics*, **34**, 3436–3440.

Dong, L., Liu, W. F. and Reekie, L. (1996), 'Negative-index gratings formed by a 193-nm excimer laser', *Optics Letters*, **21**, 2032–2034.

Dong, L., Pinkstone, J., Russell, P. S. J. and Payne, D. N. (1994), 'Ultraviolet absorption in modified chemical vapor deposition preforms', *Journal of the Optical Society of America B*, **11**, 2106–2111.

Dou, J., Li, J., Herman, P. R., Aitchison, J. S., Fricke-Begemann, T., Ihlemann, J. and Marowsky, G. (2008), 'Laser machining of micro-lenses on the end face of single-mode optical fibers', *Applied Physics A: Materials Science and Processing*, **91**, 591–594.

Douay, M., Xie, W. X., Taunay, T., Bernage, P., Niay, P., Cordier, P., Poumellec, B., Dong, L., Bayon, J. F., Poignant, H. and Delevaque, E. (1997), 'Densification involved in the UV-based photosensitivity of silica glasses and optical fibers', *Journal of Lightwave Technology*, **15**, 1329–1342.

Dragomir, A. (2002), 'Two-photon absorption properties of commercial fused silica and germanosilicate glass at 264 nm', *Applied Physics Letters*, **80**, 1114.

Dragonmir, A., McIinerney, J. G. and Nikogosyan, D. N. (2002), 'Femtosecond measurements of two-photon absorption coefficients at? = 264 nm in glasses, crystals, and liquids', *Applied. Optics*, **41**, 4365–4376.

Dragomir, A., Nikogosyan, D. N., Zagorulko, K. A., Kryukov, P. G. and Dianov, E. M. (2003), 'Inscription of fiber Bragg gratings by ultraviolet femtosecond radiation', *Optics Letters*, **28**, 2171–2173.

Du, D. (1994), 'Laser-induced breakdown by impact ionization in $SiO2$ with pulse widths from 7 ns to 150 fs', *Applied Physics Letters*, **64**, 3071.

Dubov, M., Bennion, I., Slattery, S. A. and Nikogosyan, D. N. (2005), 'Strong long-period fiber gratings recorded at 352 nm', *Optics Letters*, **30**, 2533–2535.

Dyer, P. E., Farley, R. J., Giedl, R., Byron, K. C. and Reid, D. (1994), 'High reflectivity fiber gratings produced by incubated damage using a 193nm ARF laser',. *Electronics Letters*, **30**, 860–862.

Dyer, P. E., Johnson, A.-M., Snelling, H. V. and Walton, C. D. (2001), 'Measurement of 157 nm F 2 laser heating of silica fibre using an in situ fibre Bragg grating', *Journal of Physics D: Applied Physics*, **34**, L109–L112.

Dyer, P. E., Johnson, A. M. and Walton, C. D. (2008), 'Inscription of fibre Bragg gratings in non-sensitised fibres using VUV F2 laser radiation', *Optics Express*, **16**, 19297–19303.

Dyer, P. E., Maswadi, S. M., Walton, C. D., Ersoz, M., Fletcher, P. D. I. and Paunov, V. N. (2003), '157-nm laser micromachining of N-BK7 glass and replication for microcontact printing', *Applied Physics A: Materials Science and Processing*, **77**, 391–394.

Eaton, S., Zhang, H., Herman, P., Yoshino, F., Shah, L., Bovatsek, J. and Arai, A. (2005), 'Heat accumulation effects in femtosecond laser-written waveguides with variable repetition rate', *Optics Express*, **13**, 4708–4716.

Ebeling, P., Ehrt, D. and Friedrich, M. (2002), 'X-ray induced effects in phosphate glasses', *Optical Materials*, **20**, 101–111.

Ebendorff-Heidepriem, H. and Ehrt, D. (1996), 'UV radiation effects in fluoride phosphate glasses', *Journal of Non-Crystalline Solids*, **196**, 113–117.

Ebendorff-Heidepriem, H., Riziotis, C. and Taylor, E. R. (2002), 'Novel photosensitive glasses', *Glass Science and Technology*, **75**, 54–59.

Eggleton, B. J., Westbrook, P. S., White, C. A., Kerbage, C., Windeler, R. S. and Burdge, G. L. (2000), 'Cladding-mode-resonances in air-silica microstructure optical fibers', *Journal of Lightwave Technology*, **18**, 1084–1100.

Eggleton, B. J., Westbrook, P. S., Windeler, R. S., Spalter, S. and Strasser, T. A. (1999), 'Grating resonances in air-silica microstructured optical fibers', *Optics Letters*, **24**, 1460–1462.

Ehrt, D., Carl, M., Kittel, T., Müller, M. and Seeber, W. (1994), 'High-performance glass for the deep ultraviolet range', *Journal of Non-Crystalline Solids*, **177**, 405–419.

Ehrt, D., Ebeling, P. and Natura, U. (2000), 'UV Transmission and radiation-induced defects in phosphate and fluoride-phosphate glasses', *Journal of Non-Crystalline Solids*, **263–264**, 240–250.

Endert, H., Pätzel, R. and Basting, D. (1995), 'Excimer laser: A new tool for precision micromachining', *Optical and Quantum Electronics*, **27**, 1319–1335.

Erdogan, T. (1997), 'Fiber grating spectra', *Journal of Lightwave Technology*, **15**, 1277–1294.

Fertein, E., Przygodzki, C., Delbarre, H., Hidayat, A., Douay, M. and Niay, P. (2001), 'Refractive-index changes of standard telecommunication fiber through exposure to femtosecond laser pulses at 810 cm', *Applied Optics*, **40**, 3506–3508.

Fiori, C. and Devine, R. A. B. (1986), 'Evidence for a wide continuum of polymorphs in a-SiO$_2$', *Physical Review B*, **33**, 2972–2974.

Fisette, B. (2006), 'Three-dimensional crystallization inside photosensitive glasses by focused femtosecond laser', *Applied Physics Letters*, **88**, 091104.

Fletcher, L. B., Witcher, J. J., Reichman, W. B., Arai, A., Bovatsek, J. and Krol, D. M. (2009), 'Changes to the network structure of Er-Yb doped phosphate glass induced by femtosecond laser pulses', *Journal of Applied Physics*, **106**, 083107.

Fokine, M. (2002a), 'Formation of thermally stable chemical composition gratings in optical fibers', *Journal of the Optical Society of America B*, **19**, 1759–1765.

Fokine, M. (2002b), 'Growth dynamics of chemical composition gratings in fluorine-doped silica optical fibers', *Optics Letters*, **27**, 1974–1976.

Fonjallaz, P. Y., Limberger, H. G., Salathé, R. P., Cochet, F. and Leuenberger, B. (1995), 'Tension increase correlated to refractive-index change in fibers containing UV-written Bragg gratings', *Optics Letters*, **20**, 1346–1348.

Friebele, E. (1976), 'Drawing-induced defect centers in a fused silica core fiber', *Applied Physics Letters*, **28**, 516.

Frocht, M. M. 1941 (1949), *Photoelasticity*, New York: John Wiley.

Fu, L. B., Marshall, G. D., Bolger, J. A., Steinvurzel, P., Magi, E. C., Withford, M. J. and Eggleton, B. J. (2005), 'Femtosecond laser writing Bragg gratings in pure silica photonic crystal fibres', *Electronics Letters*, **41**, 638–640.

Gagné, M. and Kashyap, R. (2010), 'New nanosecond Q-switched Nd:VO4 laser fifth harmonic for fast hydrogen-free fiber Bragg gratings fabrication', *Optics Communications*, **283**, 5028–5032.

Gattass, R. R. and Mazur, E. (2008), 'Femtosecond laser micromachining in transparent materials', *Nature Photonics*, **2**, 219–225.

Geernaert, T., Nasilowski, T., Chah, K., Szpulak, M., Olszewski, J., Statkiewicz, G., Wojcik, J., Poturaj, K., Urbanczyk, W., Becker, M., Rothhardt, M., Bartelt, H., Berghmans, F. and Thienpont, H. (2008), 'Fiber Bragg gratings in germanium-doped highly birefringent microstructured optical fibers', *Photonics Technology Letters, IEEE*, **20**, 554–556.

Glezer, E. N., Milosavljevic, M., Huang, L., Finlay, R. J., Her, T. H., Callan, J. P. and Mazur, E. (1996), 'Three-dimensional optical storage inside transparent materials', *Optics Letters*, **21**, 2023–2025.

Gower, M. (2001), 'Excimer laser microfabrication and micromachining', *RIKEN Review*, **32**, 50–56.

Gower, M. C. (2000), 'Industrial applications of laser micromachining', *Optics Express*, **7**, 56–67.

Griscom, D. L. (2011), 'On the natures of radiation-induced point defects in GeO2-SiO2 glasses: Reevaluation of a 26-year-old ESR and optical data set', *Optical Materials Express*, **1**, 400–412.

Grobnic, D., Mihailov, S. J. and Smelser, C. W. (2006), 'Femtosecond IR laser inscription of Bragg gratings in single- and multimode fluoride fibers', *Photonics Technology Letters, IEEE*, **18**, 2686–2688.

Grobnic, D., Mihailov, S. J., Walker, R. B., Smelser, C. W., Lafond, C. and Croteau, A. (2007), 'Bragg gratings made with a femtosecond laser in heavily doped Er–Yb phosphate glass fiber', *Photonics Technology Letters, IEEE*, **19**, 943–945.

Groothoff, N. and Canning, J. (2004), 'Enhanced type IIA gratings for high-temperature operation', *Optics Letters*, **29**, 2360–2362.

Groothoff, N., Canning, J., Buckley, E., Lyttikainen, K. and Zagari, J. (2003), 'Bragg gratings in air-silica structured fibers', *Optics Letters*, **28**, 233–235.

Groothoff, N., Canning, J., Ryan, T., Lyytikainen, K. and Inglis, H. (2005), 'Distributed feedback photonic crystal fibre (DFB-PCF) laser', *Optics Express*, **13**, 2924–2930.

Gusarov, A., Vasiliev, S., Medvedkov, O., Mckenzie, I. and Berghmans, F. (2007), 'Stabilization of fiber Bragg gratings against gamma radiation. Radiation and Its Effects on Components and Systems', RADECS 2007. 9th European Conference, 10–14 September 2007, pp. 1–8.

Haeberle, S. and Zengerle, R. (2007), 'Microfluidic platforms for lab-on-a-chip applications', *Lab on a Chip*, **7**, 1094–1110.

Hand, D. P. and Russell, P. S. (1990), 'Photoinduced refractive-index changes in germanosilicate fibers', *Optics Letters*, **15**, 102–104.

He, F., Cheng, Y., Xu, Z., Liao, Y., Xu, J., Sun, H., Wang, C., Zhou, Z., Sugioka, K., Midorikawa, K., Xu, Y. and Chen, X. (2010), 'Direct fabrication of homogeneous microfluidic channels embedded in fused silica using a femtosecond laser', *Optics Letters*, **35**, 282–284.

Hehlen, B., Courtens, E., Yamanaka, A. and Inoue, K. (2002), 'Nature of the Boson peak of silica glasses from hyper-Raman scattering', *Journal of Non-Crystalline Solids*, **307**, 87–91.

Hill, K. (1993), 'Bragg gratings fabricated in monomode photosensitive optical fiber by UV exposure through a phase mask', *Applied Physics Letters*, **62**, 1035.

Hill, K. O. (2000), 'Photosensitivity in optical fiber waveguides: From discovery to commercialization', *IEEE Journal of Selected Topics in Quantum Electronics*, **6**, 1186–1189.

Hill, K. O. and Meltz, G. (1997), 'Fiber Bragg grating technology fundamentals and overview', *Journal of Lightwave Technology*, **15**, 1263–1276.

Hill, K. O., Fujii, Y., Johnson, D. C. and Kawasaki, B. S. (1978), 'Photosensitivity in optical fiber waveguides: Application to reflection filter fabrication', *Applied Physics Letters*, **32**, 647.

Hill, K. O., Malo, B., Bilodeau, F., Johnson, D. C. and Albert, J. (1993), 'Bragg gratings fabricated in monomode photosensitive optical fiber by UV exposure through a phase mask', *Applied Physics Letters*, **62**, 1035–1037.

Hosono, H. (1999), 'Effects of fluorine dimer excimer laser radiation on the optical transmission and defect formation of various jours of synthetic SiO2 glasses', *Applied Physics Letters*, **74**, 2755.

Hosono, H., Kawazoe, H. and Nishii, J. (1996), 'Defect formation in SiO_{2}:GeO_{2} glasses studied by irradiation with excimer laser light', *Physical Review B*, **53**, R11921.

Huy, M. C. P., Laffont, G., Frignac, Y., Dewynter-Marty, V., Ferdinand, P., Roy, P., Blondy, J. M., Pagnoux, D., Blanc, W. and Dussardier, B. (2006), 'Fibre Bragg grating photowriting in microstructured optical fibres for refractive index measurement', *Measurement Science and Technology*, **17**, 992.

Hwang, B. C., Jiang, S., Luo, T., Watson, J., Honkanen, S., Hu, Y., Smektala, F., Lucas, J. and Peyghambarian, N. (1999), 'Erbium-doped phosphate glass fibre amplifiers with gain per unit length of 2.1 dB/cm', *Electronics Letters*, **35**, 1007–1009.

Ihlemann, J., Wolff, B. and Simon, P. (1992), 'Nanosecond and femtosecond excimer laser ablation of fused silica', *Applied Physics A: Materials Science and Processing*, **54**, 363–368.

Inamura, Y., Arai, M., Kitamura, N., Bennington, S. M. and Hannon, A. C. (1997), 'Intermediate-range structure and low-energy dynamics of densified SiO2 glass', *Physica B: Condensed Matter*, **241–243**, 903–905.

Jackson, S. R., Metheringham, W. J. and Dyer, P. E. (1995), 'Excimer laser ablation of Nd: YAG and Nd: glass', *Applied Surface Science*, **86**, 223–227.

Jin, L., Wang, Z., Fang, Q., Liu, B., Liu, Y., Kai, G., Dong, X. and Guan, B.-O. (2007), 'Bragg grating resonances in all-solid bandgap fibers', *Optics Letters*, **32**, 2717–2719.

John, S., Soukoulis, C., Cohen, M. H. and Economou, E. N. (1986), 'Theory of electron band tails and the urbach optical-absorption edge', *Physical Review Letters*, **57**, 1777–1780.

Jürgen, I., Malte, S.-R. and Thomas, F.-B. (2007), 'Micro patterning of fused silica by ArF- and F-2 laser ablation', *Journal of Physics: Conference Series*, **59**, 206.

Kakarantzas, G., Birks, T. A. and Russell, P. S. J. (2002), 'Structural long-period gratings in photonic crystal fibers', *Optics Letters*, **27**, 1013–1015.

Kakarantzas, G., Diez, A., Cruz, J. L., Markos, C., Andres, M. V. and Vlachos, K. (2011), 'Direct Bragg grating writing in a hybrid PDMS/Silica photonic crystal fiber', in CLEO/Europe and EQEC 2011 Conference Digest, OSA Technical Digest (CD) (Optical Society of America, 2011), paper CE9_3.

Kakarantzas, G., Ortigosa-Blanch, A., Birks, T. A., Russell, P. S. J., Farr, L., Couny, F. and Mangan, B. J. (2003), 'Structural rocking filters in highly birefringent photonic crystal fiber', *Optics Letters*, **28**, 158–160.

Kalachev, A. I., Nikogosyan, D. N. and Brambilla, G. (2005), 'Long-period fiber grating fabrication by high-intensity femtosecond pulses at 211 nm', *Journal of Lightwave Technology*, **23**, 2568–2578.

Kalli, K., Simpson, A. G., Zhou, K., Zhang, L., Birkin,. D., Ellingham, T. and Bennion, I. (2006), 'Spectral modification of type IA fibre Bragg gratings by high-power near-infrared lasers', *Measurement Science and Technology*, **17**, 968–974.

Kashyap, R. (1999), *Fiber Bragg gratings*. San Diego and London: Academic Press.

Kashyap, R. (2010), *Fiber Bragg gratings*. London: Academic Press.

Kazansky, P. G. (2007), '"Quill" writing with ultrashort light pulses in transparent materials', *Applied Physics Letters*, **90**, 151120.

Kerbage, C., Eggleton, B., Westbrook, P. and Windeler, R. (2000), 'Experimental and scalar beam propagation analysis of an air-silica microstructure fiber', *Optics Express*, **7**, 113–122.

Kim, C.-S., Han, Y., Lee, B. H., Han, W.-T., Paek, U.-C. and Chung, Y. (2000), 'Induction of the refractive index change in B-doped optical fibers through relaxation of the mechanical stress', *Optics Communications*, **185**, 337–342.

Klini, A., Loukakos, P. A., Gray, D., Manousaki, A. and Fotakis, C. (2008), 'Laser induced forward transfer of metals by temporally shaped femtosecond laser pulses', *Optics Express*, **16**, 11300–11309.

Knight, J. C., Birks, T. A., Russell, P. S. J. and Atkin, D. M. (1996), 'All-silica single-mode optical fiber with photonic crystal cladding', *Optics Letters*, **21**, 1547–1549.

Kondo, Y., Nouchi, K., Mitsuyu, T., Watanabe, M., Kazansky, P. G. and Hirao, K. (1999), 'Fabrication of long-period fiber gratings by focused irradiation of infrared femtosecond laser pulses', *Optics Letters*, **24**, 646–648.

Konstantaki, M., Schuster, K. and Pissadakis, S. (2011), Unpublished data.

Kühnlenz, F., Bark-Zollmann, S., Stafast, H. and Triebel, W. (2000), 'VUV absorption tail changes of fused silica during ArF laser irradiation', *Journal of Non-Crystalline Solids*, **278**, 115–118.

Kukushkin, S. A. (2007), 'Type IIA photosensitivity and formation of pores in optical fibers under intense ultraviolet irradiation. *Journal of Applied Physics*, **102**, 053502.

Ky, N. H., Limberger, H. G., Salathé, R. P. and Cochet, F. (1998), 'Effects of drawing tension on the photosensitivity of Sn–Ge- and B–Ge-codoped core fibers', *Optics Letters*, **23**, 1402–1404.

Ky, N. H., Limberger, H. G., Salathé, R. P., Cochet, F. and Dong, L. (1999), 'Hydrogen-induced reduction of axial stress in optical fiber cores',. *Applied Physics Letters*, **74**, 516–518.

Ky, N. H., Limberger, H. G., Salathé, R. P., Cochet, F. and Dong, L. (2003), 'UV-irradiation induced stress and index changes during the growth of type-I and type-IIA fiber gratings', *Optics Communications*, **225**, 313–318.

Lancry, M., Poumellec, B., Beugin, V., Niay, P., Douay, M., Depecker, C. and Cordier, P. (2007), 'Mechanisms of photosensitivity enhancement in OH-flooded standard germanosilicate preform plates', *Journal of Non-Crystalline Solids*, **353**, 69–76.

Lanin, A. V., Butov, O. V. and Golant, K. M. (2007), 'H2 impact on Bragg gratings written in N-doped silica-core fiber', *Optics Express*, **15**, 12374–12379.

Lemaire, P. (1995), 'Thermally enhanced ultraviolet photosensitivity in GeO2 and P2O5 doped optical fibers', *Applied Physics Letters*, **66**, 2034.

Lemaire, P. J., Atkins, R. M., Mizrahi, V. and Reed, W. A. (1993), 'High pressure H_2 loading as a technique for achieving ultrahigh UV photosensitivity and thermal sensitivity in GeO_2 doped optical fibres', *Electronics Letters*, **29**, 1191–1193.

Li, J., Dou, J., Herman, P. R., Fricke-Begemann, T., Ihlemann, J. and Marowsky, G. (2007), 'Deep ultraviolet laser micromachining of novel fibre optic devices', *Journal of Physics: Conference Series*, **59**, 691.

Li, L., Schülzgen, A., Temyanko, V. L., Qiu, T., Morrell, M. M., Wang, Q., Mafi, A., Moloney, J. V. and Peyghambarian, N. (2005), 'Short-length microstructured phosphate glass fiber lasers with large mode areas', *Optics Letters*, **30**, 1141–1143.

Li, M., Cheng, G., Zhao, W., Fan, W., Wang, Y. and Chen, G. (2008), 'Inscription high-fringe visibility Fabry-Perot etalon in fiber with a high numerical aperture objective and femtosecond laser', *Laser Physics*, **18**, 988–991.

Limberger, H. G., Ban, C., Salathé, R. P., Slattery, S. A. and Nikogosyan, D. N. (2007), 'Absence of UV-induced stress in Bragg gratings recorded by high-intensity 264 nm laser pulses in a hydrogenated standard telecom fiber', *Optics Express*, **15**, 5610–5615.

Limberger, H. G. and Violakis, G. (2010). Formation of Bragg gratings in pristine SMF-28e fibre using CW 244-nm Ar(+)-laser. *Electronics Letters*, **46**, 363-U4890.

Lindner, E., Canning, J., Chojetzki, C., Brückner, S., Becker, M., Rothhardt, M. and Bartelt, H. (2011), 'Thermal regenerated type IIa fiber Bragg gratings for ultra-high temperature operation', *Optics Communications*, **284**, 183–185.

Lindner, E., Chojetzki, C., Brückner, S., Becker, M., Rothhardt, M. and Bartelt, H. (2009), 'Thermal regeneration of fiber Bragg gratings in photosensitive fibers', *Optics Express*, **17**, 12523–12531.

Liou, C. L., Wang, L. A., Shih, M. C. and Chuang, T. J. (1997), 'Characteristics of hydrogenated fiber Bragg gratings', *Applied Physics A: Materials Science and Processing*, **64**, 191–197.

Liu, S., Jin, L., Jin, W., Wang, D., Liao, C. and Wang, Y. (2010), 'Structural long period gratings made by drilling micro-holes in photonic crystal fibers with a femtosecond infrared laser', *Optics Express*, **18**, 5496–5503.

Liu, X., Du, D. and Mourou, G. (1997), 'Laser ablation and micromachining with ultrashort laser pulses', *IEEE Journal of Quantum Electronics*, **33**, 1706–1716.

Liu, Y., Williams, J. A. R., Zhang, L. and Bennion, I. (2002), 'Abnormal spectral evolution of fiber Bragg gratings in hydrogenated fibers', *Optics Letters*, **27**, 586–588.

Livitziis, M. and Pissadakis, S. (2008), 'Bragg grating recording in low-defect optical fibers using ultraviolet femtosecond radiation and a double-phase mask interferometer', *Optics Letters*, **33**, 1449–1451.

Machavaram, V. R. (2006), 'Micro-machining techniques for the fabrication of fibre Fabry-Perot sensors' [Online], Cranfield: Cranfield University.

Machavaram, V. R., Badcock, R. A. and Fernando, G. F. (2007), 'Fabrication of intrinsic fibre Fabry–Perot cavities in silica optical fibres via F-2 laser ablation', *Measurement Science and Technology*, **18**, 928.

Maiman, T. H. (1960), 'Stimulated optical radiation in ruby', *Nature*, **187**, 493–494.

Malinauskas, M., Gaidukevičiūtė, A., Purlys, V., Žukauskas, A., Sakellari, I., Kabouraki, E., Candiani, A., Gray, D., Pissadakis, S., Gadonas, R., Piskarskas, A., Fotakis, C., Vamvakaki, M. and Farsari, M. (2011), 'Direct laser writing of microoptical structures using a Ge-containing hybrid material', *Metamaterials*, **5**, 135–140.

Malo, B., Hill, K. O., Bilodeau, F., Johnson, D. C. and Albert, J. (1993a), 'Point-by-point fabrication of micro-Bragg gratings in photosensitive fibre using single excimer pulse refractive index modification techniques', *Electronics Letters*, **29**, 1668–1669.

Malo, B., Johnson, D. C., Bilodeau, F., Albert, J. and Hill, K. O. (1993b), 'Single-excimerpulse writing of fiber gratings by use of a zero-order nulled phase mask: Grating spectral response and visualization of index perturbations', *Optics Letters*, **18**, 1277–1279.

Marshall, G. D., Kan, D. J., Asatryan, A. A., Botten, L. C. and Withford, M. J. (2007), 'Transverse coupling to the core of a photonic crystal fiber: The photo-inscription of gratings', *Optics Express*, **15**, 7876–7887.

Marshall, G. D., Williams, R. J., Jovanovic, N., Steel, M. J. and Withford, M. J. (2010), 'Point-by-point written fiber-Bragg gratings and their application in complex grating designs', *Optics Express*, **18**, 19844–19859.

Martelli, C., Canning, J. and Groothoff, N. (2005), 'Bragg grating in a Fresnel fibre with a water-core', *Proceedings of the 18th Annual Meeting of the IEEE Lasers Electro-Optics Society*, pp. 864–865.

Martinez, A., Dubov, M., Khrushchev, I. and Bennion, I. (2004), 'Direct writing of fibre Bragg gratings by femtosecond laser', *Electronics Letters*, **40**, 1170–1172.

Martinez, A., Khrushchev, I. Y. and Bennion, I. (2006), 'Direct inscription of Bragg gratings in coated fibers by an infrared femtosecond laser', *Optics Letters*, **31**, 1603–1605.

Meltz, G., Morey, W. W. and Glenn, W. H. (1989), 'Formation of Bragg gratings in optical fibers by a transverse holographic method', *Optics Letters*, **14**, 823–825.

Michelakaki, I. and Pissadakis, S. (2009), 'Atypical behaviour of the surface hardness and the elastic modulus of a phosphate glass matrix under 193 nm laser irradiation', *Applied Physics A: Materials Science and Processing*, **95**, 453–456.

Mihailov, S. J., Grobnic, D., Huimin, D., Smelser, C. W. and Jes, B. (2006), 'Femtosecond IR laser fabrication of Bragg gratings in photonic crystal fibers and tapers', *Photonics Technology Letters, IEEE*, **18**, 1837–1839.

Mihailov, S. J., Smelser, C. W., Grobnic, D., Walker, R. B., Ping, L., Huimin, D. and Unruh, J. (2004), 'Bragg gratings written in all-SiO$_2$ and Ge-doped core fibers with 800-nm femtosecond radiation and a phase mask', *Journal of Lightwave Technology*, **22**, 94–100.

Mihailov, S. J., Smelser, C. W., Lu, P., Walker, R. B., Grobnic, D., Ding, H., Henderson, G. and Unruh, J. (2003), 'Fiber Bragg gratings made with a phase mask and 800-nm femtosecond radiation', *Optics Letters*, **28**, 995–997.

Monat, C., Domachuk, P. and Eggleton, B. J. (2007), 'Integrated optofluidics: A new river of light', *Nature Photonics*, **1**, 106–114.

Niay, P., Bernage, P., Legoubin, S., Douay, M., Xie, W. X., Bayon, J. F., Georges, T., Monerie, M. and Poumellec, B. (1994). Behaviour of spectral transmissions of Bragg gratings written in germania-doped fibres: writing and erasing experiments using pulsed or cw uv exposure. *Optics Communications*, **113**, 176–192.

Nishii, J., Fukumi, K., Yamanaka, H., Kawamura, K. I., Hosono, H. and Kawazoe, H. (1995), 'Photochemical reactions in GeO$_2$-SiO$_2$ glasses induced by ultraviolet-irradiation: Comparison between HG lamp and excimer-laser', *Physical Review B*, **52**, 1661–1665.

Nishikawa, H., Shiroyama, T., Nakamura, R., Ohki, Y., Nagasawa, K. and Hama, Y. (1992), 'Photoluminescence from defect centers in high-purity silica glasses observed under 7.9-eV excitation', *Physical Review B*, **45**, 586–591.

Obata, K., Sugioka, K., Akane, T., Midorikawa, K., Aoki, N. and Toyoda, K. (2002), 'Efficient refractive-index modification of fused silica by a resonance-photoionization-like process using F2 and KrF excimer lasers', *Optics Letters*, **27**, 330–332.

Oh, S., Lee, K. R., Paek, U.-C. and Chung, Y. (2004), 'Fabrication of helical long-period fiber gratings by useof a CO2 laser', *Optics Letters*, **29**, 1464–1466.

Oosterbroek, R. E. and Berg, A. V. D. (2003), *Lab-on-a-chip: Miniaturized systems for (bio)chemical analysis and synthesis.* Amsterdam and London: Elsevier.

Othonos, A. and Kalli, K. (1999), *Fiber Bragg gratings: Fundamentals and applications in telecommunications and sensing.* Boston, MA and London: Artech House.

Othonos, A. and Lee, X. (1995), 'Novel and improved methods of writing Bragg gratings with phase masks', *IEEE Photonics Technology Letters*, **7**, 1183–1185.

Paek, U. C. and Weaver, A. L. (1975), 'Formation of a spherical lens at optical fiber ends with a CO2 laser', *Applied Optics*, **14**, 294–298.

Pappas, C. and Pissadakis, S. (2006), 'Periodic nanostructuring of Er/Yb-codoped IOG1 phosphate glass by using ultraviolet laser-assisted selective chemical etching', *Journal of Applied Physics*, **100**, 114308.

Petrovic, J. and Allsop, T. (2010), 'Scattering of the laser writing beam in photonic crystal fibre', *Optics and Laser Technology*, **42**, 1172–1175.

Phan Huy, M. C., Laffont, G., Dewynter, V., Ferdinand, P., Roy, P., Auguste, J.-L., Pagnoux, D., Blanc, W. and Dussardier, B. (2007), 'Three-hole microstructured optical fiber for efficient fiber Bragg grating refractometer', *Optics Letters*, **32**, 2390–2392.

Piao, F., Oldham, W. G. and Haller, E. E. (2000), 'The mechanism of radiation-induced compaction in vitreous silica', *Journal of Non-Crystalline Solids*, **276**, 61–71.

Pissadakis, S. and Konstantaki, M. (2005a), 'Grating inscription in optical fibres using 213 nm picosecond radiation: A new route in silicate glass photosensitivity', *Transparent Optical Networks, 2005, Proceedings of the 2005 7th International Conference*, 3–7 July 2005, Vol. 1, pp. 337–342.

Pissadakis, S. and Konstantaki, M. (2005b), 'Photosensitivity of germanosilicate fibers using 213nm, picosecond Nd: YAG radiation', *Optics Express*, **13**, 2605–2610.

Pissadakis, S. and Michelakaki, I. (2008), 'Photosensitivity of the Er/Yb-codoped Schott IOG1 phosphate glass using 248 nm, femtosecond, and picosecond laser radiation', *Laser Chemistry*.

Pissadakis, S. and Reekie, L. (2005), 'An elliptical Talbot interferometer for fiber Bragg grating fabrication', *Review of Scientific Instruments*, **76**, 066101.

Pissadakis, S., Ikiades, A., Hua, P., Sheridan, A. K. and Wilkinson, J. S. (2004), 'Photosensitivity of ion-exchanged Er-doped phosphate glass using 248nm excimer laser radiation', *Optics Express*, **12**, 3131–3136.

Pissadakis, S., Konstantaki, M. and Violakis, G. (2006), 'Recording of Type IIA Gratings in B-Ge codoped Optical Fibres Using 248nm Femtosecond and Picosecond Laser Radiation', *Transparent Optical Networks, Proceedings of the 2006 International Conference*, 18–22 June 2006, pp. 183–186.

Pissadakis, S., Livitziis, M. and Tsibidis, G. D. (2009a), 'Investigations on the Bragg grating recording in all-silica, standard and microstructured optical fibers using 248 nm, 5 ps laser radiation', *Journal of the European Optical Society – Rapid Publications*, **4**, 09049.

Pissadakis, S., Livitziis, M., Tsibidis, G. D., Kobelke, J. and Schuster, K. (2009b), 'Type IIA grating inscription in a highly nonlinear microstructured optical fiber', *IEEE Photonics Technology Letters*, **21**, 227–229.

Pissadakis, S., Livitziis, M., Violakis, G. and Konstantaki, M. (2008), 'Inscription of Bragg reflectors in all-silica microstructured optical fibres using 248nm, picosecond and femtosecond laser radiation', *Proceedings of the SPIE*, **6990**, 69900H.

Poignant, H., Boj, S., Delevaque, E., Monerie, M., Taunay, T., Bernage, P. and Xie, W. X. (1994), 'Efficiency and thermal behavior of cerium-doped fluorozirconate glass-fiber Bragg gratings', *Electronics Letters*, **30**, 1339–1341.

Poulain, M., Lucas, J. and Brun, P. (1975), 'Fluorated glass from zirconium tetrafluoride: Optical properties of a doped glass in ND3+', *Materials Research Bulletin*, **10**, 243–246.

Presby, H. M., Benner, A. F. and Edwards, C. A. (1990), 'Laser micromachining of efficient fiber microlenses', *Applied Optics*, **29**, 2692–2695.

Primak, W. (1972), 'Mechanism for the radiation compaction of vitreous silica', *Journal of Applied Physics*, **43**, 2745.

Primak, W. and Kampwirth, R. (1968), 'The radiation compaction of vitreous silica', *Journal of Applied Physics*, **39**, 5651.

Psaltis, D., Quake, S. R. and Yang, C. (2006), 'Developing optofluidic technology through the fusion of microfluidics and optics', *Nature*, **442**, 381–386.

Raine, K. W., Feced, R., Kanellopoulos, S. E. and Handerek, V. A. (1999), 'Measurement of axial stress at high spatial resolution in ultraviolet-exposed fibers', *Applied Optics*, **38**, 1086–1095.

Ran, Z. L., Rao, Y. J., Deng, H. Y. and Liao, X. (2007), 'Miniature in-line photonic crystal fiber etalon fabricated by 157 nm laser micromachining', *Optics Letters*, **32**, 3071–3073.

Ran, Z. L., Rao, Y. J., Liao, X. and Deng, H. Y. (2009), 'Self-enclosed all-fiber in-line etalon strain sensor micromachined by 157-nm laser pulses', *Journal of Lightwave Technology*, **27**, 3143–3149.

Ran, Z. L., Rao, Y. J., Liu, W. J., Liao, X. and Chiang, K. S. (2008), 'Laser-micromachined Fabry-Perot optical fiber tip sensor for high-resolution temperature-independent measurement of refractive index', *Optics Express*, **16**, 2252–2263.

Rao, Y.-J., Deng, M., Duan, D.-W., Yang, X.-C., Zhu, T. and Cheng, G.-H. (2007), 'Micro Fabry–Perot interferometers in silica fibers machined by femtosecond laser', *Optics Express*, **15**, 14123–14128.

Riant, I. and Haller, F. (1997), 'Study of the photosensitivity at 193 nm and comparison with photosensitivity at 240 nm influence of fiber tension: Type IIa aging', Journal of *Lightwave Technology*, **15**, 1464–1469.

Rodica Matei, R., Axel, S., Nasser, P., Albane, L. and Jacques, A. (2007), 'Photo-thermal gratings in Er3+/Yb3+-doped core phosphate glass single mode fibers', *Proceedings of the Bragg Gratings, Photosensitivity, and Poling in Glass Waveguides (BGPP) Conference*, Quebec City, Canada, 2 September 2007, Optical Society of America, paper no. BTuC3.

Roman, J. E. and Winick, K. A. (1993), 'Photowritten gratings in ion-exchanged glass waveguides', *Optics Letters*, **18**, 808–810.

Rothschild, M., Ehrlich, D. J. and Shaver, D. C. (1989), 'Effects of excimer laser irradiation on the transmission, index of refraction, and density of ultraviolet grade fused-silica', *Applied Physics Letters*, **55**, 1276–1278.

Rothschild, M., Horn, M. W., Keast, C. L., Kunz, R. R., Liberman, V., Palmateer, S. C., Doran, S. P., Forte, A. R., Goodman, R. B., Sedlacek, J. H. C., Uttaro, R. S., Corliss, D. and Grenville, A. (1997), 'Photolithography at 193 nm', *The Lincoln Laboratory Journal*, **10**, 19–34.

Russell, P. S. J. (2006), 'Photonic-crystal fibers', *Journal of Lightwave Technology*, **24**, 4729–4749.

Saifi, M. A., Silberberg, Y., Weiner, A. M., Fouckhardt, H. and Andrejco, M. J. (1989), 'Sensitivity of 2 core fiber coupling to light-induced defects', *IEEE Photonics Technology Letters*, **1**, 386–388.

Schaffer, C. B., Brodeur, A., Garcìa, J. F. and Mazur, E. (2001), 'Micromachining bulk glass by use of femtosecond laser pulses with nanojoule energy', *Optics Letters*, **26**, 93–95.

Schaffer, C. B., Garcia, J. F. and Mazur, E. (2003), 'Bulk heating of transparent materials using a high-repetition-rate femtosecond laser', *Applied Physics A: Materials Science and Processing*, **76**, 351–354.

Schenker, R. (1994), 'Deep-ultraviolet damage to fused silica', *Journal of Vacuum Science and Technology B*, **12**, 3275.

Senior, J. M. (1992), *Optical fiber communications: Principles and practice*. Harlow: Pearson Education/Prentice-Hall.

Shelby, J. (1979), 'Radiation effects in hydrogen-impregnated vitreous silica', *Journal of Applied Physics*, **50**, 3702.

Shen, Y. H., Xia, J., Sun, T. and Grattan, K. I. V. (2004), 'Photosensitive indium-doped germano-silica fiber for strong FBGs with high temperature sustainability', *IEEE Photonics Technology Letters*, **16**, 1319–1321.

Shujing, L., Long, J., Wei, J., Yiping, W. and Wang, D. N. (2010), 'Fabrication of long-period gratings by femtosecond laser-induced filling of air-holes in photonic crystal fibers', *IEEE Photonics Technology Letters*, **22**, 1635–1637.

Simpson, A. G., Kalli, K., Zhou, K., Zhang, L. and Bennion, I. (2004), 'Formation of type IA fibre Bragg gratings in germanosilicate optical fibre', *Electronics Letters*, **40**, 163–164.

Skuja, L. (1998), 'Optically active oxygen-deficiency-related centers in amorphous silicon dioxide', *Journal of Non-Crystalline Solids*, **239**, 16–48.

Smelser, C. W., Mihailov, S. J. and Grobnic, D. (2004), 'Hydrogen loading for fiber grating writing with a femtosecond laserand a phase mask', *Optics Letters*, **29**, 2127–2129.

Smelser, C. W., Mihailov, S. J. and Grobnic, D. (2005), 'Formation of type I-IR and type II-IR gratings with an ultrafast IR laser and a phase mask', *Optics Express*, **13**, 5377–5386.

Smith, C. M. and Borrelli, N. F. (2006), 'Behavior of 157 nm excimer-laser-induced refractive index changes in silica', *Journal of the Optical Society of America B*, **23**, 1815–1821.

Smith, C. M., Borrelli, N. F., Price, J. J. and Allan, D. C. (2001), 'Excimer laser-induced expansion in hydrogen-loaded silica', *Applied Physics Letters*, **78**, 2452–2454.

Sohn, I.-B., Kim, Y., Noh, Y.-C., Won Lee, I., Kim, J. K. and Lee, H. (2010), 'Femtosecond laser and arc discharge induced microstructuring on optical fiber tip for the multidirectional firing', *Optics Express*, **18**, 19755–19760.

Soppera, O., Turck, C. and Lougnot, D. J. (2009), 'Fabrication of micro-optical devices by self-guiding photopolymerization in the near IR', *Optics Letters*, **34**, 461–463.

Sørensen, H. R., Canning, J., Laegsgaard, J., Hansen, K. and Varming, P. (2007), 'Liquid filling of photonic crystal fibres for grating writing', *Optics Communications*, **270**, 207–210.

Sørensen, H., Jensen, J. B., Bruyere, F. and Hansen, K. P. (2005), 'Practical hydrogen loading of air silica Bragg gratings photosensitivity and poling in glass waveguides', *Trends in Optics and Photonics Series, Proceedings of the Optical Society of America*, Sydney, Australia, pp. 247–249.

Sozzi, M., Childs, P., Konstantaki, M. and Pissadakis, S. (forthcoming), 'Inscription of relief periodic structures inside microstructured optical fibres using organic liquids infiltration'. In progress.

Sozzi, M., Rahman, A. and Pissadakis, S. (2011), 'Non-monotonous refractive index changes recorded in a phosphate glass optical fibre using 248nm, 500fs laser radiation', *Optical Materials Express*, **1**, 121–127.

Sramek, R., Smektala, F., Xie, W. X., Douay, M. and Niay, P. (2000), 'Photoinduced surface expansion of fluorozirconate glasses', *Journal of Non-Crystalline Solids*, **277**, 39–44.

Stathis, J. H. and Kastner, M. A. (1984), 'Vacuum-ultraviolet generation of luminescence and absorption centres in a-SiO2', *Philosophical Magazine Part B*, **49**, 357–362.

Strasser, T. A. (1996), 'Photosensitivity in phosphorus fibers,' in Optical Fiber Communication Conference, Vol. 2 of 1996 OSA Technical Digest Series (Optical Society of America, 1996), paper TuO1.

Streltsov, A. M. and Borrelli, N. F. (2002), 'Study of femtosecond-laser-written waveguides in glasses', *Journal of the Optical Society of America B*, **19**, 2496–2504.

Stuart, B. C., Feit, M. D., Herman, S., Rubenchik, A. M., Shore, B. W. and Perry, M. D. (1996), 'Nanosecond-to-femtosecond laser-induced breakdown in dielectrics', *Physical Review B*, **53**, 1749–1761.

Stuart, B. C., Feit, M. D., Rubenchik, A. M., Shore, B. W. and Perry, M. D. (1995), 'Laser-induced damage in dielectrics with nanosecond to subpicosecond pulses', *Physical Review Letters*, **74**, 2248.

Sudrie, L., Couairon, A., Franco, M., Lamouroux, B., Prade, B., Tzortzakis, S. and Mysyrowicz, A. (2002), 'Femtosecond laser-induced damage and filamentary propagation in fused silica', *Physical Review Letters*, **89**, 186601.

Sugioka, K., Cheng, Y. and Midorikawa, K. (2005), 'Three-dimensional micromachining of glass using femtosecond laser for lab-on-a-chip device manufacture', *Applied Physics A: Materials Science and Processing*, **81**, 1–10.

Takeshima, N., Kuroiwa, Y., Narita, Y., Tanaka, S. and Hirao, K. (2004), 'Precipitation of silver particles by femtosecond laser pulses inside silver ion doped glass', *Journal of Non-Crystalline Solids*, **336**, 234–236.

Taunay, T., Niay, P., Bernage, P., Xie, E. X., Poignant, H., Boj, S., Delevaque, E. and Monerie, M. (1994), 'Ultraviolet-induced permanent Bragg gratings in cerium-doped ZBLAN glasses or optical fibers', *Optics Letters*, **19**, 1269–1271.

Taylor, A. J., Gibson, R. B. and Roberts, J. P. (1988), 'Two-photon absorption at 248 nm in ultraviolet window materials', *Optics Letters*, **13**, 814–816.

Taylor, R., Hnatovsky, C. and Simova, E. (2008), 'Applications of femtosecond laser induced self-organized planar nanocracks inside fused silica glass', *Laser & Photonics Reviews*, **2**, 26–46.

Tsai, T. E., Taunay, T. and Friebele, E. J. (1999). Stress-dependent growth kinetics of ultraviolet-induced refractive index change and defect centers in highly Ge-doped SiO2 core fibers. *Applied Physics Letters,* **75**, 2178–2180.

Tsai, T.-E., Williams, G. M. and Friebele, E. J. (1997), 'Index structure of fiber Bragg gratings in Ge–SiO$_2$ fibers', *Optics Letters*, **22**, 224–226.

Urbach, F. (1953), 'The long-wavelength edge of photographic sensitivity and of the electronic absorption of solids', *Physical Review*, **92**, 1324–1324.

Vainos, N. A., Mailis, S., Pissadakis, S., Boutsikaris, L., Parmiter, P. J. M., Dainty, P. and Hall, T. J. (1996), 'Excimer laser use for microetching computer-generated holographic structures', *Applied Optics*, **35**, 6304–6319.

van Brakel, A., Grivas, C., Petrovich, M. N. and Richardson, D. J. (2007), 'Micro-channels machined in microstructured optical fibers by femtosecond laser', *Optics Express*, **15**, 8731–8736.

Vengsarkar, A. M., Lemaire, P. J., Judkins, J. B., Bhatia, V., Erdogan, T. and Sipe, J. E. (1996), 'Long-period fiber gratings as band-rejection filters', *Journal of Lightwave Technology*, **14**, 58–65.

Violakis, G. and Pissadakis, S. (2007a), 'Bragg grating laser etching inside microstructured optical fibres using fluorinated gas infiltration'. Unpublished data.

Violakis, G. and Pissadakis, S. (2007b), 'Improved efficiency Bragg grating inscription in a commercial solid core microstructured optical fiber', *Proceedings of the Transparent Optical Networks ICTON 9th International Conference*, 1–5 July 2007, pp. 217–220.

Violakis, G., Konstantaki, M. and Pissadakis, S. (2006), 'Accelerated recording of negative index gratings in Ge-doped optical fibers using 248-nm 500-fs laser radiation', *IEEE Photonics Technology Letters*, **18**, 1182–1184.

Violakis, G., Limberger, H. G., Mashinsky, V. M. and Dianov, E. M. (2011), 'Fabrication and thermal decay of fiber Bragg gratings in Bi-Al co-doped optical fibers', *Proceedings of the Optical Communication (ECOC), 2011 37th European Conference and Exhibition*, 18–22 September 2011, pp. 1–4.

Wang, Y.-P., Bartelt, H., Becker, M., Brueckner, S., Bergmann, J., Kobelke, J. and Rothhardt, M. (2009b), 'Fiber Bragg grating inscription in pure-silica and Ge-doped photonic crystal fibers', *Applied Optics*, **48**, 1963–1968.

Wang, Y., Jin, W., Ju, J., Xuan, H., Ho, H. L., Xiao, L. and Wang, D. (2008), 'Long period gratings in air-core photonic bandgap fibers', *Optics Express*, **16**, 2784–2790.

Wang, Y., Liao, C. R. and Wang, D. N. (2010), 'Femtosecond laser-assisted selective infiltration of microstructured optical fibers', *Optics Express*, **18**, 18056–18060.

Wang, Y.-P., Wang, D. N., Jin, W., Rao, Y.-J. and Peng, G.-D. (2006), 'Asymmetric long period fiber gratings fabricated by use of CO_2 laser to carve periodic grooves on the optical fiber', *Applied Physics Letters*, **89**, 151105.

Wang, Y., Wang, D. N., Yang, M., Hong, W. and Lu, P. (2009a), 'Refractive index sensor based on a microhole in single-mode fiber created by the use of femtosecond laser micromachining', *Optics Letters*, **34**, 3328–3330.

Watanabe, W., Asano, T., Yamada, K., Itoh, K. and Nishii, J. (2003), 'Wavelength division with three-dimensional couplers fabricated by filamentation of femtosecond laser pulses', *Optics Letters*, **28**, 2491–2493.

Watanabe, Y. (2001), 'Photosensitivity in phosphate glass doped with Ag+ upon exposure to near-ultraviolet femtosecond laser pulses', *Applied Physics Letters*, **78**, 2125.

Weeks, R. A. (1956), 'Paramagnetic resonance of lattice defects in irradiated quartz', *Journal of Applied Physics*, **27**, 1376–1381.

Weeks, R. A. (1994), 'The many varieties of E' centers: A review', *Journal of Non-Crystalline Solids*, **179**, 1–9.

Weinberg, Z. A., Rubloff, G. W. and Bassous, E. (1979), 'Transmission, photoconductivity, and the experimental bandgap of thermally grown Si–O_2 films', *Physical Review B*, **19**, 3107.

Wiesmann, D., Hübner, J., Germann, R., Offrein, B. J., Bona, G. L., Kristensen, M. and Jäckel, H. (1999), 'Negative index growth in germanium-free nitrogen-doped planar SiO2 waveguides'. In Erdogan, T. F. E. and Kashyap, R. (eds.), *Proceedings of the Bragg Gratings, Photosensitivity, and Poling in Glass Waveguides (BGPP) Conference*, Florida, 23 September 2009, Optical Society of America, paper no. CB2.

Williams, D. L., Davey, S. T., Kashyap, R., Armitage, J. R. and Ainslie, B. J. (1992), 'Direct observation of UV induced bleaching of 240-nm absorption-band in photosensitive germanosilicate glass-fibers', *Electronics Letters*, **28**, 369–371.

Williams, G. M., Tsai, T. E., Merzbacher, C. I. and Friebele, E. J. (1997), 'Photosensitivity of rare-earth-doped ZBLAN fluoride glasses', *Journal of Lightwave Technology*, **15**, 1357–1362.

Williams, H. E., Freppon, D. J., Kuebler, S. M., Rumpf, R. C. and Melino, M. A. (2011a), 'Fabrication of three-dimensional micro-photonic structures on the tip of optical fibers using SU-8', *Optics Express*, **19**, 22910–22922.

Williams, R. J., Voigtlander, C., Marshall, G. D., Tunnermann, A., Nolte, S., Steel, M. J. and Withford, M. J. (2011b), 'Point-by-point inscription of apodized fiber Bragg gratings', *Optics Letters*, **36**, 2988–2990.

Xie, W. X., Niay, P., Bernage, P., Douay, M., Bayon, J. F., Georges, T., Monerie, M. and Poumellec, B. (1993), 'Experimental evidence of two types of photorefractive effects occuring during photoinscriptions of Bragg gratings within germanosilicate fibres', *Optics Communications*, **104**, 185–195.

Yamada, K., Watanabe, W., Toma, T., Itoh, K. and Nishii, J. (2001), 'In situ observation of photoinduced refractive-index changes in filaments formed in glasses by femtosecond laser pulses', *Optics Letters*, **26**, 19–21.

Yang, R., Yu, Y.-S., Chen, C., Chen, Q.-D. and Sun, H.-B. (2011), 'Rapid fabrication of microhole array structured optical fibers', *Optics Letters*, **36**, 3879–3881.

Yiping, W., Shujing, L., Xiaoling, T. and Wei, J. (2010), 'Selective fluid-filling technique of microstructured optical fibers', *Journal of Lightwave Technology*, **28**, 3193–3196.

Yliniemi, S., Albert, J., Wang, Q. and Honkanen, S. (2006a), 'UV-exposed Bragg gratings for laser applications in silver-sodium ion-exchanged phosphate glass waveguides', *Optics Express*, **14**, 2898–2903.

Yliniemi, S., Honkanen, S., Ianoul, A., Laronche, A. and Albert, J. (2006b), 'Photosensitivity and volume gratings in phosphate glasses for rare-earth-doped ion-exchanged optical waveguide lasers', *Journal of the Optical Society of America: Optical Physics*, **23**, 2470–2478.

Yuen, M. J. (1982), 'Ultraviolet absorption studies of germanium silicate glasses', *Applied Optics*, **21**, 136–140.

Zagorulko, K. A., Kryukov, P. G., Larionov, Y. V., Rybaltovsky, A. A. and Dianov, E. M. (2004), 'Fabrication of fiber Bragg gratings with 267 nm femtosecond radiation', *Optics Express*, **12**, 5996–6001.

Zeller, M., Lasser, T., Limberger, H. G. and Maze, G. (2005), 'UV-induced index changes in undoped fluoride glass', *Journal of Lightwave Technology*, **23**, 624–627.

Zengling, R., Yunjiang, R., Jian, Z., Zhiwei, L. and Bing, X. (2009), 'A miniature fiber-optic refractive-index sensor based on laser-machined fabry–perot interferometer tip', *Journal of Lightwave Technology*, **27**, 5426–5429.

Zhang, H., Eaton, S. M. and Herman, P. R. (2006), 'Low-loss Type II waveguide writing in fused silica with single picosecond laser pulses', *Optics Express*, **14**, 4826–4834.

Zhang, L., Liu, Y., Bennion, I., Coutts, D. and Webb, C. (1999). Fabrication of Bragg and long-period gratings using 255nm UV light from a frequency-doubled copper vapour laser. In Erdogan, T. F. E. and Kashyap, R., (eds.), Optical Society of America, BB4.

Zhou, K., Lai, Y., Chen, X., Sugden, K., Zhang, L. and Bennion, I. (2007), 'A refractometer based on a micro-slot in a fiber Bragg grating formed by chemically assisted femtosecond laser processing', *Optics Express*, **15**, 15848–15853.

Zhu, X. and Peyghambarian, N. (2010), 'High-power ZBLAN glass fiber lasers: Review and prospect', *Advances in OptoElectronics*, **2010**, 1–23.

Index

453

CPSIA information can be obtained at www.ICGtesting.com
Printed in the USA
LVOW100220180712

290534LV00008B/4/P